Geographical Thought

Geographical Thought

An Introduction to Ideas in Human Geography

Anoop Nayak and Alex Jeffrey

Newcastle University

LONDON AND NEW YORK

First published 2011 by Pearson Education Limited

Published 2013 by Routledge
2 Park Square, Milton Park, Abingdon, Oxon OX14 4RN
711 Third Avenue, New York, NY 10017, USA

Routledge is an imprint of the Taylor & Francis Group, an informa business

ISBN 13: 978-0-13-222824-4 (pbk)

British Library Cataloguing-in-Publication Data
A catalogue record for this book is available from the British Library

Library of Congress Cataloging-in-Publication Data
Nayak, Anoop.
Geographical thought : an introduction to ideas in human geography / Anoop Nayak and Alex Jeffrey.
p. cm.
Includes bibliographical references and index.
ISBN 978-0-13-222824-4 (pbk.)
1. Human ecology. I. Jeffrey, Alexander Sam. II. Title.

GF41.N42 2011
304.2--dc22

Typeset in 10/13 pt Minion by 73

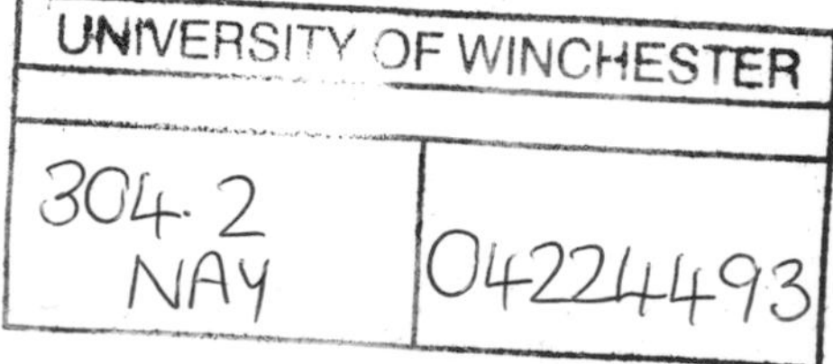

Brief contents

Contents

Authors' acknowledgements

To our students past and present

This book would not have been possible without the help and support of a number of people. This includes loved ones, family members and friends. At vital times they have lightened our moods and lifted our spirits. Special thanks go out to Andrew Taylor who was the original editor for this book. It is no exaggeration to say that your passion and clear thinking kept this project alive. We would also like to thank Rufus Curnow and Sinon Lake for bravely taking up the baton of editorship and faithfully seeing us through the final stages. Julian Brigstoke did a sterling job chasing references and responding to queries. In particular we feel privileged to have had such experienced reviewers, who painstakingly provided rich and detailed comments on each and every chapter. Those that stayed the course of the journey proved reliable anchors in a turbulent sea. We hope we have done justice to your remarks.

Anoop would like to thank Andy Pike for his rigorous comments on Chapter 4. He is indebted to friends, teachers and colleagues from the former Centre for Contemporary Cultural Studies, University of Birmingham, notably for the dialogues generated in Chapter 5. He would like to thank Peter Hopkins for his positive remarks on Chapter 8 and for providing materials for Chapter 9. A special mention goes to Mary Jane Kehily for her lucid insights on Chapter 9 and the wilful attempt to reign-in the sloppy referencing throughout his work. Your love and support kept the words flowing. Thanks go to Raksha Pande for being there to share a mutual awareness of the post-colonial present; this influence, I think, can be found in Chapter 11. Andy Law was a sounding board for all sorts of weird and wonderful conversations. Hannah MacPherson and Rachel Pain also offered kind and encouraging remarks to arrest some of the unruly tendencies creeping into Chapter 12. Alastair Bonnett is an inspiration in showing how the very best geography is often that which refuses to be named as such, or be confined within disciplinary boundaries. Finally, appreciation is given to all Newcastle University colleagues whose work is drawn upon throughout as a mark of approval. Of course, any errors, omissions and misunderstandings are ours alone – peace and love.

Alex would like to thank all his colleagues at Newcastle University for support, help and guidance with the writing of this book. In particular, he benefitted from the valuable assistance of Peter Hopkins who was always on hand to discuss ideas and point out relevant materials. Alex has benefitted enormously from teaching political geography alongside Nick Megoran; his scholarship and activism are a constant inspiration. Thanks also to Alison Williams and Fiona McConnell for discussing ideas, in particular on critical geo-politics. In addition Alison Stenning, Stuart Elden, Colin McFarlane, Alex Vasudevan, Matt Bolton, Joe Painter, Craig Jeffrey and Ewan Jeffrey have all been supportive over the course of the writing process. Most of all he would like to thank Laura and Rufus Jeffrey for their constant support, love and friendship.

Publisher's acknowledgements

We are grateful to the following for permission to reproduce copyright material:

Figures

Figure 1.1 from *The rise and fall of the British Empire*, Little, Brown and Company (James, L. 1994) p. 352; Figure 2.3 from Featured graphic: wars, massacres, and atrocities of the 20th century, *Environment and Planning A* 41, pp. 1779–1780 (Dorling, D. and Coles, P. 2009); Figure 5.1 from *Production of Culture/Cultures of Production*, Open University Press (du Gay, P. 1997)

Text

Extract on page 46 from Qualified Quantitative Geography, *Environment and Planning D: Society and Space*, 19, pp. 555–572 (Doel, M. 2001); Extract on page 143 from Situating knowledges, *Progress in Human Geography*, 21(3), pp. 305–320 (Rose, G. 1997); Extract on page 147 from Home page, www.pygyrg.org

Photographs

The publisher would like to thank the following for their kind permission to reproduce their photographs:

Alamy Images: Alan Novelli 98, Classic Image 4, David Pearson 72, Lordprice Collection 171, Network Photographers 277, Philip Hympendahl 128, Red Cover 88, Slick Shots 256, World Pictures 261; **Bridgeman Art Library Ltd:** Private Collection / Alinari Private Collection / Alinari / © DACS, © The Andy Warhol Foundation for the Visual Arts / Artists Rights Society (ARS), New York / DACS, London 2011. 217; **Fotolia.com:** Adam Radosavlievic 283, Art Jazz 295; **John Frost Historical Newspapers:** 176, 183; **Getty Images:** 131, India Today Group 232; **GNU Free Documentation License: Wikipedia:** 40, 41, 178; **Ingrid Pollard:** 188; **iStockphoto:** Chris Hepburn 221; **Christian Nold:** 31; **Pearson Education Ltd:** Photodisc 52; **Press Association Images:** Cuccuru / Milestone / Empics Entertainment 160, Mark J Terrill 150; **Reuters:** Erik de Castro 93; **Rex Features:** Giovanni Canitano 202; **Royal Geographical Society, with IBG:** 14, 237

In some instances we have been unable to trace the owners of copyright material, and we would appreciate any information that would enable us to do so.

Introduction

This book starts from a simple premise: that all ideas come from somewhere. Thinking about where ideas have come from helps us to understand how they are shaped by their historical and geographical context, how they have changed over time and how they relate to each other. Exploring the production of ideas also illuminates the process by which certain ideas are adopted and celebrated, while others are ignored and forgotten. These differences are not necessarily an outcome of whether they are 'true' or 'false' ideas; they are often a product of the relative authority of the individual or group making an assertion. But ideas are not simply a product of their context; they also have to be persuasive, in order to convince others that they offer an alternative to existing thought or practice.

When we start to examine the nature of ideas, one of the first structuring devices we come upon is academic disciplines. The boundaries between academic disciplines are often blurred and appear quite arbitrary. While geography as an academic pursuit has been oriented towards questions concerning space, place and location, there is no definitive set of issues that constitutes scholarly geography. Instead we can trace over time, attempts by individuals and groups to assert what geography *should be*: to present particular ideas as central to the practice of academic geography. We agree with David Livingstone's (1992) assertion that we should think of geography less as a discipline (since this suggests some fixed boundaries) and instead as a *tradition*: a set of practices and ideas that extend beyond those who self-identify as geographers.

This point is supported by the nature of academic geography in higher education. If you look at the variety of topics covered within different university geography programmes, you will soon see that, while there are some shared themes, the discipline is marked by a wide diversity of topics. On the one hand this reflects the innate breadth of a discipline that seeks to examine the relationship between people and their environment. On the other hand it also illustrates the rapidly changing research interests in geography, where academics that were trained 20 years ago may have different theoretical approaches to those who have just begun a career in research and teaching. As we will see over the course of the following 12 chapters, over the last century geography has been a 'magpie discipline' that has drawn on the work of economists, psychologists, sociologists and cultural theorists (amongst others) in order to develop our understanding of the world around us.

The diversity of ideas within geography makes it a hard task to reduce this to a concise introductory text. Thinking through what we have said above, when we divide up the history of geographical thought into 12 chapters we are not simply describing a neatly ordered history of the work of geographers. Rather we are presenting a particular story of the history of geography that suits our research interests. This is inevitably a *critical* history: it does not simply describe the ideas of others, but it weighs them up, presents criticisms and assesses their legacy. This is not just a consequence of our own interest in these debates; as we will explain in the book, we believe that there is no neutral perspective from which these ideas can be described. Our account reflects our own histories and geographies.

The format of the book

This book is divided into 12 chapters across three parts. Part 1, 'Foundations', explores some of the most significant ideas that emerged over the course of the late 19th- and early 20th-century geography. Here we explore how the institutionalisation of geography was shaped by processes of European imperialism, the shift to quantitative approaches in the mid-20th century and the turn to more radical theories of humanism and Marxism in the 1970s. But writing about the emergence of specific aspects of geographical thought requires a critical engagement not simply with ideas of geography, but also with concepts of history. In setting out the history of early geographical thought there is a risk of presenting unanimity where there was disagreement, or suggesting a linear path for the discipline from one 'stage' to the next, where in fact there were multiple simultaneous perspectives conducted under the banner of geography. As you will see, the prominence given to certain theoretical or methodological perspectives is not always a reflection of consensus, but rather of the power of certain individuals and institutions to define what counts as geography.

Part 2, 'Geographies of difference', charts how geographical scholarship has sought to examine and challenge the marginalisation of groups on the basis of sex, sexuality or 'race'. Over the course of the last 30 years, geographers have been at the forefront of thinking about and confronting social inequality based on ideas of identity. Much of this work has been stimulated by feminist thought: work that sought to challenge the ingrained privileging of male perspectives within academic scholarship and political life. Initially this work sought to 'write women in' by including the voices of women and the spaces women inhabit within research projects. But over time a more theoretically ambitious project has emerged that seeks to challenge the notion of gender (masculinity/femininity) as an organising structure within social life. Rather than espousing an identity on politics based on the marginalisation of women, this work seeks to question the very category of 'women'. This may seem a little abstract, but it illustrates the way in which the scholarship exploring geographies of difference is not simply about 'adding in' ignored social groups, but confronting and transforming the way in which ideas about the world are produced.

Part 3, 'Representation and post-representation', considers how post-modern ideas have forever unsettled what had once seemed as the solid truths of modernity. For some researchers this may well feel like the 'end of the world as we know it', but for others it has offered a breath of fresh air – an opportunity to think and do geography 'otherwise'. Inspired by these openings recent work on critical geo-politics, post-colonialism, and approaches concerned with embodiment and performance at once remind us of the power of representation as well as its limits. If we take the example of post-colonial geographies, the problem of representation – who speaks for whom, under what conditions and why – suggests that representation is at best an incomplete, contested and unstable affair. Representation conceals as much as it reveals. The threads of this argument are interwoven through the material and discursive approaches found in work on critical geo-politics and post-colonial geographies. Some performative accounts have even sought to develop 'non-representational' modes of expression, paying attention to the ways in which bodily sensation, affect and feeling offer new ways for sensing and interpreting our earthly surroundings. Throughout these chapters we signal some of the ways in which geographers are working between the lines of representation and in so doing are pulling apart, unfolding and reassembling our geographical imagination.

How to use this book

We have written this book as an introductory text that presents geographical theory in a palatable and easily digested fashion. We have strived to use an uncluttered and a jargon-free tone throughout, although this is not always avoidable so we have included a glossary at the end of the book where key terms are explained. Wherever we can, we illustrate our arguments using examples, either from contemporary media or from the history of the discipline. As we have already explained, this is not written as a definitive account of geographical ideas, but rather as a text that we hope will encourage further engagement with these ideas. To that end we have included further reading sections throughout each chapter, in addition to the final bibliography. In particular we would encourage students to engage with the original sources if possible.

As you read these accounts think about the types of arguments you find persuasive and ask yourself the question: what type of geographer would I like to be? Use this text as a starting point for thinking about the types of theoretical and social problems you feel are most relevant to the discipline, and how these can be addressed using geographical ideas. If you get the chance, try and develop your own empirical project, perhaps through your dissertation, that allows you to generate your own ideas. And these could go on to shape future understandings of geography: because all ideas come from somewhere.

The chapters can be read as stand-alone pieces on the specific themes, although in places they follow on closely from one another. This reflects the legacy of particular ideas in shaping future interventions and the way in which certain ideas emerged as a reaction *against* previous approaches. It also should be repeated that although there is a rough chronology structuring the organisation of the chapters, this is not supposed to suggest a progression from the dark past of empire through to the enlightenment of post-representation. Instead, we can see through this account the cyclical nature of the emergence of what we understand as present-day geography, dependent as it is on the recovery of ideas from the past. *Geographical Thought* is not just an opening into these debates, but an aperture with no attempt at closure. To this extent it is up to the reader to wrestle with these ideas: to champion, demolish or elaborate upon the approaches outlined here. For it is only through this type of close critical engagement that geographical thought and practice can proceed.

Foundations

Geographies of empire: the imperial tradition

Learning objectives

In this chapter we will:

- Provide a working definition of empire, imperialism and colonialism.
- Explore the role of imperial expansion in shaping the content of academic geography and introduce the ideas of Halford Mackinder, Ellen Churchill Semple and Petr Kropotkin.
- Examine the theoretical frameworks emerging from the relationship between imperialism and geography.
- Trace the legacies of these theories and approaches for contemporary geography.

Introduction

In this chapter we explore how geographical thought was shaped by, and shaped, the age of European imperialism. In confronting these relationships we are exploring an extremely important issue from geography's past. However, following the argument of geographers Cloke *et al.* (1991: 4), we are not engaging with **imperialism** and geography out of some 'antiquarian' concern for the discipline's history, but rather because many of the underlying themes and discussions that emerged in this period continue to shape current geographical debates and institutional priorities. This legacy is often difficult to discern, since it has shaped research agendas and institutional

The 1847 Indian Mutiny
Source: Alamy Images/Classic Image

relationships in complex ways. These will be explored later in this chapter, but it is important to emphasise at the outset that imperialism is not a discrete phenomenon that can be confined to the distant past and is unrelated to the present. Rather, imperial institutions, practices, assumptions and imaginaries continue to shape the content of geography and the conduct of geographers.

As we shall see throughout this chapter, the 18th and 19th centuries were a time of imperial expansion, as powerful states, particularly those in Western Europe, used their military and scientific advances to claim territories in Africa, South America and Asia from their indigenous inhabitants (see photo). At the time, many explained this action as a 'civilising mission', bringing order to barbarous parts of the world that lacked the supposed civility of the European states. Although certain commentators, such as Harvard University's Professor Niall Ferguson, have sought to retain this celebration of benevolent imperialism, others have drawn attention to the violence, racism and plunder on which empire thrived. The position of geography within this story is not straightforward. Certainly, geographical skills of cartography and exploration were vitally important to imperialist objectives; knowing about the world and how to traverse its surface allowed European powers to lay claim to disparate territories. But the merits of imperial expansion were the focus of sustained scholarly debate at the time. While we cannot talk of a single intellectual position among geographical scholars, there is much evidence of the ideas and theories that geographers were developing to help justify the process of imperial expansion. As geography institutionalised, it produced a range of theories that helped to explain why it was morally just that imperialism was occurring, thus entering the realm of normative theorisation (how the world *should* be) as opposed to simply positive theorisation (how the world *is*). Theories such as '**environmental determinism**' suggested that imperial powers were predisposed to rule over colonial territories on account of their climate and topography. Therefore, for geographers Neil Smith and Anne Godlewska (1994: 2), empire was 'quintessentially a geographical project'.

In this chapter we explore this contention over five sections. In the first we explore what is meant by 'imperialism'. While we predominantly examine Western European imperial conquests in this section, we must be cognisant that, just as there are *many* geographies, there are *many* imperialisms. As we have mentioned, we cannot view the age of imperialism as a discrete epoch confined to the past, but rather we can trace imperial relationships and practices in the present day. The second section explores the institutionalisation of geography over the 19th and 20th centuries. It is important to note that this process is not restricted to the emergence of geography as a scholarly discipline in universities and schools, but rather as a practical pursuit documented in museums and exhibitions, and undertaken by geographical societies across Western Europe and North America. In the third section we explore the early content of geography as it was shaped by the demands of empire, in particular drawing out the reliance on the natural sciences and the primacy of **objectivity**. This position, encapsulated by the theory of environmental determinism, was thoroughly discredited in later years, although its influence in geography can still be felt. But we must not assume that there was unanimity in the adoption of any single theory or school of thought. The fourth section outlines the contemporary dissent at the nature of geography, inspired in particular by the work of Russian-born geographer Petr Kropotkin. In the final section we document some of the lingering practical and theoretical implications of geography's entanglement with empire.

Empire, imperialism and colonialism

A brief glance at the available disciplinary histories of geography gives an indication of a long association between geography and the militarised attempts to claim territory on behalf of a particular imperial project. But if we examine these accounts more closely a number of initial – and very serious – questions come to light.

First, what is meant by the discipline of geography? There are no clear boundaries to any academic discipline, since intellectual activity takes place in a diverse range of settings: universities, schools, professional societies, public associations and so on. As the geographer David Livingstone (1992) has persuasively argued, we should not attempt to locate the boundaries of the discipline of geography – this would be pointless – and instead appreciate that geography has meant different things at different times and in different places. This means that we need to understand and explore geography in context and understand that what counts as geographical inquiry may mean different things to different people. In addition, the discipline of geography has always been characterised by debate and disagreement, so it is impossible to suggest that it displays a single political or intellectual perspective. It happens that much of the institutionalisation of geography as an academic discipline took place in the latter part of the 19th century, which coincided with a period of European imperial expansion. But this should not be taken as evidence of an obvious link between geography and empire; rather, it should lead us to question the nature of this relationship and explore the range of different perspectives on imperialism voiced within the geographical discipline.

The second question that immediately comes to the foreground when reading existing accounts of the foundation of geography in the age of empire is, what is meant by terms such as imperialism, **colonialism** and empire? These terms are explored throughout the chapter but, as an initial comment, we should say that we cannot understand these as simple descriptors, since there were many different practices of imperialism and colonialism that differed greatly in their form and substance. Just as we need to be careful not to assume a singular definition of geography, we must be similarly attentive to accounts of imperialism and colonialism that reduce these to simple characterisations. Geographers such as Livingstone, Mike Heffernan, James Sidaway and Klaus Dodds have provided rich accounts of imperial practices in a range of different places that illustrate the shared traits but also the considerable divergences between these processes.

The third question that comes to mind is to what extent was geography alone as a discipline that is inextricably linked to the age of European imperialism? This is a crucial question, since some accounts of the relationship between geography and empire present this as an occurrence unique to this discipline. Instead, we can trace similar processes in the institutionalisation of disciplines such as anthropology, biology and medicine. As we will see, the common thread between these endeavours is a desire to utilise the analytical strength and disciplinary prestige afforded by advances in scientific practices and theories. Processes of imperial expansion necessarily drew on these same scientific practices of measurement and calculation, and thus connections may be made between these disciplines and imperialism. But we need to be careful in our unproblematic presentation of science as a benevolent form of rationality desired by many different academic disciplines. Often scientific ideas were transformed or truncated in their adoption by different disciplines and therefore their meaning was changed. As we will see, one of the key examples of this troubled adoption is Charles Darwin's theory of evolution. The geographer David Stoddart (1966) has provided a detailed account of the dangers of piecemeal adoption of Darwin's ideas, where certain facets of his theory were ignored and others accentuated.

Defining terms

Empire, imperialism and colonialism are terms that are used to describe how people and territory are ruled. As we have discussed, definitions of these terms are necessarily of limited use, as it is in practice that we see specific differences. But as a starting point, it is important that we grasp the basic differences implied by these terms.

Imperialism is derived from the Latin word *imperium*, which roughly translates as power or ability to command. Therefore imperialism relates to the practice of enacting power over a particular group of people or territory. In conventional usage an empire is created through the successful deployment of imperial power. An empire can therefore be defined as an unequal territorial relationship between states often based on economic exploitation. This definition is somewhat problematic in the cases of ancient empires, such as the Roman Empire, where individual states may not have been discernable, but the economic system was still clearly structured around the extraction of wealth, or *tribute*, from peripheral areas back to the imperial centre in Rome.

In contrast to the idea of unequal imperial relationship, the Berkeley geographer Michael Watts describes **colonialism** as the 'establishment and maintenance of rule, for an extended period of time, by a sovereign power over a subordinate and alien people that is separate from the ruling power' (2000: 93). Thus, while imperialism focuses on an unequal relationship between states, colonialism draws our attention to the physical settlement of territory for the material or military advantage of a colonial sovereign power. As these definitions would suggest, we cannot divorce the exercise of colonial or imperial power from the expansion of the capitalist system and the emergence of a global division of labour. Therefore we would draw attention to three aspects of empire:

1 The establishment and maintenance of rule by a sovereign power over a range of separate territories.
2 There is a fundamental economic imperative underscoring these imperial or colonial practices.
3 We can trace relations of exploitation from a discernible imperial centre towards a colonial periphery.

In order to understand the emergence of geography as an academic discipline and its entanglement with colonial practices, we need to identify some general historical points. In identifying such a historical narrative it is easy to rely on generalisations of large and complex systems of rule, for example to identify overriding characteristics of empires that covered large tracts of the Earth's surface for several hundred years. In short: we need to remember that imperialism and colonialism have always been diverse and dynamic practices of rule, that are best confronted and understood through empirical examples.

Portuguese and Spanish empires

Let us take the example of European imperialism from the 15th century onwards to explore the central tenets of colonial expansion. Although preceded by Portuguese colonial exploration, we could isolate the voyages of Christopher Columbus between 1492 and 1504 and Vasco de Gama in 1498 as starting points of the era of European colonial expansion. Columbus was originally from Genoa (in present-day Italy) although he was largely funded by King Ferdinand and Queen Isabella of Spain. His voyages across the Atlantic to the Caribbean, Central America and the coast of South America ensured personal enrichment and paved the way for Spanish colonial expansion. At this early stage, geographical and navigational knowledge was vital, although not always well developed. For example, Columbus famously never accepted that he had landed on a 'new' continent for European explorers; he remained certain that he had sailed to the eastern coast of Asia. Vasco de Gama similarly assisted Portugal in its colonial ambitions, serving as the first European to round the Cape of Good Hope at the southern tip of Africa and sail to India. These early explorations allowed Spain and Portugal to establish extensive colonial possessions. From the 15th to the 16th centuries, Portugal earned great wealth through the spice trade between Asia and Europe. At a similar time, Spain enjoyed a 'golden era' as its *conquistadores* (conquerors) decimated the Inca, Maya and Aztec populations in Central and South America. These acts of violence established access to lucrative reserves of silver and gold across their fledgling colonies, while the creation of sugar plantations in the Caribbean provided further revenue streams. In addition, both the Portuguese and Spanish empires were sustained through the trade and exploitation of slave labour.

At a straightforward level these examples illustrate how geographical knowledge, and particularly navigational skills of cartography, combined with military might to help

European powers project their power and economically exploit distant territories. But in the example of Portugal the geographers James Sidaway and Marcus Power (2005) have pointed to a more complex picture that underpinned processes of imperial expansion. In a line of argument that we will see repeated in later parts of the chapter, these scholars suggest that the effects of imperial expansion should not simply be traced in terms of militarism and economic exploitation. They examine how processes of imperialism have fostered forms of cultural transformation in Portugal that are still felt in the present day, arguing that imperial experiences are central to the country's culture:

> This is a state whose 'national anthem' *begins* with the words 'herois do mar, nobre povi' (heroes of the sea, noble people), whose flag features at its *center* the navigational sphere, and whose coins and banknotes (before they were subsumed into the euro) featured maps of southern Africa and portraits of explorers. (Sidaway and Power, 2005: 528)

Consequently the authors argue that the imperial experience is central to Portuguese understandings of national identity, in particular centring on Portugal as the metropolitan centre of an expansive empire. The authors are making an important geographical point here. They are suggesting that rather than seeing imperialism as a process of territorial expansion (literally the amount of land ruled by one state) it is, in fact, a process of identity formation. The process of imperial expansion shifts how people think and act about themselves and others.

British Empire

The combination of geography, militarism and exploitation exhibited in the Portuguese and Spanish examples was repeated on a greater scale in the case of the British Empire. Initially established in the Caribbean and North America in the 17th century, the roots of the British Empire can be found in England's superior naval power and geographical knowledge. The initial colonies established on the coastal fringes of present-day America and the Caribbean led to the establishment of significant trade flows in sugar, tobacco, cotton and rice as well as greater numbers in the slave trade. But the empire did not remain static; the colonies in the present-day United States of America were lost after the War of Independence from 1775 to 1782 and Britain outlawed the use of slave labour in 1807. But perhaps the most significant shift occurred as a consequence of the Industrial Revolution in the United Kingdom over the course of the 18th and 19th century. The profound economic and social changes that took place over this period transformed the pace and scope of British colonial expansion. The rise of new mechanised industries, driven by steam power, acted as a catalyst to emergent circuits of global capital, structured around a need for new markets, new access to raw materials and new opportunities to invest profit and surplus. Nowhere is this economic imperative more evident than in the case of India. British colonisation of India had begun in 1600 through the creation of the British East India Company as a focal point of the trade of spices, tea and opium. Despite these corporate beginnings, the process of colonising India over the 19th and early 20th century was often explained by the protagonists as part of Britain's civilising mission. Historians, geographers and economists have since questioned such altruistic motives, drawing attention to the fact that by the mid-19th century India had become a key export market for British products, absorbing goods that had previously been sold to continental Europe. As the historian Lawrence James (1994: 219) suggests, 'the process of modernising India, which gave so much satisfaction to the Victorians, was vital for balancing the books at home'.

In addition to greater British control in India, advances in explorations and heightened demand for raw materials led to a 'scramble for Africa' between the Western European powers over the 1880s and 1890s. France, the United Kingdom, Spain, the Netherlands, Italy, Germany and Portugal divided the African continent into colonies by the outbreak of World War I in 1914. This process encapsulated the nexus of geography (exploration and cartography), violence (colonial warfare and suppression) and capitalist economics (need for markets, investment opportunities and unprocessed raw materials). Following the entrenchment of British colonial power over India and parts of Africa at the end of the 19th century, the British Empire reached its greatest extent, covering a quarter of the world's land surface and comprising a population of 425 million, of which 316 million lived in India (see Figure 1.1).

European colonial expansion was a protracted and varied process that is difficult to summarise in a brief historical narrative. We would draw attention to a number of common themes prior to exploring the emergence of geography as an academic discipline and practical pursuit.

First, as we have seen, the process of colonial expansion required the intersection of scientific knowledge, military might and economic exploitation. Some of this scientific knowledge is directly related to geographical inquiry, such as: navigational competence, cartography, environmental and meteorological skills. The combination of science and militarism provided the tools necessary to establish (often exploitative) rule over distant lands.

Second, this form of imperialism was dependent on the establishment of colonies; that is, the transfer of populations to settle in newly appropriated territories in order to exert authority over the indigenous population. This was a vital process, since at this time the technology did not exist to rule at a distance. This process of resettlement may be of crucial significance to the endurance of colonial expansion. In the case of French imperial expansion in the mid-19th century, the historians Andrew and Kanya-Forstner (1988) suggest that the impetus for colonial expansion did not come from political elites in Paris; they were actually rather unconvinced of the need for the colonies. Rather, the drive to continue the expansion of the French Empire came from the military officers stationed in the colonies. As Heffernan (1994b: 95) explains, these 'military empire-builders ran roughshod over the wishes of Parisian politicians and were effectively beyond the control of their civilian "masters" in France'. The process of colonisation, then, in some instances, created the conditions for the continuation of imperial policy.

Third, we should not restrict the processes of colonialism to the sites that were colonised: in the imperial core (whether London, Paris, Lisbon, Madrid or Berlin) the policies of imperial expansion needed to be legitimised. As we have seen in the case of the Portuguese Empire, the creation of 'exotic' and essentially *different* categories of peoples and places was a vital part of colonial rationality. Consequently, imperial 'discoveries' constituted important parts of museum exhibits, international expositions and theatrical shows. For example, Sidaway and Power (2005: 535) describe the mechanisms through which 1940 Exposition of the Portuguese World sought to position Portugal on a global scale:

> . . . Lisbon's Jardim do Ultramar [imperial gardens] and other public spaces and parks in the capital would house a cocktail of African peoples assembled and presented in their

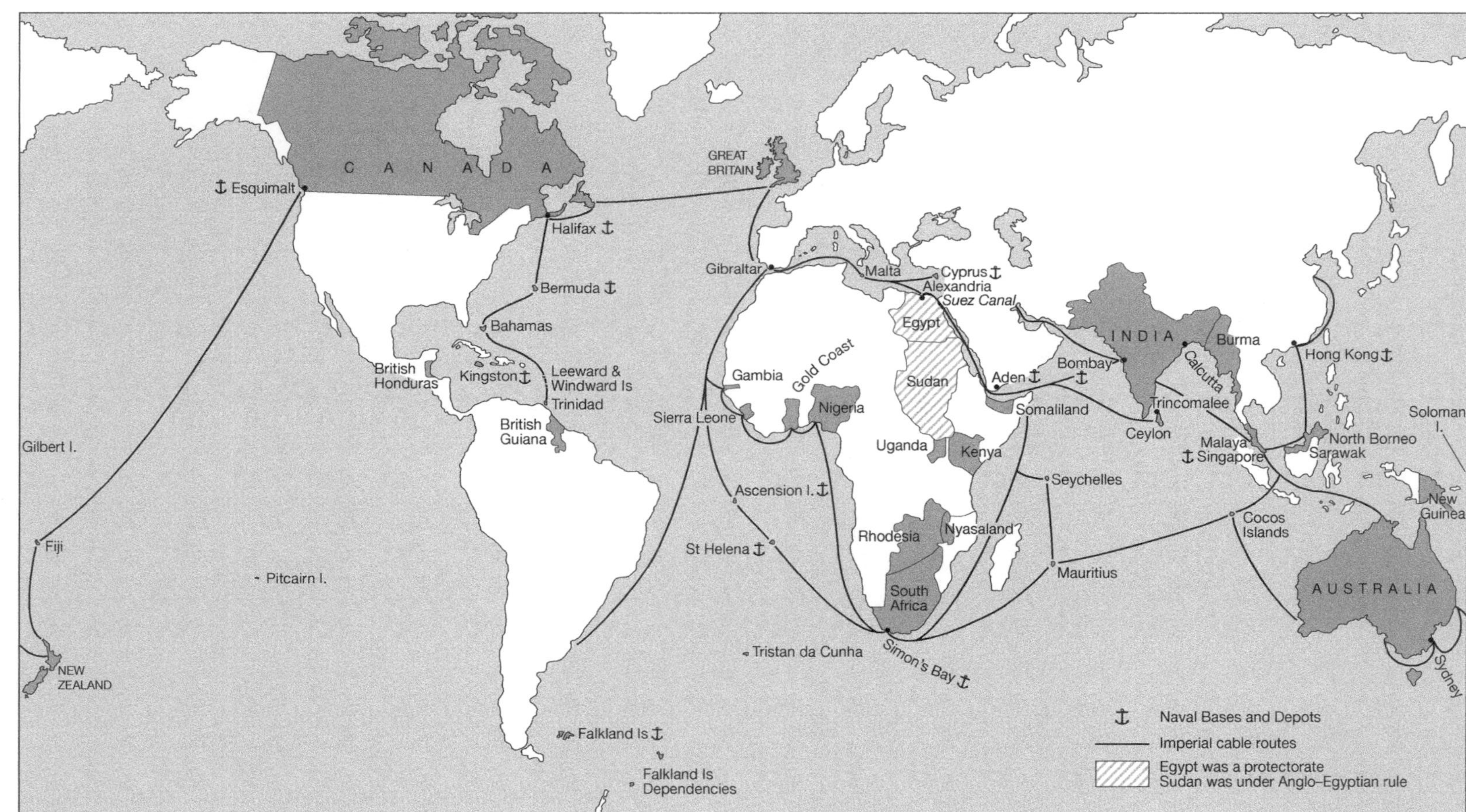

Figure 1.1 **The territorial extent of the British Empire in 1914**

Adapted from James, 1994: 352

'authentic' and 'original' habitats in order to disseminate a vision of Portugal and its peoples as part of a protected space, at the centre of empire, occupying a privileged position from which to imagine and define its sense of national identity and belonging.

Finally, the process of British colonialism produced a central contradiction between the professed ideals of liberalism (that is, individual freedoms) and the subjugation of colonial populations. This required sustaining an imagined division between the virtues of the colonisers over the perceived moral, social or political failings of the colonised. Although the formal colonies of the European powers have (largely) been broken up, the legacy of colonial relationships remains to the present day.

The institutionalisation of geography

> [A] geography may be worked out which will satisfy at once the practical requirements of the statesman and merchant, the theoretical requirements of the historian and scientist, and the intellectual requirements of the teacher. (Mackinder, 1887: 159)

When we consider the status of geography as a popular scholarly pursuit of the 20th and 21st centuries, it is hard to imagine a time when it was merely emerging as an academic discipline. To adequately trace this process we need to engage in more detail with the content and methodologies of geographical inquiry. A number of historians of the geographical discipline have found that the Ukrainian-born author Joseph Conrad is helpful in this regard (see Box 1.1). Conrad is best known for his fictional work, which often engaged with issues of imperialism and geographical imagination. Perhaps his most famous example of this **genre** is *Heart of Darkness* (1924b, originally serialised in 1899), where Conrad draws on his own maritime experiences to chart the voyage of a sailor named Marlow up the River Congo in Africa to search for a British Colonial agent named Kurtz. The themes of exploration, geography and empire that permeate *Heart of Darkness* were developed by Conrad in a non-fiction essay entitled 'Geography and Some Explorers', which was published in the *National Geographic* (1924a). In this brief essay Conrad isolates three stages of geography's development: Geography Fabulous, Geography Militant and Geography Triumphant. For Conrad, Geography Fabulous refers to a phase of imaginative and speculative geography where maps consisted of 'pictures of strange pageants, strange trees, strange beasts, drawn with amazing precision in the midst of a theoretically conceived continents' (Conrad, 1924b, cited in Driver, 1992: 23). In contrast, Conrad described the epoch of Geography Militant as an era of science and exploration, where the 'only object was the search for truth' and the explorers 'devoted themselves to the discovery of facts in the configurations and features of the main continents' (Conrad, 1924b, cited in Hampson, 1995: xiii). Conrad is explicit in connecting this new exploratory zeal with an 'acquisitive spirit', that exploration and scientific measurement are prompted by 'the desire to trade or the desire of loot, disguised in more or less fine words' (*ibid.*). Conrad identifies Captain Cook as the epitome of this intersection of exploration, science and accumulation. Finally, Geography Triumphant refers to the expansion of European geographical knowledge across the globe, to the point where space has succumbed to the dominion of science.

Box 1.1
Joseph Conrad (1857–1924)

Joseph Conrad (born Jozef Teodor Konrad Korzeniowski) was born in the Ukraine in 1857. His parents died when he was still young and he was subsequently raised by his uncle. From an early age, Conrad had a desire to go to sea, and this was fulfilled in 1878 when he joined the French Merchant Navy. Nearly ten years later he joined the British Merchant Navy and, in 1890, he made a formative journey up the Congo River, a key inspiration for *Heart of Darkness* (1924b). In 1894 Conrad left his life at sea and settled in Kent, England, where he produced a series of important novels challenging prevailing notions of morality and politics. He is often considered one of the greatest fictional writers in the English language, although his views on imperialism and race have been the focus of sustained literary criticism. In particular, Chinua Achebe (1977), the acclaimed Nigerian writer, is critical of what he perceives as Conrad's lack of awareness of the racism on which colonial practices were founded.

Conrad's three-part framework should not be mistaken for an unquestioned truth; instead it must be analysed with a critical eye. Conrad's account has been criticised for being Eurocentric, meaning that it is a history written from the point of view of a European and therefore makes assumptions about the 'correct' path of progress. Conrad's approach is also reliant on a form of thinking that emerged from the Enlightenment, structured around the triumph of science, rationality and reason over supposed superstition and myth. Conrad's dependence on this transition fosters a belief that the introduction of new scientific procedures and techniques can purge geographical insights of imaginative and irrational supposition. As we will explore later in the chapter, this belief in the possibility of a neutral and detached scientific observer has been widely criticised as a device through which Western ideas and interests could be promoted over other non-Western forms of knowledge and rationality. In addition, Conrad's model suggests that geographical inquiry can be divided into distinct epochs progressing to a final end point of geographical enlightenment. This concept, developed in Thomas Kuhn's (1996) conception of the '**paradigm**', underemphasises the overlapping nature of eras of scientific thought and the debates that existed as to what constitutes geographical inquiry.

From 'fabulous' to 'militant' geography

We would therefore view Conrad's framework as a *situated* account of the values of late 19th- and early 20th-century geography. By this we mean that it is best understood within the context that it was produced and as a set of guidelines rather than an unquestioned scientific truth. With this in mind, the aspect of Conrad's framework that we will look to here is the supposed shift from a 'fabulous' to a 'militant' style of geographical knowledge production. To understand this process we need to focus on the importance of geographical knowledge to the twin historical processes of expanding trade and the strengthening of the internal capacities of Western European states. Although at first sight these may appear disconnected from the dispassionate practice of scholarly inquiry, we will argue in line with Halford Mackinder's opening quote to this section, that geography has never simply been a pursuit of interest simply to the scholar or teacher, but it has also long served the interests of the 'trader and the statesman'. Knowledge of places, distances, measurements and map-making were crucial in the emergence of mercantile exchange in Western Europe over the 16th century. This process is described by Miles Ogborn (1998) in his account of the increased infrastructural and bureaucratic capacity of the English state at this time. In Ogborn's account he traces the significance of the increased regulation of beer trade in England and Wales in the late 17th century, analysing in particular the state's increased capacity to govern trade through new instruments of regulation and the production of a uniform instrument of measurement: the beer cask. This process of producing 'calculable spaces' through careful measurement of distances and trading quantities proved vital to increasing tax income which, in turn, funded imperial expansion in the following two centuries.

Considering the mercantile and military requirements for geographical knowledge it is understandable that skills such as cartography and exploration began to be taught in British universities between the 16th and 19th centuries. The exact date is open to debate, since there is disagreement over what counts as 'geography' in early universities, with some scholars suggesting that we cannot claim the emergence of a geographical discipline until the appointment of Halford Mackinder as a Reader in Geography at Oxford University in 1887. But as historical geographers Withers and Mayhew (2002) have suggested, universities of the 16th and 17th centuries did not have the disciplinary frameworks we have become used to, so we would not expect to find a clearly defined 'discipline of geography' at this stage. They explain:

> The early modern 'university' still held to the aim its etymology suggests: to provide an education in the whole world (*universitas*) of learning, not a particular discipline, this to be recognised in the form of a degree in the arts; both humanities and sciences in modern parlance. As such we should not expect the early modern university to teach geography separately, this being fundamentally at odds with its ideal of universal learning. (Withers and Mayhew, 2002: 13)

While methods of geographical inquiry were emerging within the curricula of British universities, there was a more concerted institutionalisation of geography outside the academy. As we have detailed, geography was both a practical and popular pursuit that captured an emerging public interest in exploration, travel and the empire. The Travellers' Club, founded in London in 1819, encapsulates the popular and adventurous nature of early geographical institutionalisation, where membership was granted to those who had 'travelled out of the British Isles to a distance of 500 miles from London in a straight line' (see Bonnett, 2008). Over the 19th century, museums, public fairs, lectures and international expositions sought to present the products of exploration and Britain's expanding colonial possessions.

The Royal Geographical Society

But geography was more than a public diversion; scholars and politicians sought to establish geographical societies to promote geography as a scientific endeavour. First Paris (1921), then Berlin (1928) and then London (1930) established geographical societies to capture the public and governmental interest in exploration, cartography and knowledge of the world. Later, further societies emerged in the UK in the cities of Manchester (1884), Newcastle upon Tyne (1887), Liverpool (1891) and Southampton (1897). This institutionalisation serves as an important example of the close proximity of geographical inquiry to the interests of imperial expansion. We will take the London Geographical Society, which later gained a Royal Charter and subsequently became the Royal Geographical Society

The Royal Geographical Society, London
Source: Royal Geographical Society, with IBG

(RGS), as an example of this complex relationship. From its foundation through to World War II, the RGS served as a focal point for public and scholarly debate regarding the content and direction of geography. The society emerged out of a gentlemen's dining club on Savile Row, and moved its present location on Kensington Gore in 1913 (see photo).

The move to Kensington allowed for an increase in the fellowship of the RGS to 5,300 by 1914, ensuring that the RGS remained the pre-eminent institutionalisation of geography in the UK. The activities and archives of the RGS provide an unparalleled set of insights into the relationship between this phase of geography's institutionalisation and practices of British imperialism. We must be careful to avoid characterisations: the RGS was not simply 'imperialistic' even if some of its most vocal representatives were supportive of the British Empire. The fellowship of the RGS reflected wider UK society in containing individuals who both supported and criticised imperial practices. For example, regular speakers at the RGS's public lectures criticised either the practice of imperialism or the specific practices of agents of the British Empire, as we will see in the case of Petr Kropotkin, later in this chapter. But the organisation of the RGS grants us insights into the close relationship that was often forged between geographical inquiry and British imperial practices. We will focus on two in particular: the role of gender in the RGS constitution and the attempts to forge links between geography and the natural sciences.

Women fellows at the RGS

Tracing the connections between geography and imperialism within the RGS requires an understanding of the marginalisation of women in the early years of the society. As we will see throughout this book, the role of women in the history of geographical inquiry is often marginalised on account of the male-dominated nature of academic societies and universities. As Avril Maddrell's (2009) book *Complex Locations: Women's Geographical Work in the UK 1850–1970* makes clear, many histories of geographical thought have consequently ignored women and provided a male-oriented account of geographical inquiry. While certain scholars have challenged the need for a feminist historiography (see Stoddart, 1991), Maddrell's account illustrates that a better understanding of women's contribution to geographical scholarship provides a more detailed picture of the relationship between the discipline and imperial practices. First, the inclusion of women's accounts can challenge an over-emphasis on the militarisation of the discipline, since it is often through military or ex-military service personnel that we engage with early geographical studies. Following this line of argument we find that the inclusion of more of the silenced women's voices may provide accounts of alternative, or anti-imperial, approaches to geographical thought. Second, and in contrast, a study of the writing of excluded women can support close links between geography and imperialism on account of the position of eminent female geographers such as wives or relatives of colonial or military figures. In part this is a reflection of the economic barriers to travel for many: it was only affordable to the wealthy, who may be less disposed to criticising a system that is supporting their relative advantage. Therefore we cannot make a generalisation as to the effect of greater attention to the voices of women geographers to our understanding of the relationship between geography and imperialism: we must look at the process of exclusion in practice.

The geographers Morag Bell and Cheryl McEwan (1996) provide such a practical account through their study of the debate surrounding women's exclusion from the RGS. Until 1913 women were barred from becoming full fellows of the RGS. Despite the fact that this reflected the wider unequal nature of British society, other geographical societies had allowed membership on an equal basis between the sexes for many years: Manchester since 1884, Tyneside since 1887 and Liverpool since 1891. The decision to exclude women seems to have been based on male assumptions relating to the relationship between gender and competence at scientific inquiry. This is encapsulated in the comments made in *The Times* newspaper by Lord Curzon, future Viceroy of India, UK Home Secretary and President of the RGS (1911–1914):

> We contest *in toto* the general capacity of women to contribute to scientific geographical knowledge. Their sex and training render them equally unfitted for exploration, and the genus of professional globe-trotters with which America has lately familiarised us is one of the horrors of the nineteenth century. (Cited in Bell and McEwan, 1996: 298)

Beyond this assumption of the incapacity of women, other commentators at the time tied the possibility of women's fellowship into the wider public debate in the UK concerning political enfranchisement. As we will see in Chapter 7, the first part of the 20th century saw the rise of the Suffragette movement: a group of political activists seeking the right for women to vote in parliamentary elections. Indeed Bell and McEwan note that correspondence to the RGS in 1913 included letters that emphasised the need to exclude what were termed 'militant Suffragettes' who should not be eligible for fellowship of the RGS.

By the early decades of the 20th century these concerns no longer reflected the majority opinion within the RGS. In 1913 a vote was held, where around 75 per cent of fellows voted to allow full women fellows. While this provided the necessary institutional context for women to become fellows, individual applicants still needed to demonstrate a contribution to the advancement of geographical knowledge to be made a fellow. Bell and McEwan note that this may have has the consequence of strengthening links between geographical practice and imperialism:

> For those [fellowship applicants] who lacked professional authority, knowledge of the empire and imperial connections formed the basis for a convincing case. Indeed 'lived in India' appeared frequently on the nomination forms. (Bell and McEwan, 1996: 306)

Bell and McEwan (p. 302) suggest that over half the 163 women fellows in 1913 were travel writers and explorers such as Mary Hall, the first woman to cross Africa from north to south, and Charlotte Cameron who had twice circumnavigated the globe.

From this discussion of gender and the RGS we would like to draw a number of points. First, and perhaps most clearly, we need to be attentive to the role of women in the foundation of geography, a role that is often erased through a preoccupation with formal intellectual voices which were predominantly male. Second, this process of incorporation does not necessarily challenge the link between geography and imperialism: as we have seen, the eventual inclusion of women in the RGS in some ways strengthened this link through the value placed by the RGS committee on imperial ties and experiences. Third, and most fundamentally, we need to be attentive to the contexts within which geographical knowledge is produced and select our methodology of study accordingly. In order to understand the role of women in the history of geographical thought, feminist geographers have developed skills in biographical and archival methodologies in order to draw out the contextual factors that shape the production of knowledge.

Science and militarism

The second consequence of the centrality of the RGS was the desire among prominent members of the society to position geography as a scientific endeavour with clear utility to the British imperialism. Studying the formal proceedings and the make up of the fellowship illustrates that the early preoccupations of the RGS were closely structured around the interests of the British Empire. This orientation is evident in the initial prospectus of the society in 1831 which made an explicit reference to the utility of geographical knowledge 'to the welfare of a maritime nation like Great Britain, with its numerous and extensive foreign possessions' and in 'conferring just and distinct notions of physical and political relations of our globe' (cited in Livingston, 1992: 167). The geographer David Livingstone suggests that early proclamations of the RGS make plain 'the imperialistic undergirding of the institution's entire project and thereby [reveal] that Victorian geography was intimately bound up with British expansionist policy overseas' (1992: 167). Felix Driver uses the RGS presidency of Sir Roderick Murchison (president for four terms during the mid-19th century) to highlight the close intertwining of geography and empire. Driver's (2001: 43) account of Murchison's influence on the British military expedition to Abyssinia (present-day Ethiopia) highlights these interconnections:

> The Abyssinian force, totalling thirteen thousand men from Britain and India under the command of Sir Robert Napier, was despatched to force the release of British consular officials and other Europeans held captive by the Abyssinian King, and more generally to restore British influence both in the region and beyond. Murchison used his political influence to secure the appointment of a scientific team to accompany the expedition, including a botanist, a meteorologist, a geologist and a geographer, Clements Markham (working in the India Office).

Although there was some resistance within the RGS to the close connection between militarism and science, the example of the Abyssinian campaign marks a crucial public recognition of geography's service to imperial expansion. This relationship was also evident in British government financial support for the Society: after 1854, Sir Roderick Murchison secured an annual government subsidy for the Society's map room. In addition the RGS presidents and membership strengthened the connection, both material and imagined, between the RGS and imperial expansion:

> The first two presidents of the RGS were colonial ministers, and many of their successors were career diplomats. Moreover, army and naval officers constituted around one-fifth of the 460 founding members of the RGS, and this proportion was to remain remarkably stable throughout the next seventy years. (Driver, 2001: 41)

Historians of geography often cite Halford Mackinder as a key figure in the connection between the RGS and state imperialism (see Box 1.2).

Box 1.2

Sir Halford Mackinder (1861–1947)

Mackinder was a key figure in the institutionalisation of geography and an advocate for the close connection between imperialism and geographical study. His work has enjoyed a revival in interest in recent years, becoming the focus of geographical debate (see Dodds and Sidaway, 2004; Heffernan, 2000).

Mackinder originally studied zoology at Oxford (1880–1883), gaining an interest in the natural sciences that he incorporated into his later geographical work. Mackinder was the first dedicated scholar of geography at a UK university when he was appointed as a Reader at Oxford University in 1887. But he did not restrain his geographical practices to scholarly pursuits. In 1899 Mackinder rekindled geography's age of exploration when he led an expedition to climb Mount Kenya, the highest mountain in Kenya. There can be no doubt of Mackinder's motivation to lead this difficult expedition, as Brian Blouet (2004: 323) has remarked: '[t]he Mount Kenya expedition was part of imperial expansion'.

Mackinder's scholarship centred on attempts to map the opportunities and threats facing Britain in what he termed a 'post-Columbian era' – a reference to the end of the 'Colombian age' of European exploration and expansion, lasting from the 15th century to the late 19th century. He was a proponent of the incorporation of Darwin's ideas of competition and evolution into the social realm. Such social Darwinism highlighted the world as a stage for competition between races and between nations (Kearns, 2004: 341). This incorporation of the laws of natural science into political geography emphasised the role of the climate and topography on the social development of human populations. For example, in 'The Geographical Pivot of History' (1904: 423) Mackinder identifies the topographic and climatic differences between Russia and Western Europe:

> From a physical point of view, there is, of course, a like contrast between the unbroken lowland of the east and the rich complex of mountains and valleys, islands and peninsulas, which together form the remainder of this part of the world. At first sight it would appear that in these familiar facts we have a correlation between natural environment and political organization so obvious as to hardly be worthy of description, especially when we note that throughout the Russian plain a cold winter is opposed to a hot summer, and the conditions of human existence are thus rendered additionally uniform.

Mackinder (1904, 1919) identified the Eurasian landmass as the 'geographical pivot' (later renamed the 'heartland'), a phrase he used to imply the centrality of this region to world power. Mackinder's thesis sought to frame the potential power of different territories according to their topography and climate. As Klaus Dodds and James Sidaway (2004: 294) suggest:

> Mackinder suggested that the resources, railways and the remoteness of the heartland would prove an irresistible force in the future. Britain and other sea-based powers such as the United States would have to respond to the challenge posed by the heartland power and defensive readiness and financial reform were a necessity.

Box 1.2 continued

Mackinder's vision thus posed a clear warning to the powerful states and empires of the early 20th century, such as Britain. In *Democratic Ideals and Reality* (1919) he condensed his thesis of threat:

> Who rules East Europe commands the Heartland;
> Who rules the Heartland commands the World-Island;
> Who rules the World-Island commands the world.

Mackinder would therefore seem to epitomise geography's connection to empire: he emulated the imperial exploration in his sojourn to Mount Kenya, while he remained committed to securing British military pre-eminence. Geographer Michael Heffernan (2000: 348) agrees with this view describing his Geographical Pivot paper as:

> part of an imperial, Eurocentric planetary consciousness. This was a masculinist ex-cathedra vision of a dangerous world viewed from the commanding heights of governmental and academic institutions.

But in tracing the connections between Mackinder and imperialism we need to be careful to reduce his complex affiliations and motivations to stark stereotypes. It is certain that across Mackinder's varied career choices (from lecturer, to MP, to his chairmanship of the Imperial Economic Committee) he expressed a complex relationship with imperialism. A number of scholars have pointed to Mackinder's conviction that the British Empire could act as a benevolent structure, supporting free trade that would 'allow the economy to grow, British financial institutions would invest capital in developing areas to produce long-term income and prosperity' (Blouet, 2004: 328). Nick Megoran (2004: 356) draws attention to Mackinder's democratic instincts, citing Mackinder's commitment to the 'universal ideals of freedom'. While recognising this paradox between imperial domination and commitment to ideals of free trade and expression, we must not erase Mackinder's support for British imperialism, an unjust and oppressive structure of material accumulation and dispossession.

Further reading

Kearns, G. (2004) The Political Pivot of Geography, *The Geographical Journal*, 170(4): 337–346.

Mackinder, H. (1904) The Geographical Pivot of History, *The Geographical Journal*, 23(4): 421–444.

Megoran, N. and Sharapova, S. (2005) Mackinder's "Heartland" Help or Hindrance in Understanding Central Asia's International Relations? *Central Asia and the Consensus*, 4(34): 8–21.

The contradictions we can see in Mackinder's views, where imperialist and democratic outlooks seem to be combined, serve to illustrate the broader contradictions in the emerging geographical discipline. Just as Mackinder's long career cannot be reduced to a simple imperial essence, so too the RGS represented a range of social, political and scientific interests. As Driver (2001: 47) explains '[i]t is difficult to characterize a body which finds room for missionaries, anti-slavery campaigners, roving explorers, mountaineers, antiquarians, geologists and naturalists'. We must, then, be wary of reducing the RGS to the single voice of its president and erasing from the historical record the many dissenting or divergent intellectual and political movements that the society encompassed. But in doing so it is important to recognise the tension between an appreciation of the complexity of an organisation such as the RGS and a disciplinary concern for its manifest links with British imperialism.

The Societé de Géographie de Paris (SGP)

The danger of reducing the nature of geography to a single imperial essence is further illustrated in the case of the foundation of the SGP in France. The geographer Mike Heffernan (1994a, 2005) has provided a series of insights into the connections between imperial expansion and the institutionalisation of geography in France. His investigations of the foundation of the SGP and, in particular, his examination of the intellectual positions of its key members, provide a picture of the complex relationship between geography and empire building over the 19th and 20th centuries. Instead of drawing a link between geography and imperialism, Heffernan's account directs attention to the significance of the military defeat suffered by the French in the Franco–Prussian War in 1870–1871. He argues that this event, more than any other, stimulated a desire amongs the French people to reform education as a means of 'national rejuvenation' (1994b: 97). Geography's engagement with the physical and social aspects of the French state, as well as documenting France's imperial possessions in Africa, was considered by senior political and educational figures as crucial to bolstering this sense of national rejuvenation. The membership of the SGP reflects this increased interest in geography, as it rose from around 600 in 1870 to 1,353 in 1875 (Heffernan, 1994b: 98).

Heffernan's (1994b) account illustrates the different positions towards imperialism that existed within French geographical debates in the 19th century. He identifies five interlinked perspectives within the geographical movement: utopian imperialism, cultural imperialism, economic imperialism, opportunistic imperialism and anti-imperialism. It is worth examining these in a little more detail, as they illustrate the considerable diversity of opinion as to the relative merits of imperial expansion.

Utopian imperialism

The first perspective identified by Heffernan may come as a surprise, since imperialism is so often connected with processes of militarism and economic exploitation. He suggests that, in fact, many geographers, including prominent figures within the SGP, supported French imperialism on the grounds that it had the potential to bring more utopian forms of society into being. This utopian perspective was linked to a religious movement in the 19th century known as Saint-Simonianism. This group, founded on the thinking of social theorist Henri de Saint-Simon, was based on broadly socialist principles that stressed the importance of work, collaboration and science to advancing human society. In relation to imperialism, advocates of this mode of thought believed in the emancipatory power of technology and industrial development to bridge divides between religious and cultural communities (Heffernan, 1994b: 100). Therefore imperial expansion was supported on the basis that it allowed new infrastructures of trade and industry to develop, therefore increasing the productive capacity and collaborative networks between different groups of people. While Heffernan notes that this utopian perspective was rarely voiced within military circles, it was a popular intellectual thread within geographical communities in France.

Cultural imperialism

Heffernan's second perspective is drawn from the role of geography in bolstering senses of France's cultural and national prestige. He suggests:

> While the carefully directed study of France's rich and varied regional geography was to be a central educational component in fostering the spirit of patriotism and devotion to the Republic, the expansion of France overseas was interpreted as crucial to the nation's future survival as a vibrant culture and civilization. (1994b: 102)

This point shares aspects of Sidaway and Power's (2005) observation of the centrality of concerns of national identity and prestige within Portuguese imperialism (see above). But this was not simply a point about national aggrandisement: geography teaching sought to set France's imperial practices and motives apart from other European colonies. For example, the British Empire was presented as more mercantile and money-oriented than the more benevolent and culturally sensitive French imperial practices (thereby overlapping with the concept of utopian imperialism).

Economic imperialism

Despite the prevalence of utopian and cultural perspectives on imperialism, economic motives continued to play a significant role within geographical debates on imperialism. Just as today, there was no consensus as to whether imperialism served to benefit the French economy; many saw the expenditure on military action to outweigh the benefits of cheap raw materials. One important process that Heffernan highlights is the role of economic motives in the rise of provincial geography societies in France, stimulated by a desire to better utilise the economic opportunities afforded by imperial expansion. The distinction between economic and cultural motives led to marked differences in the economic role of imperialism within French life:

> the economic imperialists sought to demonstrate the commercial advantages of overseas colonies, while many cultural imperialists spent their time celebrating the costs of empire as evidence of France's commitment to the cultural advancement of the colonies.
>
> (Heffernan, 1994b: 106)

We need to be wary of seeing these positions on imperialism as absolutes: people would often hold both cultural and economic perspectives and voice different opinions at different times. But it remains important to see the variety of positions on the economic question, which can often be overlooked due to the exploitative nature of imperial relations.

Opportunist imperialism

As we have stressed throughout this chapter, we cannot understand geography's relationship with imperialism outside of the specific historical context within which it emerged. To some extent this will always involve an aspect of opportunism: that particular events stimulated the possibility for new forms of imperial expansionism. In the French context this is particularly appropriate in the case of World War I. Over the course of the war many French geographers were engaged in secret negotiations to ensure the survival – and expansion – of the French Empire. This 'opportunistic imperialism', as Heffernan calls it, saw the possible break-up of the Ottoman Empire and the seizure of German colonies in Africa and the Pacific. Similar processes of secret negotiation are found in other geographical movements: the geographer Neil Smith (2000) has documented the US President Woodrow Wilson's establishment of a group of intellectuals and policymakers during World War I was tasked with exploring (amongst other things) how a possible peace agreement could suit US economic interests. As we discuss in Chapter 10, the geographer Isaiah Bowman played a prominent role in this group. For Heffernan the events of World War I acted to stimulate French public interest in greater imperial expansion:

> Although clearly informed by the earlier debates about the nature and form of French imperialism, the attempt to expand the French empire by acquiring German colonies or Ottoman territory as a result of the war was motivated, at least in part, by the general spirit of revanchism that gripped France in 1916 and 1917. This persuaded even those who would otherwise have doubted the long-term benefits of this dramatic expansion of the French empire, that such a policy was necessary in order to weaken Germany and its allies.
>
> (Heffernan, 1994b: 111)

Anti-imperialism

The final perspective examined in Heffernan's work is that of anti-imperialism: forms of resistance to the French imperial expansion from within the geographical community. Interestingly he finds no real organised criticisms of French imperialism from geographers, perhaps reflecting the dominant political perspectives of the time. Certainly, we see within French geography resistance to extreme nationalistic views or the forms of racial and environmental determinism by prominent thinkers such as Vidal de la Blanche (see Heffernan, 2005). But the absence of a distinct anti-imperial project reflects the persuasive powers of the utopian, cultural, economic and opportunist arguments set out above.

Environmental determinism: climate and race

> The idea that climate stamped its indelible mark on racial constitution, not just physiologically, but psychologically and morally, was a motif that was both deep and lasting in English-speaking geography. (Livingstone, 1992: 224)

In this section we explore geography's relationship with empire in more detail by examining some of the key theories and models that geographers were using over this period. In particular, we will focus on the incorporation of ideas from the natural sciences, such as biology and climatology, into geography. This work, grouped under the heading of 'environmental determinism', has been the focus of intense criticism in the intervening years, in particular since it has been accused of producing racist categorisations and legitimising imperial conquest. Rather than simply documenting the location of spatial arrangements, geographers sought to construct over-arching explanatory frameworks of the patterns of occupation on the Earth's surface (see Johnston and Sidaway, 2004: 46). One of the earliest proponents of the incorporation of the natural sciences into geographical theory can be found in the work of the German political geographer Friedrich Ratzel (1844–1904). Ratzel's training in the natural sciences had proved influential in his incorporation of Darwin's theory of evolution (published in *On the Origin of Species* in 1859) into understandings of the formation of political communities. In particular, Ratzel oriented his thinking to political geography, suggesting that the state can be understood as a living organism, striving with and against others to grow and develop. In doing so, Ratzel described the state's imperative for *Lebensraum* (living space), arguing that stronger states should expand territorially into areas that were not exploited efficiently by their current residents (see Agnew, 2002: 58–60).

We must be careful in our reading of Ratzel's work and in particular his use of Darwinian thinking. As the geographer David Stoddart (1966) has pointed out, Ratzel himself was only selectively drawing on Darwin's ideas, in particular his notion of evolution and then, somewhat crudely, applying this to the practice of states. Ratzel drew on Darwin's ideas and terminology since they granted his theory scientific legitimacy. But being granted scientific legitimacy does not mean that the theory itself was scientific, in the sense of being objective. In particular, Stoddart (1966: 696) examines how ideas of random variation, central to Darwin's philosophy, were omitted in geographical uses of his ideas:

> Darwin's theory made a clear distinction between the way in which evolution was effected, and the course of evolution itself: geography seized on the latter and ignored the former.

The partial adoption of Darwin's ideas leads to deterministic understandings of evolutionary change: it is a process that is presented as preset and unchangeable, rather than an outcome of apparently random variations. Criticising this form of evolutionary thinking should not be confused with criticisms of either the natural sciences or of attempting to apply scientific models to geography. Rather, it identifies the limitations of incomplete adoptions of scientific ideas and the social and political implications this may cause.

Ratzel's reading of Darwin was extremely influential to the development of the theory of environmental determinism. This theory represents an early attempt by geographers to generalise the geographical processes and distributions observed in the late 19th century. The objective was to establish a single model through which human distribution and territorialisation could be explained. The earliest mentions of environmental determinism can be found in the work of Baron de Montesquieu in 1748, where he suggests that the social habits of European, Asian and 'Barbarian' cultures were connected to global climate patterns; human nature was thus a construct that could be related to the environment by scientific laws (Coombes and Barber, 2005: 303). The conception that human social, cultural and moral characteristics were determined by their geographical environment became extremely popular in both Western Europe and North America. In the US, Ellen Churchill Semple constituted a key figure in this field, in particular in her major work *The Influences of the Geographic Environment* (1911) (see Box 1.3). As we see from Semple's work, the climate and topography of a given environment was deemed to affect the entire population in uniform ways, leading geographers to feel confident in pronouncing on the racial characteristics of given populations. The domination colonial populations were thus explained by the racial inferiority in comparison to European or American populations.

Box 1.3

Ellen Churchill Semple (1863–1936)

Ellen Churchill Semple was a leading figure in geography in the early 20th century, lecturing at the University of Chicago before being granted a chair in Anthro-Geography at Clark University in 1921. She was also the first woman to chair the Association of American Geographers. Like Ratzel, Semple sought to strengthen understandings of the development and distributions of human societies through the application of theories of evolutionary biology. Reflecting the views of her contemporaries, Semple felt that previous analysis of the relationship between the environment and human development had lacked scientific rigour and empirical testing. As a corrective, Semple produced a series of lengthy publications to validate emerging theories, the most influential of which was *The Influences of the Geographic Environment* (1911). Semple opens this classic text with a statement of the role of the environment in determining human behaviour and development:

> Man is a product of the earth's surface. This means not merely that he is a child of the earth, dust of her dust; but that the earth has mothered him, fed him, set him tasks, directed his thoughts, confronted him with difficulties that have strengthened his body and sharpened his wits, given him his problems of navigation or irrigation, and at the same time whispered hints for their solution. She has entered into his bone and tissue, into his mind and soul.
>
> (Semple, 1911: 1)

There is a lyricism to Semple's account of the role of environmental context on the social, cultural and, in the following excerpt, religious development of human societies:

> Up on the wind-swept plateaus, in the boundless stretch of the grasslands and the waterless tracts of the desert, where he roams with his flocks from pasture to pasture and oasis to oasis, where life knows much hardship but escapes the grind of drudgery, where the watching of grazing herd gives him leisure for contemplation, and the wide raging life a big horizon, his ideas take on a certain gigantic simplicity; religion becomes monotheism, God becomes one,

Box 1.3 continued

> unrivalled like the sand of the desert and the grass of the steppe, stretching on and on without break or change. (Semple, 1911: 1)

In adopting an explicitly scientific approach, Semple was keen to dispel what she felt was the irrationality and simplicity in previous attempts to connect climate and **race**. Despite drawing on the work of earlier scholars such as Montesquieu, Buckle and von Treitschke, she distanced herself from a simple linear relationship between climate and the moral, social or political characteristics of a given population. She argued that 'Environment influences the higher, mental life of a people chiefly through their economic and social life; hence its ultimate effects should be traced through the latter back to the underlying cause' (Semple, 1911: 22). In doing so, Semple produced a model of environmental determinism that sought to trace the social and cultural effects of a range of environmental factors (such as topography, fluvial systems and climate). For example, she had a particular concern for the role of mountain ranges on the psychological and cultural development of human groups. She felt that '[m]ountain regions discourage the budding of genius because they are areas of isolation, confinement, remote from the great currents of men and ideas that move along river valleys' (Semple, 1911: 20). In the following quote she applies this thesis to the Indian sub-continent:

> The scientific geographer, grown suspicious of the omnipotence of climate and cautious of predicating immediate psychological effects which are easy to assert but difficult to prove, approaches the problem more indirectly and reaches a different conclusion. He finds that geographic conditions have condemned India to isolation. On the land side, a great sweep of high mountains has restricted intercourse with the interior; on the sea side, the deltaic swamps of the Indus and Ganges Rivers . . . have combined to reduce its accessibility from the ocean. (Semple, 1911: 18–19)

Although such observations regarding India's topology may seem reasonable, when mapped onto the development of the India population, Semple turns to the established racist tendencies of environmental determinism. She resorts to generalisations as to the character and capacity of the Indian population, suggesting the effect of such isolation is 'ignorance, superstition and the early crystallization of thought and custom' (Semple, 1911: 19). This example neatly illustrates the broader paradox at the heart of much of the scholarship under the banner of environmental determinism: while it deployed scientific approaches to cartography, surveying and demographics, its analysis often resorted to intangible and prejudiced claims that labelled entire populations with negative assumptions regarding morality, spirituality or culture.

Further reading

Frenkel, S. (1992) Geography, Empire and Environmental Determinism. *Geographical Review*, 82(2), 143–153.

Semple, E.C. (1911) *The Influences of the Geographic Environment*, London: Constable.

Environmental determinism and the Panama Canal

Although the racial stereotypes of environmental determinism bolstered the imperial policies of powerful states, taken to its limits this thesis also served to undermine certain colonial practices. This potential paradox can be seen in the account of the construction of the Panama Canal and surrounding colonial outposts in the work of Stephen Frenkel (1992). Frenkel explores how American policymakers drew on the apparent scientific basis of environmental determinism to explain and justify their imperialism in Panama in the late 19th century. Frenkel remarks that 'Panama was popularly perceived as environmentally dangerous, a place of snakes, malarial mosquitoes, and rank, dank vegetation running riot' (1992: 146). By representing Panama as a barbarous place, American policymakers established what they termed a 'mandate from civilisation' to enter Panama, appropriate resources and oversee the construction of the Panama Canal. In doing so, the construction process relied on crude racial categorisations to justify the poorer wages and conditions suffered by West Indian labourers in comparison with white American overseers. This hierarchy depended on an image of West Indians as 'slow in thought and action' (Huntington, 1915) and thus capable only of manual labour. The white population, coming from a temperate climate, presented themselves as 'bold and strong' (Frenkel, 1992: 149). Frenkel continues:

> Environmental determinism rationalized a theoretical position by which Americans considered it natural that different races should be treated differently and that races could be ranked according to environmentally based biological differences. (Frenkel, 1992: 149)

The categorisations and hierarchies of race, evoked through the Panama Canal example, highlight the colonial consequences of the framework of environmental determinism. At the level of daily exertion, Frenkel observes how white Americans could present themselves as endowed with strength and courage yet simultaneously declare themselves unsuited to the exertion of manual labour. Proponents of environmental determinism explained this apparent inconsistency by stating that those with white skin were out of their accustomed climatic context and therefore could only conduct office-based skilled or supervisory roles. Mirroring the examples in the work of Ellen Churchill Semple, this explanation extended beyond simple physical discomfort or illness to extend into the moral sphere of the lives of white Americans sent to work in the Panama Canal area. The fear among American policymakers was that the environment in Panama would not only exhaust their white workers but could also corrupt their moral character. Such rhetoric led to the organisation of Christian societies and the encouragement of constructive pastimes through the construction of racially segregated sports facilities. But, perhaps more significantly, the potential physical, mental and moral risk posed to the white population was used to justify pay discrepancies between different racial groups on account of the supposed 'enervating and debilitating influence of an extended stay in the tropics' (Frenkel, 1992: 151).

The perceived danger of those with white skin working outside their accustomed climatic context led to a second, broader, consequence of environmental determinism. While the theory seemed to offer a mandate to subject colonial territories to their control and domination, the dangers of white colonists living in these environments seemed to prevent them populating such areas. The issue of acclimatisation, the possibility of white races adapting to different climatic contexts, thus became a crucial area of geographical, scientific and political debate. As David Livingstone shows (1992: 232–241), this debate left uncontested the central thesis connecting moral aptitude and climatic conditions, with opinion instead divided between whether white colonisers had the ability to acclimatise to tropical environments. Livingstone cites Emory Ross (1919) who is seemingly indicative of an anti-acclimatisation position when he pronounced on the effect of Liberia's climate on British colonists:

> The odors, the mists, the sights, the sounds get on the nerves; the heavy, drooping, silent, impenetrable green forest everywhere shuts one in like a smothering grave; the mind grows sick, the body follows. For these reasons, largely mental, no one should stay on the West coast longer than eighteen months. (Ross, 1919: 402, cited in Livingstone, 1992: 236)

Countering this **representation** of the tropics, a less-influential group of scholars sought to temper what was termed the 'superstition' surrounding the acclimatisation debate. David Livingstone draws attention to the work of Italian scholar Luigi Westenra Sambon who used an address to the RGS to dismiss the supposed scientific basis for arguments against acclimatisation. For Sambon, arguments in favour of acclimatisation lacked scientific rigour and had instead risen to prominence through their legitimisation of colonial labour relations:

> The truth about the labour problem is that white men will not work; they go with a fixed resolve to gain wealth by coloured labour, which only too often is another word for slave labour.
> (Sambon, 1898: 594, cited in Livingstone, 1992: 239)

This critique was not well received at the RGS. Perhaps this is not surprising given the history and affiliations of the society (see earlier section). But Sambon's assertion draws attention to more than simply the fact that geographical ideas were used to support practices of colonial occupation and dispossession. His work also signals what became a growing critique of environmental determinism over the early 20th century: that it attempted to juxtapose a scientific attachment to rationality, rigour and neutrality with moral judgements of the populations of colonial territories. Where geographers had attempted to position the discipline as a scientific pursuit that made a distinct break from its 'fabulous' past (to use Joseph Conrad's phrase) in making claims to environmental determinism they relied upon unverifiable supposition and conjecture.

A number of brief points should be made in summary regarding the adoption of environmental determinism within the geography. First, this example highlights the importance of historical context in understanding the main theoretical elements of the geographical discipline. We cannot study geography divorced from the social, political and economic context within which it is situated. Second, although the broad tenets of environmental determinism have been widely criticised, the racial categories and stereotypes on which they were based continue to exert an influence both within geography and in wider policymaking circuits. Such categorisations and stereotypes are easily deployed to explain xenophobic or racist understandings of the world. Third, we should not view environmental determinism as the only theoretical or political movement within the emerging geographical discipline. In the following section we draw out a number of dissenting scholars and commentators, individuals who imagined a different future for the geographical discipline than adding scientific legitimacy to colonial practices.

Criticisms and dissent

In tracing the links between imperialism and geography we must be careful that we do not silence or ignore voices of dissent and criticism. It is relatively easy to construct a historical narrative that neatly allies geography with imperial power, and highlight the very visible manifestations of this link (such as the RGS or the theory of environmental determinism). While this is an important central theme, it underplays the voices of opposition and dissent to colonialism that emerged in the late 18th century, both within the academy and outside the geographical discipline.

But articulating these two positions of critique is not a straightforward task. On the one hand we need to identify key figures within geography who offered alternative narratives of geographical change and organisation to the dominant theories that, as we have seen, legitimised colonial conquest. On the other hand, we need to look to a broader range of writers and commentators who were marginalised at the time of the emergence of the geographical discipline but whose ideas have since been 'recovered' through scholarly inquiry. Such a task highlights the distinction between geography as a recognised discipline and geography as a set of ideas about space and place that are negotiated and debated in people's everyday lives. In the following discussion we see that while individuals within the geographical discipline may have made limited attempts to critique colonial practices (with notable exceptions), individuals outside the discipline, speaking from what are often termed 'marginal positions', offered robust alternative visions of how the world could be.

As we have seen, the emergence of geography as an academic discipline was intimately connected to imperial concerns, economies and practices. To this extent, finding dissenting voices is difficult, since the emerging institutions carefully screened who could be heard and seen within its debates (for example through the exclusion of women's membership to the RGS). As we have discussed earlier, in the French context, geographer Michael Heffernan suggests that 'anti-imperial' geography was 'virtually non-existent', in the sense that there may have been privately held concerns over specific imperial practices but these did not translate to sustained arguments against French colonial expansion. Instead, Heffernan notes that 'the most consistent and vociferous opponents to French colonialism were economists associated with the conservative Right' (1994b: 112). Such opposition was structured around an argument that colonialism was a waste of French resources, valuable commodities that should be redirected to other threats facing the French nation (at that time specifically its neighbour, Germany). Therefore the morality of the violent and oppressive nature of colonial practices went uncritiqued; rather, they were seen as economically disadvantageous to the French nation.

It is almost as difficult to find dissenting voices in other geographical contexts. Even in the US, which, despite its activities in Panama, did not have an empire on the scale of those established by European powers, critiques of colonial practice did not come from geographers. If we look at the work of one of the most prominent US geographers at the time, Isaiah Bowman, his commentary on British colonialism written in 1928 only serves to highlight the problems of governing disparate and populous colonial possessions and confronting the challenges to economic trade. He writes, 'Colonial diversity is owing in part to the fact that Great Britain's colonial empire is wholly tropical and subtropical, and is made up of populations mixed and in the main non-European and largely primitive' (1928: 74). Thus, in his analysis he only serves to reproduce the notion of essential differences between 'civilised' colonisers and the 'primitive' colonised. Despite Bowman's repeated claims to the contrary, such characterisations have no basis in scientific research, but are rather hierarchies that serve to legitimise conquest on the basis of biological superiority.

In the context of the UK the key figures in the discipline posed little direct challenge to the strength of the British Empire. One individual who, although not born in the UK, went on to present a series of influential works critiquing the role of geography within imperial power was Russian émigré Petr Kropotkin (see Box 1.4).

Box 1.4
Petr Kropotkin (1842–1921)

Petr Kropotkin was born in Moscow in 1842 to an aristocratic family with close ties to the Russian military. In 1861, Kropotkin chose to serve a military tour as an officer in Siberia, a move motivated less by a commitment to armed service and more by curiosity about the physical geography of the region and the conditions of its large prisoner population. Kropotkin was reportedly disheartened with what he witnessed in Siberia, in particular the failure of the central Russian government to improve the conditions of the peasantry. This experience motivated a lifetime of interest in geographical issues, taking him far beyond the borders of Russia. He travelled across Asia and Europe and his observations extended beyond mere descriptions of the physical landscape of his destinations to include insights regarding the social and economic conditions of their inhabitants. Seeing the plight of peasant communities across Europe, Kropotkin developed anarchist political sensibilities whereby he questioned the imposition of state and imperial power over oppressed peasantry and colonial subjects. Such sentiments, perceived as radical and unwanted within certain circles of power, led to his arrest and imprisonment in Russia and France over the latter years of the 19th century. Finally, Kropotkin moved to London where he was a regular visitor at the RGS in London. Kropotkin was subsequently asked by Sir James Scott Keltie to contribute to a symposium on the future of geographical education. It was through these arenas that Kropotkin managed to articulate dissent to processes of imperialism.

Further reading

Breitbart, M. (1981) Petr Kroptkin, the Anarchist Geographer, in D.R. Stoddart, (ed.), *Geography, Ideology and Social Concern*, Oxford: Blackwell, pp. 134–153.

Kearns, G. (2004) The Political Pivot of Geography. *The Geographical Journal*, 170(4): 337–346.

Like many others, Kropotkin drew on Darwinian ideas in constructing his theorisation of historical and geographical development. But, unlike scholars such as Halford Mackinder and Ellen Churchill Semple, Kropotkin structured his arguments around the importance of *cooperation* rather than *competition*. In *Mutual Aid* (1914), his most influential publication, he argued that the representation of competitive struggles in nature was more a caricature than a representation of real life. As Breitbart (1981: 137) suggests, 'Kropotkin's historical research indicated that struggles for existence were carried on not by individuals, but by groups of individuals cooperating with one another'. Whereas geographers such as Freidrich Ratzel had emphasised the image of the state as a self-interested individual struggling with others, Kropotkin's work drew attention to the importance of mutuality, cooperation and community. But in doing so Kropotkin did not reject the structuring role of the environment in determining the social and political characteristics of a given society, as this excerpt from his 1885 paper 'What Geography Ought to Be' attests:

> [Geography] must show that the development of each nationality was the consequence of several great natural laws, imposed by the physical and ethnical characters of the region it inhabited; that the efforts made by other nationalities to check its natural development have been mere mistakes; that political frontiers are relics of a barbarous past; and that the intercourse between different countries, their relations and mutual influence, are submitted to laws as little dependent on the will of separate men as the laws of the motion of planets.
>
> (Kropotkin, 1885: 942)

Therefore, although Kropotkin criticised the politics of conventional environmental determinism (i.e. of competition, struggle, invasion and domination) he did not question the role of the environment on living phenomena. Despite these political differences, Kropotkin was a regular speaker at the RGS and wrote 'What Geography Ought to Be' at the invitation of the society's then Inspector of Geographical Education, Sir James Scott Keltie (see Kearns, 2004). It was in this paper that he outlined in most precise terms the moral that:

> [There is another task for geography:] that of dissipating the prejudices in which we are reared with regard to the so-called 'lower races' – and this precisely at an epoch when everything makes us foresee that we soon shall be brought into a much closer contact with them than ever. When a French statesman proclaimed recently that the mission of the Europeans is to civilise the lower races by the means he had resorted to for civilising some of them- that is, by bayonets and Bac-leh massacres – he merely raised to the height of a theory the shameful deeds which Europeans are doing every day. (Kropotkin, 1885: 943)

Thus Kropotkin's work stands as an important beacon of anti-imperialist scholarship at a time when others were looking to enrol Darwinian metaphors in order to justify violent territorial gains. Although Kropotkin may not have been celebrated in the same way as Sir Halford Mackinder (he certainly wasn't knighted) he stimulated a groundswell of popular support for his egalitarian geographical imagination. This may be evidenced by the fact that at his funeral in Russia in 1921, it is reported that over 100,000 mourners attended to pay their respects.

Conclusions

This chapter has examined the relationship between imperialism and the emerging geographical discipline in the late 19th and early 20th centuries. The historical focus of this chapter should not be taken to suggest that geography has successfully exorcised its colonial past from its intellectual present. The themes and methodologies emerging in the late 19th and early 20th centuries endure in the discipline, despite sustained effort to distance geographical inquiry from oppressive and unjust practices of imperial expansion. In many places through the rest of the book, although in particular in Chapter 11 on post-colonial geographies, we explore how scholars have sought to challenge the theoretical and practical legacies of imperialism. By way of conclusion we would draw out three points that are central to our discussion of geography and militarism.

First, we need to be attentive to the origins of key ideas within human geography. The theories that you cover in your geographical studies did not emerge from nowhere, they were developed in particular intellectual and political contexts. In the case of geography's institutionalisation as a discipline, this was a context where one of the dominant preoccupations of European powers was the establishment and retention of global empires. The ideas that geographers developed, in particular environmental determinism, reflected these imperial desires.

Second, colonial geographies required the production of racial and spatial categorisations. In order to justify processes of imperial subjugation and exploitation, geographical ideas were often founded on the absolute difference between 'white' and 'indigenous' peoples, as we saw in the case of the Panama Canal example. This reduces difference to 'essences'; that is, to fundamental attributes that individuals hold that may not be altered. Of course, there were scholars who sought to critique this approach and who instead stressed notions of common humanity, such as Petr Kropotkin. In Chapter 11 we consider how geographers have sought to challenge and deconstruct these processes of differentiation and the racist assumptions upon which they are founded.

Third, while the events we have discussed may seem in the distant past, their legacy can be keenly felt in the present day. Indeed, the practice of geography as a subject continues to reflect these founding experiences. The importance granted to fieldwork and field experiences is a consequence of the importance placed on travel and exploration in the early years of geography's institutionalisation. This continues to set human geography apart from other disciplines, such as political science, that do not value field experience. Furthermore, the close links between geography and development studies can be traced back to these early preoccupations with global concerns. In many ways the following chapters reflect the reverberations of imperial pasts as they are reflected in geography's present and future.

Summary

- The institutionalisation of geography shaped, and was shaped by, the process of European colonialism.
- We need to think of the theoretical tools that allowed for such geographical imaginings: in particular the racist connections between physical and racial characteristics encapsulated in environmental determinism.
- In doing so we must be careful not to simplify complex scientific theories: it was a simplistic reading of Darwin's ideas that led Ratzel to advocate forms of environmental determinism.
- The examples of the exclusion of women from RGS fellowships and the construction of the Panama Canal illustrate the importance of understanding these processes in context.
- As we will see in Chapter 11, geographers have recently been at the forefront of critiquing the discipline's imperial past and considering its effects on how we think and act as geographers.

Further reading

Bell, M. and McEwan, C. (1996) The Admission of Women Fellows to the Royal Geographical Society, 1892–1914: The controversy and the outcome. *The Geographical Journal*, 162(3): 295–312.

Godlewska, A. and Smith, N. (eds) (1994) *Geography and Empire*, Oxford: Blackwell.

Kearns, G. (2009) *Geopolitics and Empire: The legacy of Halford Mackinder*, London and New York: Oxford University Press.

Livingstone, D. (1992) *The Geographical Tradition*, Oxford: Blackwell.

Maddrell, A. (2009) *Complex Locations: Women's geographical work in the UK 1850–1970*, Oxford: Wiley-Blackwell.

Semple, E.C. (1911) *The Geographic Influences of the Environment*, London: Constable.

Sidaway, J.D. and Power, M. (2005) 'The Tears of Portugal': Empire, identity, 'race', and destiny in Portuguese geopolitical narratives. *Environment and Planning D: Society and Space*, 23: 527–554.

Withers, C.W.J. and Mayhew, R.A. (2002) Re-thinking Disciplinary History: Geography in British universities, c. 1580–1887. *Transactions of the Institute of British Geographers*, 21(1): 11–29.

2 The quantitative revolution

Learning objectives

In this chapter we will:

- Examine the intellectual and political origins of the 'quantitative revolution', examining the role of World War II and an emerging desire for a more scientific approach to geography.
- Outline the key aspects of the quantitative revolution in the 1950s and 1960s, identifying the important individuals and publications.
- Explore the criticisms of quantification that have emerged both at the time and in the intervening years.
- Identify the legacy that has been left by quantitative approaches on the discipline of geography.

Introduction

In 2004 the artist, designer and educator Christian Nold invented what he termed a 'bio-mapping device' (see photo). This comprised a global positioning system (on the right) and a biometric sensor (on the left) capable of measuring changes in levels of sweat in the wearer, registered through finger monitoring straps. The purpose of this device was to monitor how changes in emotion are shaped by the experience of the wearer moving through different landscapes. After each volunteer had worn the bio-mapping device, he or she would examine the read-out of the biometric sensor and talk

A 'bio-mapping device' created by Christian Nold
Source: Christian Nold

through the emotional journey. The perception of the wearer differed from that of the technical read-out, as Nold (2004: 5) explains:

> While I would see just a fairly random spiky trail, they saw an intimate document of their journey, and recounted events which encompassed the full breadth of life: precarious traffic crossings, encounters with friends, meeting people they fancied, or the nervousness of walking past the house of an ex-partner. Sometimes people who walked along the same path would have spikes at different points, with one commenting on the smells of rotting ships, while another being distracted by the CCTV cameras.

The point of this bio-mapping is not, then, to quantify emotion but rather to use GPS and bio-sensory technology to provide a starting point for attempting to visualise human emotional responses to space and place. The researchers have scaled up this technique and coupled it with the mapping tools of Google Earth to visualise collective emotional responses to a wide range of cityscapes, including London, Paris and San Francisco. While this project is participatory in its desire to include research participants in the design, execution and analysis of research outcomes, it is underpinned by advanced location technology that places all of the data in spatial sequences. This project illustrates the fundamental importance of quantification and spatial analysis to understanding human geographies. As we will see in this chapter, the initial incorporation of quantitative approaches often sought to produce laws concerning the spatial relations of humans and environments. While elements of this perspective still exist, recent adoptions of quantitative approaches are more concerned with examining localised processes and using quantitative techniques to illustrate trends and patterns. The quantitative geographer A. Stewart Fotheringham (2006: 240) has recently commented on the significance of quantitative approaches to understanding human agency: 'quantitative geographers increasingly recognise that spatial patterns resulting from human decisions need to account for aspects of human decision-making processes'. In this chapter we will examine how practices of quantification emerged within geographical study, and how their application and utility have transformed over the past 50 years.

The spread of quantitative techniques within geography after World War II changed the direction of the discipline. For many years geographers had embraced the 19th-century European geography of regionalism, attempting (not always successfully) to leave behind the environmental determinism that characterised much geographical scholarship in the late 19th and early 20th century. This shift in focus had multiple origins, from a desire to provide more 'useful' research to wartime governments, to a need to raise the prestige of geography, which had, in some quarters, been presented as an 'easy' discipline of listing attributes of particular places. The spread of quantification stimulated some ground-breaking work in geography, incorporating complex mathematical techniques to analyse spatial distributions. Key figures emerged working in centres such as the University of Washington (Seattle, USA) and Cambridge University (UK) that shaped this new set of tools for geographers. The work stimulated fierce debates between the 'old' regionalism and the 'new' quantifiers, between those who sought to look at specific geographical localities and those who sought to produce mathematical laws.

The decline in prominence of quantitative methodologies in the years since the early 1970s should not be taken to mean they were surpassed or should be relegated to the discipline's past. Instead we need to understand how quantitative techniques reshaped the methods and philosophies of geographical inquiry, and how these remain prevalent and important to the present and future study of geography. Following the broad approach of this book, we will investigate in this chapter how and why quantitative techniques were adopted by human geographers over the 1950s and 1960s. This shift in emphasis did not emerge simultaneously across the world; it diffused from specific universities and was guided by the interests of a group of influential scholars. In doing so, we are keen to illustrate how this shift in geographical thought was a consequence of wider political, social and economic contexts within which the study of geography is embedded.

In order to explore these processes this chapter is divided into four sections. In the first we examine the origins of the quantitative revolution, drawing attention in particular to the specific intellectual and political contexts from which quantitative techniques emerged. The second section explores some of the themes and thinkers of the quantitative revolution. The third section looks at key criticisms of the quantitative revolution, rooted as they are in the perceived attachment of quantitative study to the philosophy of **positivism**. That is not to say that quantifiers were also philosophers, but rather that the practice of quantifying geographical phenomena often implicitly relied upon a positivist philosophy. The final section examines the legacy of quantification, examining the continued reliance on aspects of positivist thinking, the increasing prominence of quantitative techniques within geographical information systems (GIS) and the opportunities for collaboration across human and physical geography offered by quantitative analysis.

The origins of the quantitative revolution

Chapter 1 examined the nature of geographical thought during European colonialism at the end of the 19th and beginning of the 20th centuries. As we saw, leading figures in the emerging discipline of geography, such as Ellen Churchill Semple, Halford Mackinder and Friedrich Ratzel, based their ideas (to differing extents) on theories of environmental determinism, where the environment is considered to determine cultural, political and social traits. Perhaps the most important aspect of these works was its illumination of the relationship between geographical ideas and the prevailing political contexts. The concept of environmental determinism served to legitimise European colonial ambition, as it rendered racist categorisations in natural terms. That is, indigenous populations were presented within this work as naturally inferior to white settlers on account of their specific climate and topography. As we saw, this was an attempt to make a universal law that supported a hierarchy of 'races' on the basis of environmental factors. The service provided by geographical knowledge to the furthering of colonial ambition was the subject of fierce criticism, both at the time and in more recent years. Environmental determinism was largely rejected by European and North American geographers over the early decades of the 20th century, and replaced by a study of the nature and properties of regions of the world. Rather than theorising on a grand scale, geographers sought to produce inventories of regions, harnessing the technological advances of travel to what had previously been considered remote parts of the world, and map, log and classify the topography and landscape. As an example, this conceptual shift from environmental determinism to regionalism is illustrated in the career of Isaiah Bowman (see Chapter 10).

In place of the grand (or even global) theory of environmental determinism, regional geography in the early 20th century 'rested on scrupulously recorded observations of a lone scholar, and tended towards classification, even the encyclopaedic' (Barnes and Farish, 2006: 807). This, however, did not equate with a complete break from generalisation and often carried with it some of the assumptions that underpinned environmental determinism. The methodological approach of regionalism was influenced by two earlier figures from the European geography: Alexander von Humboldt (1769–1859) and Karl Ritter (1779–1859). These scholars sought to advance knowledge about the world through explanation of the environmental characteristics of specific regions, gathered together into encyclopaedic volumes of observational material coupled with theoretical reflection. While there are variations between von Humboldt's and Ritter's scholarship, their work was founded on belief in the importance of registering the differences between regions of the world and basing these on parallel observations of environmental factors (see Johnston and Sidaway, 2004: 49).

Regional geography, then, became the basis of geographical teaching within European and North American universities following the shift away from environmental determinism. As Tim Unwin (1992: 107) notes, the outbreak of World War II 'marked a turning point in the practice of geography'. He continues:

> The war that spanned the globe from Europe to eastern Asia provided a rare opportunity for geographers once again to satisfy the task that Strabo had allocated them almost two millennia previously, that of providing military intelligence.

We can then see the outbreak of World War II as another moment that highlights the relationship between geographical knowledge and the prevailing political context. Military strategy required knowledge of cartography and detailed understanding of the locations where battles were being fought. We can see a difference of experience between the UK and US contexts. In the UK there was a surge in funding for a variety of different geography-related research projects, for example examining the nature of agricultural production in different areas of the UK or exploring the character of wave and sea patterns in order to perfect amphibious landings (Unwin, 1992: 107). In the US, however, the reaction of politicians to the possibilities of geographical knowledge were quite different, perceiving the discipline to lack substantive or sufficient knowledge of different parts of the world and little understanding of global physical processes (see Ackerman, 1945; Unwin, 1992). The demands for military intelligence saw the enrolment of geographers into government service within the UK and US, and a shift in focus away from regional descriptions into a more scientific and process-based form of spatial analysis.

In the decade following the Allied victory in World War II a discernable shift took place in the study of geography within Western Europe and North America. The geographers Gauthier and Taaffe (2002) suggest that the increased travel of servicemen and women across the world stimulated greater interest in understanding spatial processes and patterns. They observe that 'a number of veterans, many who had worked in intelligence and related fields, were attracted to graduate studies in geography' (p. 511). Rather than relying on a regional perspective, new approaches that embraced mathematical modelling sought to understand geography as a practice of spatial analysis. The geographer Rob Kitchin (2006: 22) has described this as a move from 'an ideographic discipline (fact gathering) focusing on regions and places to a nomothetic (law producing) science focused on spatial arrangement'. Scholars have suggested a number of reasons for this shift in approach and emphasis, which we will explore under the headings of intellectual and political reasons.

Intellectual reasons

For many practitioners the regional geography of the early 20th century had 'degenerated into a plodding, enumerative exercise lacking both the intellectual vigour and the moral zest of its earlier champions' (Livingstone, 1992: 310). Over the course of the late 1940s and early 1950s we see repeated calls by prominent geographers to shift the intellectual basis of the discipline from a descriptive enterprise and embrace mathematical techniques. For example, in 1945 in the US, Edward Ackerman made 'a particularly strong appeal for a greater stress on topical work in geography, arguing that the material prepared by geographers for wartime intelligence was weak in content and of limited usefulness' (cited in Gauthier and Taaffe, 2002: 511). Figures such as the German émigré to the US, Kurt Schaefer, sought to challenge the regionalism of pre-war geography and focus intellectual attention on the scientific study of spatial arrangements (see Bunge, 1979 and Box 2.1).

This characterisation of geography as a moribund intellectual backwater had profound implications for the future direction of geography as a university discipline. Many departments sought to recruit staff with competence at quantification, while others reacted against what was often considered an 'ungeographical' turn towards mathematics and statistics. In many ways we can see this tension played out in the corridors of North America's

Box 2.1

F. Kurt Schaefer (1904–1953)

Schaefer was born in 1904 in Berlin. His father was a metal worker and it was the expectation that Kurt would follow in his footsteps. Consequently, in 1918 he became an apprentice metal worker. As William Bunge (1979: 128) explains: 'He was a man of his times and German metal workers [. . .] were political radicals.' This is an important contextual point, as it shaped his political commitments throughout his life. Bunge suggests that although his interests in geographical scholarship were genuine, they seemed to come from 'a frustration of his political life'. Schaefer attended the University of Berlin from 1928 to 1932, immersing himself in the study of political geography, mathematics and statistics. This combination was to be important for his future research, as well as the direction of geographical study. Following university, the rise of the Nazi party posed a serious threat to Schaefer and his political allies on the left. Bunge (1979: 129) describes his conditions:

> The Nazis came to his neighbours' houses, first on the one side and then on the other, with sirens screaming in the middle of the night. He was under constant police surveillance.

Under this threat and suspicion, Schaefer fled Germany, first to England and then later to the US. He undertook a number of jobs and roles, finally ending up at the University of Iowa as an instructor in geography. Over the late 1940s, Schaefer dedicated himself to the study of geography, and in particular sought to incorporate the spatial analysis of August Lösch and Walter Christaller into the discipline of geography. Although little of Schaefer's writings remain, his paper 'Exceptionalism in Geography: A methodological examination' in 1953 in the *Annals of the Association of American Geographers* was a watershed moment in the adoption of quantitative spatial analysis. As Tim Unwin (1992: 113) has argued, 'Schaefer's essay was an unbridled attempt to launch geography firmly into the mainstream of positivist science.' Indeed, as we shall see later in the chapter, Schaefer was influenced by his friendship with the prominent member of the Vienna School of Logical Positivists, Gustav Bergmann. At the heart of Schaefer's 1953 paper was a desire to shift the attention of geographers away from describing unique events and instead look for patterns and laws:

> Geography, thus, must pay attention to the spatial arrangement of the phenomena in an area and not so much to the phenomena themselves. Spatial relations are the ones that matter in geography, and no others. Nonspatial relations found among the phenomena in an area are the subject matter of other specialists such as the geologist, anthropologist or economist. (p. 228)

Schaefer's engagement with the spatial analysis of Christaller and Lösch was rejected by many who see this as a turn to mathematics, rather than geographical inquiry. It took the development of these ideas in the late 1950s and 1960s by other geographers, particularly William Garrison at the University of Washington, Seattle, to extend and develop Schaefer's approach. The article was to prove to be Schaefer's great work, although he died in 1953 from a heart attack, prior to its publication. William Bunge (1979) suggests that it was pressure from the changing political context in the US that influenced his health problems. In particular, Schaefer was sickened by the witch hunt for supposed communist sympathies spearheaded by the Republican Senator Joseph McCarthy. Bunge remarks that 'McCarthyism repulsed and sickened Schaefer in a way that those of us who have not seen Hitler come to our native land cannot understand' (Bunge, 1979: 132). Schaefer's untimely death meant that he did not live to see the profound influence his adoption of quantitative techniques would have on the discipline of geography over the following half century.

most prestigious university: Harvard. Until 1948 Harvard had a geography programme and appointed some high-profile quantitative geographers. As Neil Smith (2003: 439) explains, leading scholars at Harvard:

> were beginning to explore the application of systems theory and mathematics to location analysis, conceiving human geography in positivist terms. Their wartime work quickened this

process, and they were widely seen as among the most innovative young researchers of a new generation, on the cutting edge of a more sophisticated human geography.

But disciplinary rivalry (specifically between geography and geology) sought to undermine geography at Harvard, fuelled by a concern from those outside geography that university resources should not be directed to a discipline of 'dubious intellectual worth' (Smith, 2003: 239–240). In 1948 the geography programme at Harvard was closed, marking a sudden withdrawal of support for the discipline. This move was highly symbolic and in the words of French geographer Jean Gottmann: 'it was a terrible blow for American Geography [from which] it has never completely recovered' (cited in Smith, 1987: 155). In the wake of the Harvard move, David Livingstone (1992) suggests that geographers were seeking a means of bolstering the prestige of a subject that risked being marginalised within academic circles. As Ian Burton (1963: 157) remarked at the time, to be accepted and accorded an honoured place in society, geography needed 'to acquire demonstrable value as a predictive science without a corresponding need to control, restrict, or regiment the individual'. The adoption of scientific techniques provided such an opportunity.

Political reasons

But the motivations for shifting the focus of geographical inquiry onto quantitative spatial analysis are perhaps more diverse than simply an attempt to bolster disciplinary prestige. The use of geographical ideas, metaphors and thinkers within the Nazi war effort (discussed in Chapter 10) seemed to imprint an indelible connection between geography and militarism. Where the absence of geographical data and analysis was criticised within the war effort in North America, in Germany it was the excess of geographical theory that has received critical attention. The most commonly cited connection between the violence of World War II and geography was the adoption by the Nazi party of Friedrich Ratzel's (1844–1904) concept of *Lebensraum* (living space) to legitimise the appropriation of the territory of neighbouring states. In his text *Politische Geographie* (1897), Ratzel applied Darwinian concepts of evolution and succession to the behaviour of states. In doing so this theory suggested that stronger states should expand into neighbouring states if the territory was not being exploited effectively by its current residents. Of course, this style, linking the natural and social sciences, had been discredited through environmental determinism, but it remained a potent political force through its association with Nazism.

After the war, geographers sought to deliberately break with Ratzel's thinking, criticising its support for aggressive territorial expansion. But this was not the only political context that shaped the nature of geographical study. The newly emerging Cold War rivalries between the capitalist US and the communist Soviet Union presented a new political dynamic that shaped the nature of geographical inquiry, particularly in the US. David Harvey (1984: 5) draws attention to the McCarthy era of the 1950s, where the fear of political suspicion of communist sympathies 'led many progressive geographers thereafter to express their social concerns behind the supposed neutrality of "the positivist shield"'. Of course, the extent to which the adoption of positivist approaches equates to neutrality is open to debate, and this became a central point of critique among many of those who rejected quantification in geography. But, importantly these debates turned critique inwards towards the philosophical assumptions of geography, rather than using geographical ideas to critique prevailing political practice.

The quantitative revolution

> It was a time of great intellectual excitement, the sort of excitement that can only come from seeing new paths opening up, new connections being made, and real challenges being met. There was a sense of discovering and forging, of breaking out of the banal, factual boxes erected by old men, and a sense of reaching out to scholars in fields to which we had never properly been introduced, but which seemed friendly enough if you were prepared to learn.
> (Gould, 1979: 139)

In the above excerpt, the geographer Peter Gould evokes the 'revolutionary' nature of the turn to quantitative approaches in geography in the 1950s and 1960s. Reflecting on the nature of geographical scholarship under the banner of regionalism, Gould (1979: 143) suggests that 'geography had stagnated for decades, without tools, without methodological insight and development, without principles and constructs that gave coherence and order to observations, and without concepts to guide their gathering'. The crucial point to make about the quantitative revolution is that it was not an immediate adoption of new techniques but a slow and uneven process, both temporally and spatially. Some of the most important work exploring the quantitative revolution has sought to examine the adoption of quantification through the biographies of some of the leading figures and first-hand accounts from the time (see Barnes, 2001a; Haggett, 2008; Barnes and Farish, 2006). Indeed, many commentators feel the label 'revolution' is unhelpful, since it suggests a rapid overthrow of past approaches rather than a gradual shift in perspective that was certainly not universally adopted (see Burton, 1963). One of the key refrains of literature recalling the events within geography departments of the 1950s and 1960s in the US and Western Europe is the initial marginalisation of quantitative research, and the subsequent struggle to publish new scientific approaches in major geographical journals.

One of the best ways to understand the emergence of quantitative techniques is to examine the specific university settings within which they were developed. In the US two institutions were crucial: the University of Washington, Seattle, and the University of Iowa. In the UK, the University of Cambridge and the University of Bristol were similarly influential. In Sweden the University of Lund became a significant node of quantitative activity. These focal points did not emerge organically; they were constructed by passionate individuals who often brought mathematical or statistical expertise from allied disciplines. In Seattle, quantitative approaches were championed by William Garrison. Garrison was a trained meteorologist, who had previously worked for the US Air Force. The geographer Trevor Barnes (2001) has examined Garrison's role in developing quantitative geography in the late 1950s. A key moment in this history is the 1955 advanced statistical methods course that Garrison delivered, the first of its kind in a US geography department. For Barnes, what was crucial about this event was not that it simply involved enumerating, but rather it was Garrison's desire to promote 'a new theoretical sensibility'. This sensibility followed Kurt Schaefer's earlier desire (see Box 2.1), outlined in his 1953 paper, to incorporate the quantitative spatial analysis of scholars such as August Lösch and Walter Christaller into geographical study. But it was not simply the personnel at Seattle that helped to bolster a quantitative perspective. As Barnes (2001b: 552) explains, the role of new computing technology assisted in developing quantification in Seattle but also in diffusing these ideas across the US:

> In an early advertisement for the department, its head, Donald Hudson [. . .], boasted about his department's use of an IBM 604 digital computer, another national first. Also important were the large Friden desk calculators and the duplicator that allowed Berry [. . .] and others to circulate a stream of internal position papers and, in March 1958, to launch the Washington Discussion Paper series, that was sent to kindred souls around the world. Not only paper circulated and promoted the Washington message. The students themselves did so as they were hired and established their research agendas at several prestigious US universities and departments, including Chicago (Berry's institution for seventeen years), Northwestern (where Garrison and a number of his Washington 'space cadets' held positions) and the University of Michigan.

An important aspect of the diffusion of quantitative ideas was the strong links established between the University of Washington and the University of Lund. In 1959 the geographer Torsten Hägerstrand was invited to Seattle to teach a seminar concerning his work on population movement and the diffusion of innovations. This was followed up by a Lund Symposium on Urban Geography, which Richard Morrill (2005: 334) describes as 'highly influential in spreading advanced methods and theories in urban geography generally'. Morrill, a key figure himself in the rise of quantification, explains how his experience of post-doctoral work with Hagerstrand shaped his geographical imagination: 'I was convinced that the test of theory in geography would be through our ability to explain and model historical processes of development' (Morrill, 2005: 334). This desire to present geography as a predictive enterprise was epitomised by Morrill's mentor:

> Torsten had at the end of the seminar cut a circle out of a map and told the class, 'When geography has learned enough, we will be able to fill in the missing part.'

Across the late 1950s and early 1960s, quantitative approaches were becoming increasingly visible and influential within academic geography in Europe and North America. Perhaps the most explicit signal of the incorporation of quantification in the UK came in 1969, when the Quantitative Methods Study Group was established within the Institute of British Geographers. There are numerous texts that are dedicated to illuminating the detail of the styles of methodology and analytical practice that flourished during this time (see Johnston and Sidaway, 2004). In order to focus specifically on the changing nature of geographical thought, we will focus on two interlinked traits of this work: it was based on laws and models, and it was underpinned by an adherence to the tenets of positivism.

Laws and models

> Spatial relations among two or more selected classes of phenomena must be studied all over the earth's surface in order to obtain a generalization or law. Assume, for instance, that two phenomena are found to occur frequently at the same place. A hypothesis may then be formed to the effect that whenever members of the one class are found in a place, members of the other class will be found there also, under conditions specified by the hypothesis. To test any such hypothesis the geographer will need a larger number of cases and of variables than he could find in any one region. But if it is confirmed in a sufficient number of cases then

> the hypothesis becomes a law that may be utilized to 'explain' situations not yet considered. The present conditions of the field indicate a stage of development, well known from other social sciences, which finds most geographers still busy with classifications rather than looking for laws. We know that classification is the first step in any kind of systematic work. But when the other steps, which naturally follow, are not taken, and classifications become the end of scientific investigation, then the field becomes sterile. (Schaefer, 1953: 229)

In his ground-breaking paper of 1953, Kurt Schaefer makes a case for the incorporation of scientific techniques into geography. His argument, as laid out above, is that it is necessary to enable geographers to undertake socially useful research that can provide general explanations and predictions based on past observations. As Cloke *et al.* (1991: 12) have stated, the enrolment of **quantitative methods** involved the 'adoption of the view there existed spatial laws or rules, which (if only they could uncover them) would prove to be at the root of all human existence'. The shift towards generalisation and law-making necessitated greater interest in quantitative modelling. The geographers of the 1950s and 1960s were not the first to herald the significance of spatial modelling, indeed much of their work was concerned with recovering and updating modelling practices from geography's past and from allied disciplines, in particular economics and geology. One of the most prominent examples of such a model is Central Place Theory.

Although it is best known as a product of the German geographer Walter Christaller (1893–1969), Central Place Theory is a broad label for a group of different models of urban settlements. The formation of the model is based upon an economic theory of utility maximization, where locations of settlements and infrastructure are organised on account of the decision making of rational actors. Walter Christaller developed these ideas in German in the 1930s, though a translation into English was only available in the late 1950s. Johnston (2009: 76) explains:

> [Central Place Theory] was based on two concepts: the range of a good – the maximum distance that people will travel to buy it; and the threshold for a good – the minimum volume of sales necessary for a viable establishment selling that good (or a bundle of linked goods, such as groceries). In order to maximise their utilities, retailers locate establishments to be as near their customers as possible and customers visit their nearest available centre: in this way, expenditure on transport costs is minimised and spending on goods and services maximised.

This was a model, then, that sought to identify an ideal distribution of settlements of different sizes 'acting as the marketing centres of functional regions, within constraining assumptions relating to the physical environment and the goals of both entrepreneurs and customers' (Johnston and Sidaway, 2004: 77). The visual result of the interplay of decision making by these rational actors is a range of spatial models depicting settlements located within a hexagonal hinterland (see Figure 2.1).

Christaller's Central Place Theory was a key starting point for many of the thinkers within the quantitative revolution. Geographers were attracted by Christaller's move beyond description of actual places, instead seeking to construct a model that depicted the relationship between and spatial arrangement of urban centres. This allure is evident in the opening paragraph of Brian Berry and William Garrison's paper 'The Functional Bases of the Central Place Hierarchy', published in *Economic Geography* in 1958:

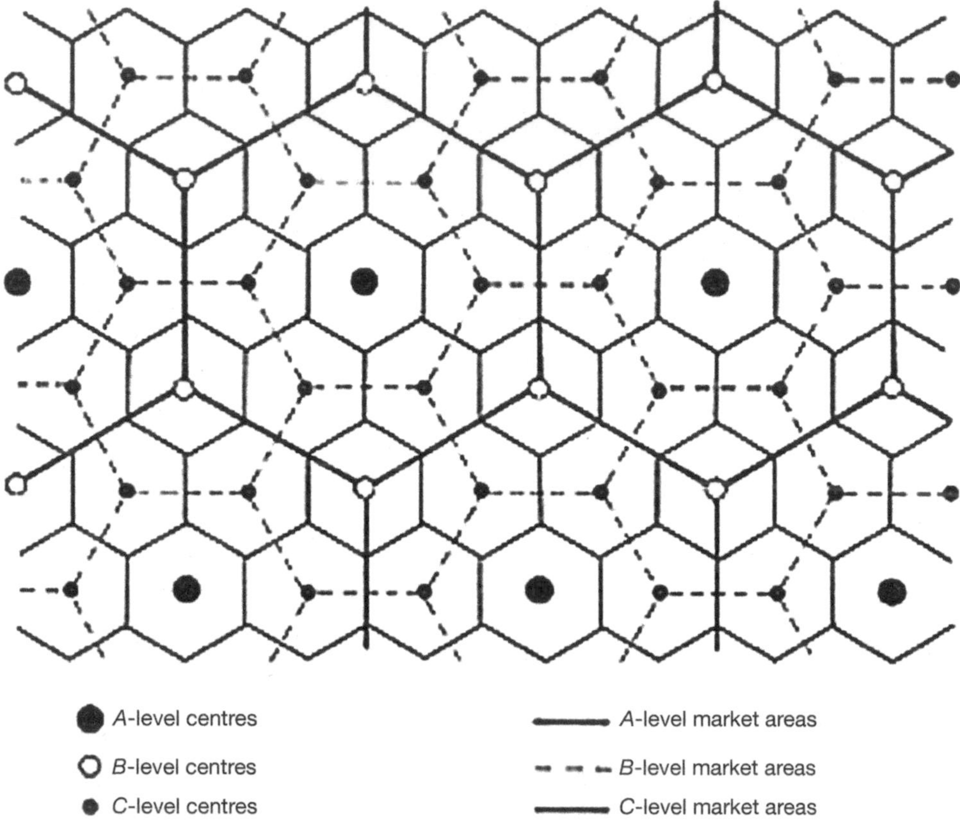

Figure 2.1 Christaller's Central Place Theory: setting out the model of interaction between a hierarchy of settlements

Source: GNU Free Documentation License: Wikipedia

> It is obvious that urban centers differ, each from others. On the intuitive level one notion of difference is that of classes of urban centers. The wealth of descriptive terms available illustrates this notion: hamlet, village, town, city, metropolis, and the like. The present study is concerned with this problem of the differentiation of centers into broad classes. In particular it provides original and urgent evidence that a system of urban center classes exists of the type identified on an intuitive level above. (Berry and Garrison, 1958: 145)

The paper goes on to provide evidence of a central place hierarchy in Snohomish County, Washington. The authors sorted the urban centres into three 'classes', based on the various services and functions that could be observed (for example the numbers of churches, petrol stations and bars). Rather than relying on intuition to grade and rank cities, Berry and Garrison used Christaller's approach to come to a scientific understanding of the relationship between urban centres of different sizes (see Figures 2.1 and 2.2). This argument illustrates a fundamental philosophical tenet of the quantitative revolution: that what had previously been guided by intuition and speculation may be assessed and analysed through observation and classification. Christaller's model provided a starting point through which phenomena could be categorised, sorted and placed in relation to one another through scientific and quantitative assessment.

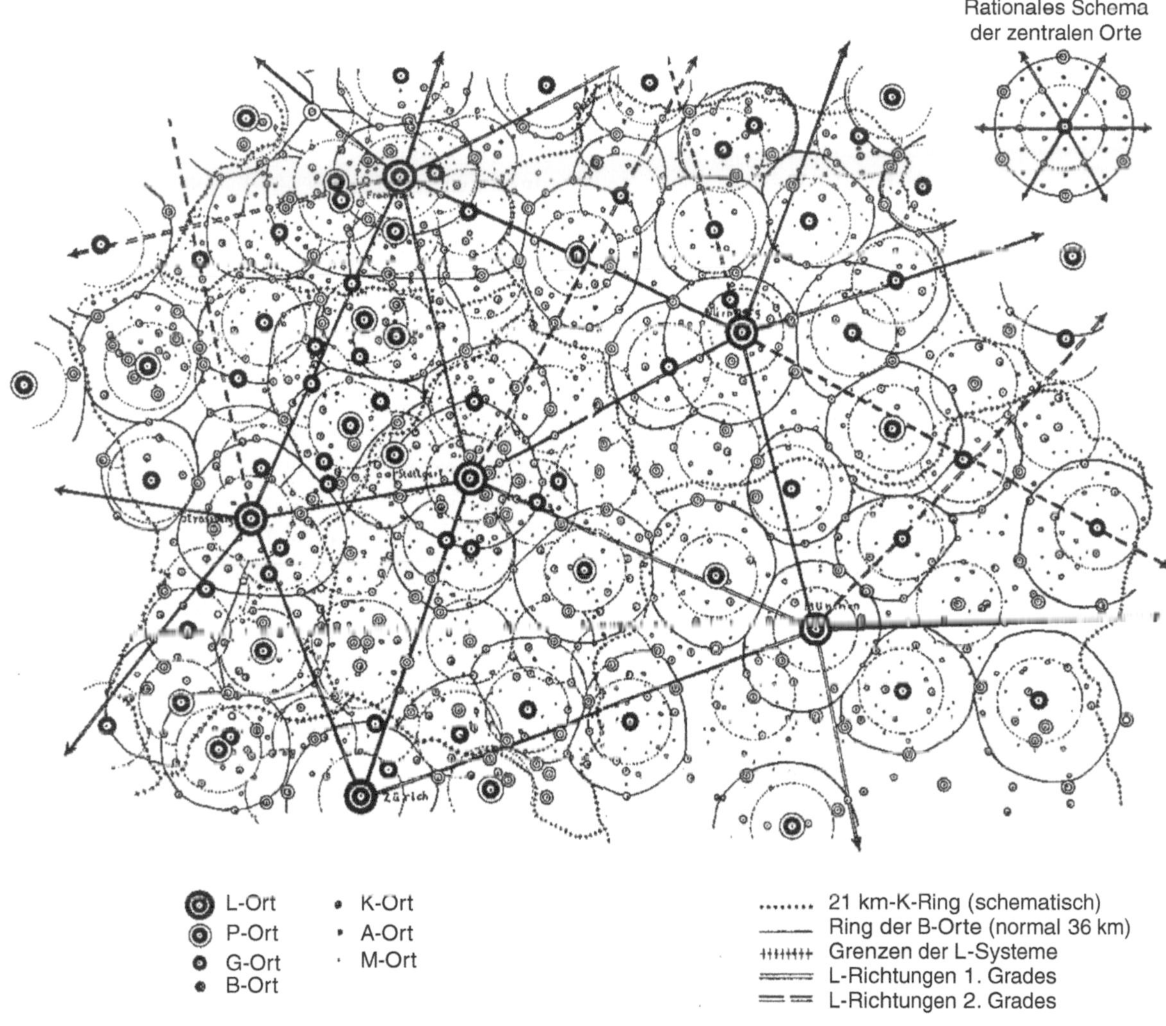

Figure 2.2 Christaller's Central Place Theory: an application of the model to Southern Germany in 1933
Source: GNU Free Documentation License: Wikipedia

Positivism

As we have seen, the adoption of quantitative techniques was informed by the philosophy of positivism. Despite the often-cited link between quantification and positivism, it is noticeable that many of the studies conducted at the time (and some of the more recent commentaries) avoid directly engaging with the philosophical tenets upon which quantitative methods rely. For example, Fotheringham (2006: 237) suggests that quantitative geographers 'do not concern themselves with philosophy' and goes on to argue that the label of 'positivism' 'has little or zero impact on the way we prosecute our work'. This is an interesting perspective that illuminates a common criticism of quantification; that it attempts to portray itself as 'non-philosophical' or 'non-ideological' while simultaneously communicating a series of philosophical and ideological assumptions. We will explore this point in greater detail under the question of criticisms of quantification.

Certainly, the label of 'positivism' was not widely used or reflected upon during the 1950s and 1960s, and it was not until the 1970s that sustained critique and exploration of quantification's philosophy began. While geographers often avoided exploring (or extending) the philosophical roots of positivism, it was certainly an important influence on early quantifiers. For example, Kurt Schaefer's work was shaped by his tuition under Gustav Bergmann, a member of the original Vienna Circle (see below) who had fled Austria in the 1930s and settled in Iowa City. The application of models such as Christaller's Central Place Theory depended on a positivist perspective, where repeated observations allow the scientist to develop knowledge about the world. But positivism extends beyond simple empiricism; it also has an intersubjective element, since it also depends upon the notion that all observations may be repeated, and it is in repeated observations that truths are uncovered (see Cloke *et al.*, 2004: 250).

While much quantitative work relied on positivist notions, we should not use the terms 'positivism' and 'quantification' interchangeably, as sometimes occurs. Positivism is an extremely broad movement that makes the claim that the only worthwhile knowledge is scientific knowledge that may be tested and verified. On the other hand, metaphysical knowledge (non-scientific reflection) is denied significance as worthless speculation. At its core the tenets of positivism simply imply 'that our ability to generate knowledge is restricted to those things that we can observe in reality' (Fotheringham, 2006: 239). The roots of this approach can be found in the work of French philosopher August Comte (1798–1851), in particular in his *Course in Positivist Philosophy* published in six volumes between 1830 and 1842 (see Barnes, 2009). This work provided the inspiration for numerous revisions and new variants of positivism, leading to a daunting array of labels for different branches on the tree of positivist thought. Perhaps the best-known example is Logical Positivism, championed by a group of thinkers called the Vienna Circle during the 1920s and 1930s. This group was concerned with 'extending the range of science over the entire gamut of systematic knowledge, and in doing so spearheaded an anti-metaphysical assault on traditional philosophy' (Livingstone, 1992: 319). David Livingstone (1992: 319) continues:

> To the logical positivists, transcendental metaphysical claims just had no meaning because there was no possible way of testing them in experience. Hence the verification principle, as it was called, became – for a short time – a methodological tool by which meaningful and meaningless sentences could be discriminated.

Grounding geographical inquiry in the language of positivism certainly helped the discipline bolster its scientific and mathematical credentials, whereby assertions may be supported by language of verification, corroboration and quantification. However, it was the adoption of positivist perspectives that became the focus of criticism of quantitative approaches.

Criticisms of quantification

Quantitative approaches required the adoption of techniques from other disciplines, in particular mathematics and economics. This interdisciplinary approach was the focus of much criticism during the 1950s and 1960s from other (often older) scholars who felt that quantification was consequently 'un-geographical'. Numerous prominent quantifiers have remarked in reviews published since that they found their work rejected by journal editorial boards and side-lined at conferences by adherents to previous geographical approaches, in particular scholars who retained a regionalist perspective. But with the publication of major texts supportive of quantification, such as David Harvey's (1969) *Explanation in Geography* and Chorley and Haggett's (1967) edited collection *Models in Geography* (see Box 2.2), we see evidence of the entrance of quantitative approaches to the mainstream of Anglo-American geography.

Box 2.2
Key text: *Models in Geography*

Edited by Dick Chorley and Peter Haggett, *Models in Geography* (1967) was a text that encapsulated the approach and spirit of the quantitative revolution. Like many major edited collections, this text has a fascinating history. It emerged from a series of summer schools organised by Chorley and Haggett at Madingley Hall, a rural retreat belonging to Cambridge University, UK. The summer schools ran for five years from 1963 and were open to around 28 geography teachers each year. Peter Haggett (2008: 343) explains the organisation of the events:

> Typically, the course began with lectures covering developments in each of the main areas of geography with an emphasis on spatial modeling. There were then practical classes covering a few main techniques (e.g. morphometric analysis, graph theory, linear programming, multivariate modeling, remote sensing applications, spectral analysis). Mid-week provided an opportunity for field work (with an emphasis on sampling designs) and visits to the Cambridge geography department. The end of the week had a specifically schools focus wrestling with practical classroom implementation problems.

This account demonstrates how the Madingley Hall events were key moments for disseminating the methods and theoretical perspectives of quantification. This is important, since the participants were schoolteachers and thus these new developments in university geography soon found their way into school curricula.

The organisers were keen to have some form of permanent record of the discussions that took place at Madingley Hall, and three edited volumes finally emerged across the late 1960s. The most substantial of these was the second volume, *Models in Geography* (which is subtitled 'The Second Madingley Lectures'). This text consists of 18 chapters in five sections by scholars from across the UK and USA. The book presented an integrated approach to geography – combined physical and human aspects of the discipline – and followed a broadly positivist approach. This wide-reaching focus allowed the book to take stock of changes that had happened in the discipline over the previous 15 years. In order to serve this function, the editors specifically sought to avoid complex mathematical equations and terminology that may be unfamiliar to a geographical audience.

The book had a significant impact on the study of geography, introducing scholars to a range of new approaches and perspectives. But its timing was significant. Published towards the end of the 1960s, it marked an era where the revolutionary nature of quantitative analysis was waning. Four years before its publication, Ian Burton had claimed 'the quantitative revolution is over'. Golledge (2006) suggests that *Models in Geography* marks both an end of an era and the beginning of an era. It gathered together work by many of the key thinkers that had pioneered quantitative approaches over the 1950s and 1960s. However, it also marks a moment when the application of positivist methods was being increasingly questioned, a new era of limits to quantification. As we will see later in the chapter, these new critical perspectives led scholars to interrogate the philosophical underpinnings of quantitative perspectives and offer alternative methods. As Golledge (2006: 112) explains, *Models in Geography*:

> was seen by many disciplinary participants to be the epitome of an attempt to change a more general and descriptive way of looking at human–environment relations to a process of scientific and analytical examination that was too artificial, and did not (because of limiting assumptions in both the theory and the analytical methods) actually represent and deal with how people thought, dealt and behaved.

But the success of quantification also brought renewed scholarly scrutiny. As we will see throughout this book, geography is a critical discipline that seeks to challenge dominant ideas and this is well illustrated in the case of quantification. At the turn of the 1970s, publications began to emerge that critiqued quantitative approaches. In particular, scholars were drawn to exposing the philosophical assumptions that undergirded quantitative approaches. As the geographer Rob Kitchin (2006: 24) has remarked, quantification over the 1950s and 1960s had 'largely operated in a philosophical vacuum: it focused on methodological form, not the deeper epistemological structure of knowledge production'. It was to these deeper questions that many scholars turned, questioning the theoretical and political assumptions upon which quantitative approaches depended. Indeed, some of the key figures from the quantitative revolution, such as William Bunge and David Harvey, were among the first to engage in such philosophical reflection and critique (see Livingstone, 1992: 329). Harvey's shift from a quantitative to a Marxist perspective is often presented as a key illustration of the wider disciplinary disaffection with quantification. Over the course of the 1970s, humanistic, feminist and Marxist approaches to geography emerged, perspectives that share a starting point of identifying the limits to quantification. While we cover these critiques in detail in the following chapters, we can draw out three central criticisms that serve as a starting point.

The assumption of neutrality

One of the key concerns of critics of quantification was the cultivation of supposed neutrality through the use of mathematical and scientific techniques. At the root of this concern is an issue of the role of the scholar. The use of mathematical language and models allowed geographers to present their work as dispassionate and neutral representations of social life. It was a language through which 'the scientific authority of their assertions could be reinforced' (Livingstone, 1992: 326). But feminist scholars began to question the production of spatial models as 'detached observation' and instead argued that they presented male-oriented understandings of economic and political practice. As Mei-Po Kwan (2002: 647) explains, conventional scientific objectivity:

> denies the partiality of the knower, erases subjectivities, and ignores the nexus of power-knowledge in its discursive practice. Feminist critics see this mode of knowledge production as masculinist.

Quantitative accounts are not 'views from nowhere', but rather they present a partial and situated form of knowledge that must be understood in its context of its production, a point we develop in Chapter 6 on feminist geographies. The work of Trevor Barnes (2001a, 2001b) has been instrumental in illuminating the social contexts within which quantitative approaches emerged. In his fascinating account of the background to Berry and Garrison's (1958) paper 'The Functional Bases of the Central Place Hierarchy', Barnes traces the biographies of the paper's authors to illuminate how the conjunction of a range of material and social networks and processes came together to enable the paper to come to fruition. His point is to challenge the notion that quantitative approaches 'inhabit some separate realm, unsullied by the lives and times of their originators and users' (2001a: 425). Instead he suggests that the use of quantitative methods was 'very much connected to their life experiences and the wider context of 1950s America' (*ibid.*). This approach brings the analysis developed in the field of the sociology of scientific knowledge (SSK) to geography and has been further developed by David Livingstone (2003). We will explore the implications of these critiques in later chapters, but they have sought to challenge the notion of quantification as a universal rationality, and rather look to embed the methods in the lives and places in which they are produced.

The absence of politics

Trevor Barnes (2010: 670) has recently commented that the 'original turn to formal theory and numbers in geography during the 1950s was driven, at least for a few of its participants, by [. . .] a critical sensibility'. At first glance this may seem a surprising statement, since within recent critiques quantification is rarely noted for its progressive social or political stance. However, this may be a reflection of the life-cycle of the quantitative revolution. In the early years the work of quantifiers sought to challenge existing frameworks of thought, to break down the existing categories and rebuild geography as a credible scientific discipline. Over the course of the 1960s we perhaps see the slow draining out of the potential for quantitative models to challenge existing forms of social inequality and political oppression. One of the central charges placed at this work is that it is over-descriptive rather than normative. That is, the energies of quantifiers were largely directed towards explaining how things are rather than how things should be. David Harvey (1972) argued that where normative considerations came into play they were largely structured around Malthusian notions of resource scarcity, rather than a more critical notion of the socially created unequal distribution of resources (in Harvey's analysis, through the capitalist economic system). This line of critique turns attention again to the mechanisms of the academy, whereby the supposed neutrality of quantification masks the uneven landscape of power through which geographical knowledge is produced.

The uniformity of human subjects

One of the distinctive features of early quantitative work was its reliance upon a uniform understanding of human subjects, suggesting that their decision making process was structured around utility maximisation. The models of early quantifiers were structured around the decision making of the so-called 'rational-economic man' (and the gender is important). The value of these models relies on an assumption of behavioural stability, whereby choices under certain environmental circumstances may be predicted from one time period and spatial location to the next. There are two choices facing scholars in the wake of such criticism. The first is an attempt to 'add in' the variation of human subjects to the existing models, to make more complex and intricate quantitative analyses. This approach has become more credible in recent years as computer modelling processes become more sophisticated and quantitative geographers have brought qualitative insights into their analysis (see the section on the legacy of the quantitative revolution below). The other option is to reject the logic of quantification, arguing that it presents an instrumental understanding of social life, where the diversity of human existence is drained out to be replaced by the organising logic of mathematics (Barnes, 1994). The reliance on mathematical formulas and equations, Barnes argues, reproduces the notion that we live in a fundamentally ordered world, where differences may be easily categorised, sorted and represented through spatial models. This more radical option has been the mainstay of post-structural critiques of quantification that have drawn on the work of philosophers such as Jacques Derrida, Alain Badiou and Jacques Lacan (see Box 2.3).

Box 2.3
Post-structural critiques of quantification

In 2001 the post-structural geographer Marcus Doel used a combination of scholarly and literary sources to mount a theoretical challenge to the concept of quantification. His particular focus of criticism in this paper is the number one. Doel opens the paper with an admission, and it is worth quoting at length:

> Let me put my cards on the table so that you know where I am coming from: I detest every one. No one in particular: just one in general. I prefer not to count on one. For me, number is a horror story. It is the most brutal of levelling devices. I look up from the screen and force myself to count: one and one and one and one and one and one and one and one and there's another one and one and one and one and another one and one and one and one and one and on and on and on and on and on and on to another one and one and one and on and on to a different one and one and one. It is an onerous task. I find it truly nauseating. Counting on one necessarily renders everything as one and one, on and on. Such is the semblance of pure positivity, as if everything were being affirmed by being counted upon. Life, the universe, and everything may be digitized. On and on, one and one. But make no mistake, by way of this operation that goes on and on interminably, the heterogeneous texture of the world is being liquidated: ambivalence reduced to equivalence. In a very little while the portion over here and the portion over there will be rendered equivalent, substitutable, and exchangeable. Each will lose its specificity, cease to exist in and of itself, and enter into general circulation. One and one, on and on. Is there anything that cannot be counted on? One world. One word. One life. One hand. One year. One tear . . . (2001: 555–556)

This challenging paper, an example of a vibrant field of post-structural critique of calculation, provides two insights. First, it illustrates a radical departure from the tenets of quantification in some parts of human geography. This work challenges the neutrality of science, seeking to expose the assumptions and power relations that lie beneath quantification. This mode of critique is unhelpful for some of the original quantifiers, who lament that (along with Marxism and post-modernism) **post-structuralism** have deflected geography 'from the cusp of scientific respectability' (Berry *et al.*, 2008: 229). The second insight that emerges from Doel's paper is the remarkable shift in the use of language from the dispassionate scientific rubric of quantification through to the more playful and rule-breaking nature of some post-structural scholarship (see Doel's repetition of 'and one', above, for example). This strategy seeks to bring to the fore the role of language in constraining thought, where the structure of grammar and the limits of vocabulary set rules. It is argued that these rules do not simply shape our ability to communicate, but actually set down habits of the mind to categorise and organise people and things in particular ways. The important aspect of language is that we reproduce these rules uncritically through the process of writing and speech. Of course, just as in the case of early quantification, this kind of critical scholarship that draws on philosophy and psychology is open to the criticism of being 'un-geographical'. Rather than being an absolute criticism, this is perhaps best understood as a reflection of the nature of geography as an area of scholarship that has always drawn on affiliated disciplines and seeks to create space to critique dominant ideas.

The legacy of the quantitative revolution

The emergence of Marxist, humanist and feminist perspectives in geography drew attention away from quantification over the course of the 1970s. The prominent human geographer Ron Johnston suggests that over the following decades two 'camps' have emerged: 'spatial analysis' (using quantitative approaches) and 'social theory' (using **qualitative methods**). He goes on to suggest that by the mid-1990s the 'two camps were no longer debating but were instead operating in separate spheres, co-existing within geography but little more' (Johnston, 2000: 129). This fragmentation saw forms of quantitative analysis develop, but often outside of the mainstream of human geography in the UK or USA. For example, quantitative work was largely published in more specialist sub-disciplinary journals such as *Geographical Analysis* rather than the main geography journals. While geographers still used quantitative methods across the 1980s and 1990s, there was rather less reflection on the underlying philosophies and neither were edited collections of spatial analysis garnering the same levels of attention, as was the case in the 1960s. Certainly (as we have seen) many of the leading figures within the quantification movement continued to pursue such methods and many turned to writing the history of the quantitative revolution. One of the key points from this work, and in particular the path-breaking work of Trevor Barnes (2001a), is that the story of the quantitative revolution is not one of a shift from the irrationality of regionalism through to the enlightenment of quantification, as some of the more conventional histories of science may suggest. Rather, quantification introduced a range of new methods and perspectives, some of which have been challenged and rejected since by large parts of the discipline (such as a commitment to logical positivism) and others have been embraced and reproduced (such as a commitment to rigour or the value of empirical testing).

The point we want to make in this final part of the chapter is that the clear division between the two 'camps' outlined by Johnston has been eroded in recent years. A range of technical, methodological and philosophical developments has seen a resurgent quantification, often coupled with qualitative techniques. This has led Trevor Barnes to announce in 2010 (p. 669) 'quantification is back'. The emergence of special issues of the journals *Environment and Planning A* and *Professional Geographer* (both in 2009) dedicated to thinking through critical quantitative geography illustrate this resurgence. In particular the development of geographical information systems (or Science, GIS) over the 1990s has bolstered the visibility and prestige of quantification. GIS is the 'mix of hardware, software and *practices* used to run spatial analysis and mapping programmes' (Schuurman, 2009: 579, emphasis in original). These powerful qualitative analytical tools have been harnessed by a range of human geographers and combined with what were previously thought to be antagonistic philosophical positions. While some scholars seem keen to retain a concept of two distinct 'camps' in human geography, others have suggested that practices of 'enumeration, statistical analysis and spatial modelling can also form part of a wider portfolio of techniques open to all human geographers' (Cloke *et al.*, 2004: 251). We think Elvin Wyly's (2009: 319) comment sums up this attempt to adopt quantification for critical purposes:

> All statistics are social constructions, but when critical scholars abandon statistics, we give up the opportunity to shape and mobilize these constructions for progressive purposes.

This point is illustrated by the scholarship of geographer Danny Dorling who uses quantitative approaches to illustrate social inequalities in the UK and beyond (see Dorling, 2010; Dorling and Pritchard, 2010). Dorling's work uses quantitative statistics and graphics to illustrate the stark spatial differences in access to education, healthcare, social welfare and housing. This critical work often challenges mainstream government representations of universal welfare coverage, and Dorling is a regular contributor to national and international media coverage of these issues. Dorling's high profile illustrates the ability of quantitative work to reach a wide audience and intervene in policy debates. For example, Figure 2.3 shows a graph drawn by Dorling and Paul Coles that represents the numbers of deaths attributable to war, massacres and atrocities in the 20th century. The 12 shaded histograms use the scale on the left to show the number of deaths in specific wars or massacres from 1900 to 1995. The stacked shaded bars to the right show the cumulative deaths attributable

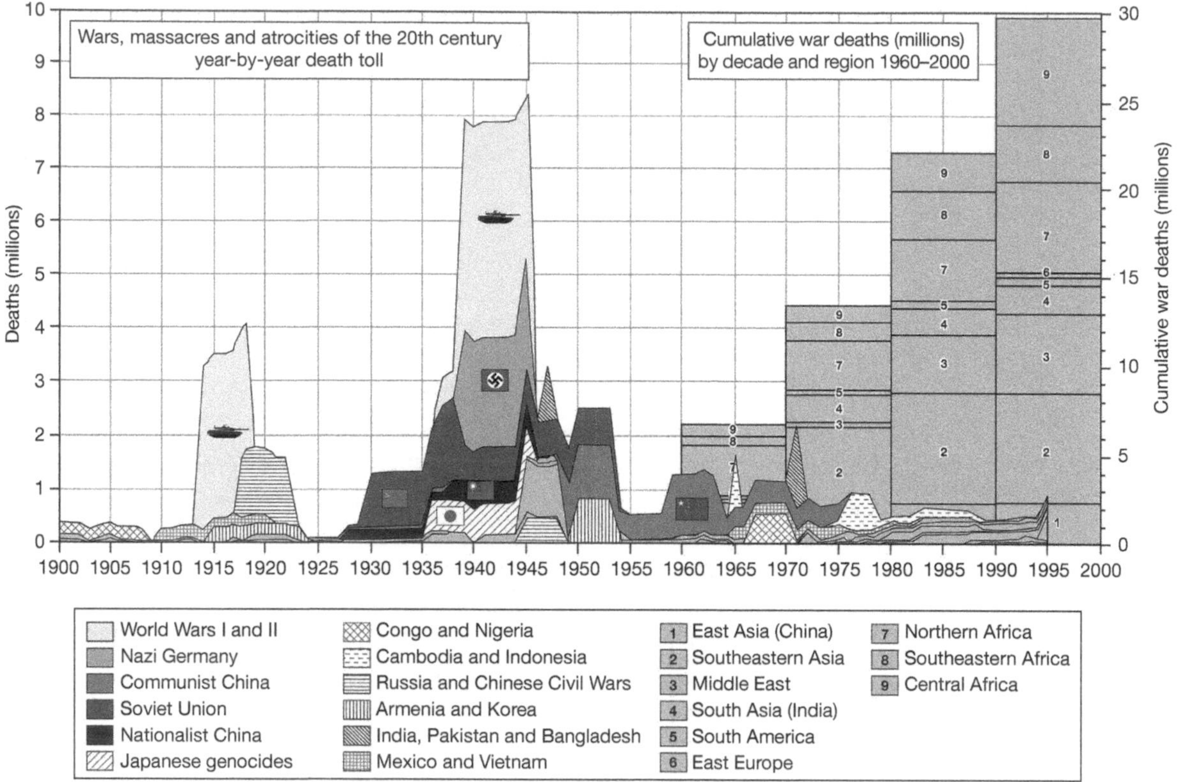

Figure 2.3 Wars, massacres and atrocities of the 20th century
(Dorling and Coles, 2009: 1779)

to war across nine geographical regions. This powerful graphic challenges the common assumption that the numbers of people dying as a result of war have been decreasing since the end of World War II. The use of quantitative data and stark graphical illustration presents a powerful argument that can be used within both policy and academic circles to encourage greater concern over 'hidden' casualties of more recent conflicts.

Reflecting this form of 'new quantification', Mei-Po Kwan (2002) has advanced a form of feminist GIS. In this framework, Kwan (2002: 650) seeks to use the power of GIS to illuminate the forms of oppression and marginalisation experienced by women:

> the purpose of using GIS in feminist geographic research is not to uncover universal truth or law-like generalizations about the world, but to understand the gendered experience of individuals across multiple axes of difference.

Part of the utility of GIS for Kwan is its ability to aggregate large data sets concerning information at a range of geographical scales, in particular the highly localised scale of the street or building. As we saw earlier in the chapter, conventional quantitative approaches had a tendency to explore phenomena on a larger scale and focused on model building. Kwan (2002: 651) argues:

> The ability of GIS to incorporate information about the geographical environment across spatial scales renders it a useful tool for feminist research. As geographic data of urban environments at fine spatial scales (e.g., at the parcel or building level) can be assembled

> and incorporated into a GIS, it is possible to link the trajectories of women's everyday lives (including activities locations and travel routes) with their geographical context at various geographical scales. This would allow a mode of analysis that is more sensitive to scale and context than are conventional methods. Further, when individual-level data are available, GIS methods can be attentive to the diversity and differences among individuals.

Therefore Kwan proposes a form of GIS that seeks to expose and challenge the marginalisation or oppression of women using quantitative techniques. For example, Kwan discusses the use of mapping technology to monitor the incidence of breast cancer in a community in Long Island, New York (see Timander and McLafferty, 1998; McLafferty, 2002). There was a fear among the community that the incidence of breast cancer was shaped by environmental hazards, and the quantitative technique could establish if this was the case and isolate specific clusters. While this movement managed to influence public policy and capture funding, it also exposed the unequal relations of power which are inherent in this kind of research. Kwan urges scholars to remain attentive to the differential access to GIS technology, in terms of both technical know-how and access to appropriate technology. This leads to inequality between social movements, but also more significantly between poorly resourced non-governmental organisations and well-resourced and networked government agencies. The introduction of statistics into the public realm always involves a degree of politics and presentation, and this is where Kwan points out that the inequality between these groups can be significant.

Conclusions

In 2006 Harvard University reintroduced a geography programme in the form of the Center for Geographic Analysis (CGA), marking a return for a subject so notably abolished from the university in 1948. But, as the name suggests, Harvard's geography is resolutely quantitative, providing 'continuous time-monitoring of the social and natural environment' using powerful GIS technology (Center for Geographical Analysis, 2010). This represents an indication of the continued prestige of scientific analysis and its tense relationship with the mainstream of human geography in the UK and North America.

While GIS has clear policy relevance and gains significant public and private funding, it is no longer at the heart of the geographical enterprise in the style of the 1950s and 1960s. The positivist approach that is often folded into quantification has been strongly criticised – and even rejected – by strands of humanistic, feminist and Marxist scholars. This point is addressed in the forthcoming chapters, in particular the following chapter examining humanistic geography. While attempts have been made recently to bridge the quantitative–qualitative 'divide', in particular through innovative uses of GIS, there remains suspicion among many human geographers as to the philosophical assumptions that underpin quantification. But we should keep in mind that quantification is not alone in facing questions regarding its philosophical and political implications. As such, we should see these as areas for debate rather than reasons for sidelining quantitative approaches. As Danny Dorling's work illustrates, if we as geographers want to reach out to public and policy audiences we need to be able to support our arguments using statistical evidence. How we can do this, while remaining sensitive to the 'human' aspects of inquiry, is explored in the following chapters.

Summary

- The emergence of quantitative approaches is closely tied to the historical and geographical context from which they emerged: post-World War II geography within Western Europe and North America.
- Quantification involved the adoption of techniques from economics and mathematics into an integrated human and physical geography in order to make laws concerning spatial phenomena.
- Quantification relied upon (often implicitly) a positivist philosophy that enshrined empirical testing and possibility of replication of such tests.
- Quantification is not a relic of geography's past, but a central strand of the methods and philosophy of the discipline since World War II. We are seeing a resurgence of quantification, often operating in tandem with sophisticated qualitative techniques.

Further reading

Barnes, T.J. (2001) Retheorizing Economic Geography: From the quantitative revolution to the 'cultural turn'. *Annals of the Association of American Geographers,* 91(3): 546–565.

Barnes, T.J. (2004) Placing Ideas: Genius loci, heterotopia and geography's quantitative revolution. *Progress in Human Geography,* 28(5): 565–595.

Berry, B. (1993) Quantitative Revolution: Initial conditions 1954–1960: A personal memoir. *Urban Geography,* 14: 434–441.

Burton, I. (1963) The Quantitative Revolution and Theoretical Geography. *The Canadian Geographer,* 7: 151–162.

Dorling, D. (2010) All Connected? Geographies of race, death, wealth, votes and births. *Geographical Journal,* 176(3): 186–198.

Gregory, S. (1983) Quantitative Geography: The British experience and the role of the institute. *Transactions of Institute of British Geographers,* 8: 80–89.

Johnstone, R. (2000) On Disciplinary History and Textbooks: Or where has spatial analysis gone? *Australian Geographical Studies,* 38(2): 125–137.

Kitchin, R. (2006) Positivistic Geographies and Spatial Science, in S. Aitken and G, Valentine (eds), *Approaches to Human Geography,* London: Sage, pp. 20–29.

Morrill, R. (2005) Hägerstrand and the 'Quantitative Revolution': A personal appreciation. *Progress in Human Geography,* 29: 333–335.

3 Humanistic geographies

Learning objectives

In this chapter we will:

- Examine how a body of humanistic work emerged in geography in the late 1960s and 1970s, and the challenge this posed to prevailing positivist approaches (see Chapter 2).
- Identify two central theoretical approaches to humanistic geography: phenomenology and existentialism.
- Focus on the contribution of Yi-Fu Tuan: one of the most influential humanistic geographers.
- Explore how humanistic ideas have been challenged and critiqued.

Introduction

Machu Picchu is an Inca city located in the Andes mountains in present-day Peru. It is a site of historical interest; scholars believe the original settlement was constructed around AD 1400. As geographers we can discern a number of facts about its location and altitude: it is situated at a height of around 2,430 metres above sea level on an elevated shelf in the Urumbamba river valley. But these scarce facts tell us little about how this site is understood by individuals living nearby, those who visit on the popular treks or who act as guides and maintenance workers on this famous tourist site. Surely, in order to grasp how Machu Picchu is understood as a place we need, first, to

Machu Picchu in Peru
Source: Pearson Education Ltd/Photodisc

visit there ourselves and, second, talk to those who live, work and visit this historical site. This approach demonstrates the importance we place on experience in knowing about the world around us. But this also poses problems: places such as Machu Picchu are not experienced in the same way, either between groups, such as tourists and workers, or even within these groups themselves. Some will find this a place of wonder and discovery, while for others it is a site of toil and banality. Humanistic approaches in geography have developed theoretical and methodological approaches to explore the significance of human experience as to how we understand the world around us. The places we inhabit are not universally understood: their meanings are being produced and reproduced all the time through people's experience. Humanistic geography sought to orientate our attention to this human process and explore what it means for how people and space interact.

In Chapter 2 we outlined the pre-eminence of spatial science within the geographical discipline over the 1960s. The use of positivist analysis of geographical problems and processes seemed to assure the discipline's prestige as a rigorous scientific pursuit. But this certainty and methodological pre-eminence did not last long. New critiques of positivism emerged in the late 1960s and early 1970s. Questions of existence and being, previously felt to be secondary to concerns of distribution and location, came to prominence, as geographers sought a more 'anthropocentric' ('human-centred') understanding of the creation of space and place. Scholars wanted to bring to the centre of attention the philosophical challenge of understanding what it was to be human and how this re-shaped our understanding of concepts of space and place. While the approaches of spatial science were adequate for *describing* geographical processes, scholars began to feel less confident that they could *understand* the meanings behind human choices, distributions and inequalities.

To provide a richer account of the motivations that underpinned human life, geographers turned to two bodies of thought. The first, which will be discussed in the next chapter, was Marxism, a theory that focuses on the material relations between humans and stressed the importance of class relations and capital ownership in shaping human history and geography. The second, humanist position, drew on philosophies of **existentialism** and **phenomenology** to draw attention to the importance of human values and perceptions to explaining spatial distributions and practices. It is this body of thought that this chapter will explore. In analysing this work we will draw attention to the new methodologies that this approach demanded, the challenges it faced and the enduring legacy it has left within geographical thought.

The chapter is divided into four sections. In the first we examine how geographers challenged positivism by drawing on humanist ideas. This account charts how scholars began to explore philosophical questions of being and consciousness within geographical studies. The following section provides a theoretical basis for this discussion by examining the two most prominent philosophies drawn upon in humanistic geography: phenomenology and existentialism. These are very challenging philosophical positions that are difficult to neatly summarise, so we try and introduce the key tenets of these ways of understanding the world. The following section looks at the work of a key proponent of humanistic geography: Yi Fu Tuan. As we will see, it is important to examine humanistic geography 'in practice' since this is a form of inquiry that seeks to avoid abstraction and focus on lived experience. Finally we examine the philosophical and political critiques of humanistic geography. The arguments in this section will be developed in later chapters of the book, since we can see the legacy of humanistic approaches in later critical, feminist and non-representational approaches to geographical thought.

Humanistic geography and the challenge to positivism

Humans are not machines. Their actions, perceptions and motivations cannot therefore be reduced to mechanistic models and detached observations. These claims, central to the thinking of humanistic geographers, served to question the purpose of seeking a verifiable scientific model of human distribution and behaviour. From the environmental determinism of Ellen Churchill Semple (see Chapter 1) to the Central Place Theory of Walter Christaller (see Chapter 2), geographers had previously sought a demonstrable theory of the spatial development of human societies. Humanistic geographers were sceptical of such objectives and their critiques of spatial science gained popularity among historical and cultural geographers over the late 1960s and early 1970s. As the name suggests, this approach involved the incorporation of concepts of **humanism** into the theories and practices of academic geographers. Although humanism is subject to a range of philosophical interpretations, at heart it advocates a philosophical reflection on what it means to be human. Therefore, humanism focuses our attention on what a human person is and can do. American geographer Nicolas Entrikin, a key figure in the development of humanistic geography, sets out the framework of the incorporation of humanism into geography:

> humanism in geography emphasizes the study of meanings, values, goals, and purposes. Within this humanist perspective concepts of traditional significance in geography are given

> existential meanings. For example, place is defined as a centre of meaning or a focus of human emotional attachment. The humanist approach is defined by its proponents in geography and in other human sciences as a reaction against what they believe to be an overly objective, narrow, mechanistic and deterministic view of man presented in much of the contemporary research in the human sciences. Humanist geographers argue that their approach deserves the appellation 'humanistic' in that they study the aspects of man which are most distinctively 'human': meaning, value, goals, and purpose. (Entrikin, 1976: 616)

From this initial critique we may draw out four specific criticisms of positivist thought offered by the newly emerging school of humanistic geography:

1 *De-humanised*: there was a feeling that quantitative models did not reflect the complexity of human life – choices were reduced to rational economic choices, which poorly reflected the values and perception of the human decision making process.
2 *De-contextualised*: spatial location and distribution were not considered in context, but were rather abstracted to present a mechanistic representation of human nature. Humanistic geography incorporated the physical, social and emotional surroundings into its geographical thought.
3 *De-particularised*: scholars such as Nicolas Entrikin (1976) felt that logical positivist approaches had emphasised abstract space over individual places – there was need to refocus on the particular instead of the generalised.
4 *De-politicised*: as a consequence of the first two limitations, scholars began to draw attention to the inadequacy of spatial science perspectives to intervene in the key political issues of the late 1960s and early 1970s – for example: the growing inequality between rich and poor expressed at a range of spatial scales; the complex spatial consequences of de-colonisation in Africa; and new conflicts fuelled by the Cold War.

Taking forward these four critiques, geographers sought to develop new methodologies and philosophical approaches that moved beyond positivist approaches. For some, the approaches of logical positivism needed extending, arguing that the quantitative approaches of positivism need further development through the use of qualitative methodologies. In this case scholars sought to *extend* the theoretical and methodological sophistication of positivist thought. In a second perspective, geographers have rejected the arguments of positivist thought, questioning the existence of a single reality and focusing attention instead on the construction of individual human experience rather than attempting to state essential truths. In this case geographers have sought to *revise* positivist thought and construct alternative philosophical schemas on which to consider human spatiality. We explore each of these approaches in more detail below.

Extension

In the case of *extension*, geographers have sought to draw on the ideas of humanistic geography to enrich the models provided by positivistic spatial science. David Ley, another key figure within humanistic geography, outlines this approach in the case of modelling incidents of crime:

> if group X 'causes' crime here, why is it that they do not 'cause' crime in other locations? Why is it that the same urban neighbourhoods now occupied by group X also tended to be high crime areas a decade ago, when they were occupied by group Y? Clearly statistical or

> cartographic analysis alone is not sufficient to provide an understanding of the social action behind the map of crime, though it may well be a useful first step; such variables, though convenient for the interests of the researcher, are not always demonstrably salient for the interests of the research. (Ley, 1978: 42)

In this case the explanatory power of a spatial scientific model is *extended* through a humanistic investigation into broader social, cultural or historical factors that could potentially shape the spatiality of crime. Here, logical positivism serves as 'a useful first step' but the more pressing issues of social and cultural complexity are left open to question and are worthy of further research. In so doing, the underlying **epistemology** of positivism, that an objective reality exists that can be subjected to rational interpretation, remains intact. It is this model of *extension* that appears most pervasive among human geography research, evidenced by the 'adding in' of qualitative insights into quantitative models of human distributions or the concept of 'triangulating' methodologies by combining quantitative and qualitative approaches. The geographer Susan Smith explains this impetus to draw on multiple methodologies to validate or generalise from fieldwork experiences:

> It is tempting . . . to 'hedge one's bets,' to conclude that case materials will not stand alone, and to advise that they are used to best effect only with the support from other 'harder' empirical data taken from surveys, censuses and similar 'objective' and supposedly representative sources. (Smith, 1984: 357)

To provide an illustration of the practice of *extension* we can stay with the issue of the geography of crime. The last two decades have seen a dramatic increase in research exploring the spatiality of crime and, perhaps more significantly, the fear of crime (see Smith, 1987; Pain, 2000; Nayak, 2003c). Much of this research seeks to analyse and extend the quantitative picture of crime distribution supplied in national surveys, such as the British Crime Survey (BCS) in the UK and the National Crime and Victimization Survey in the US. For example, the geographer Rachel Pain has been at the forefront of developing methods of surveying and analysing the spatiality of fear of crime – which she defines as the wide range of emotional and practical responses to crime and disorder individuals and communities may make (Pain, 2000: 367). Pain acknowledges the correlation between the emergence of the BCS in 1982 and an upturn in scholarly interest in the distribution of crime, often focusing on the implications for urban planners and architects. But the inspiration of this strand of research is to move beyond, or extend, the quantitative data provided by the BCS and, in doing so, develop a more nuanced understanding of the development of individual and community experiences of the fear of crime. Pain (2000: 381) identifies the importance of incorporating qualitative methodologies such as **ethnography**:

> With the help of more sensitive methodologies, researchers are beginning to forge more holistic accounts of the fear of crime: as a phenomenon that varies between individuals; which has geographical, economic, social, cultural and psychological dimensions; which is influenced by a whole range of processes and relations scaled from the global, national and local to the household and the body; and which is rooted in place and variable between places.

While this research programme seeks to explore subjective experiences of fear and the consequent emotional responses, Pain is simultaneously aware that this produces a tangible problem worthy of academic and policy attention. This research, then, serves to extend the scientific perspective of the BCS, without undermining its philosophical underpinnings of the existence of an objective reality that we may measure and understand.

Revision

But while such approaches may adopt elements of humanism, humanistic geographers have posed a more fundamental critique of the philosophical underpinnings of positivist thought. In a framework, we will term *revision*, geographers have argued that the project of spatial science is based on a flawed premise: that a separation is possible between the human observer and an external reality. What we understand as reality, these scholars argue, is so grounded in human experience that we need to engage with how individuals perceive the world around them rather than assuming that everyone interprets the world in the same way. Therefore, rather than simply 'adding in' human experiences into existing positivist approaches, this approach seeks to offer an alternative framework for understanding human geographies based on individual experience.

This can be seen as a fundamental reworking of the geographical approach: rather than seeking to *describe* distributions or spatial locations, this work seeks to understand how humans perceive the world and how this perception shapes their actions. This approach is challenging to describe, since it is based on the assessment of individual thoughts and actions, but one feature we can draw out is the sensitivity to change and difference. Humanistic scholars have argued that understanding the human condition is based upon an attempt to analyse the categories used by individuals to understand the world around them. For example, in describing a daily routine, a commuter may describe part of his or her daily routine where he or she walks through a dark and unlit car park. The car park is therefore categorised as a site of fear and apprehension. In contrast, a group of young skateboarders may view this same site as a location of recreation and play; while the security guard tasked with patrolling the car park may apprehend this space as one of work and daily routine. So we see across these three examples different perceptions of a single space, which then feeds into different actions (whether to enter the car park and what activities are undertaken once in the car park). Furthermore, the reactions of the different car park users change depending on the time of day, the weather conditions or simply their own mood: potentially a consequence of unrelated factors in the economy, politics or society.

The car park example highlights three important points regarding humanistic approaches. First we need to understand there will be a diversity of perspectives, but perhaps through critical reflection we can begin to see common themes and perceptions. Second, these perspectives will change over time, although, again, we may be able to trace certain routines and rhythms to individual attitudes. The third issue in some senses works in opposition to these first two points: we need to simultaneously try and construct some generalised points about 'the human condition'. So on the one side we have an approach that revels in diversity, while simultaneously seeking to try and uncover some general points about how humans interpret and act in the world around them. The geographers David Ley and Marwyn Samuels used the concept of 'cognitive categories' as a means for trying to comprehend this relationship between diversity and **universalism**. They argued that the human condition is based on cognitive categories, referring to the mental frameworks that people create through which to order their lives. For example, cognitive categories may be employment, family, sport and so on. While these categories may endure, Ley and Samuels argue that their validity and meaning may change with 'changing social, economic, political, intellectual, broadly historical circumstances' (1978: 19).

So, by placing perception at the very forefront of inquiry, some humanistic scholars have rejected the idea of positivist approaches acting as a necessary 'first step' and sought to revise the methodology through which geographical inquiry is practised. But this was more than simply a change in the way that field research is conducted, highlighting the value of methods such as participant observation, extended interviews and, later, focus groups. In order to understand human perceptions, scholars sought to bring philosophical reflection into geographical inquiry. In particular the humanistic approach requires an engagement with some of the most significant philosophical questions that have faced humanity: what is meant by *being* and *consciousness* (Samuels, 1978: 23). We should be careful not to overstate the novelty of this turn to philosophical reflection. As we saw in Chapter 1, much of early geographical scholarship related to imperialism was framed in terms of individual perceptions and often drew on travel writing and other creative sources through which to draw out geographical arguments. What was different about this humanistic approach was its reticence to develop laws of spatial distribution, as we saw in the case of environmental determinism and, in more rigorously scientific frameworks, in the discussions of modelling and theory in Chapter 2.

The division between *extension* and *revision* is presented here as a means through which to illustrate the variety of ways that scholars have adopted humanistic approaches. Of course, as you read humanistic accounts they often combine both positions by including wider quantitative studies and questioning the applicability of positivist analysis. And, as we will see in future chapters, humanistic geographers were not alone in critiquing positivist perspectives; we will see similar arguments emerging in the discussions of the **cultural turn**, feminist geography and **non-representational theory**. In a sense this is a reflection of the wide impact of humanistic approaches: the language of experience and perception has been pervasive in human geography over the last 20 years. Consequently we need to engage with the underlying theoretical perspectives of humanistic geography in order to clarify its specific contribution.

Phenomenology and existentialism

This section explores two of the key underlying philosophies that have been used by humanistic geographers: phenomenology and existentialism. These are very challenging terms that are often used in different ways by different scholars. They are also terms that are, in other ways, trying to grasp what it is to be human. This makes providing a neat synthesis of these ideas a difficult task and it may be that on first read this seems like a rather abstract set of issues for geographers. It would be a good idea to explore the original texts to develop these ideas; a list is provided in 'Further reading', at the end of this chapter.

Phenomenology

Seamon and Sowers (2009) have recently defined phenomenology as: a philosophical approach that examines and describes phenomena (that is, things and experiences) as human beings experience those things or experiences. This seems remarkably straightforward and it is worth remembering this definition as further philosophical complexity is brought into the discussions in this chapter. Uses of phenomenology in geography often draw on the work of Edmund Husserl, a 19th- and 20th-century philosopher born in what is now the Czech Republic. Husserl was interested in developing a philosophical method which would allow the scholar to go 'back to things themselves' (Entrikin, 1976: 617). So, at its heart, phenomenology has been concerned with challenging abstraction and thinking about the lived experience of human life. In terms of scholarly practice, this approach advocates empirical inquiry and the importance of the humans at the centre of this process. But in getting 'back to things themselves' Husserl was also advocating setting aside all preconceptions of the nature of human experience and, instead, examining the word afresh.

Husserl's phenomenological approach is therefore both straightforward (it focuses on human experiences of things and events) and radical (it seeks to challenge existing scientific categories and explore instead how new categories are formed through experience). What lies at the heart of both these points is that we can only understand human phenomena in practice. Let us take an example of a phenomenon that has been the focus of much attention in geography: language. Phenomenologists argue that we can only understand language when observed in practice. The anthropologist Simon Charlesworth (2000) illustrates this point within his phenomenology of working-class lives in Rotherham, UK. Charlesworth uses personal testimonies given by residents of Rotherham to examine forms of marginalisation and oppression of the working class. Since the author is adopting a phenomenological approach it is important that individual narratives are re-created using the terms, speech styles and pronunciations that were experienced over the course of the research. But this presents challenges when it comes to the silences within the research transcripts, as Charlesworth (2000: 4) explains:

> . . . what is impossible to represent here is the persuasive silence which surrounds instances of testimony and which resultant transcription only apparently contradicts. [This inarticulation] . . . does pervade some of the transcription and is a reason why I tried so hard to preserve the verbal form of speech, with all its inarticulate mumblings and broken lapses, but the form of the work cannot capture the bleak darkness of the invisibility of these people's lives to themselves.

We would like to draw out two important points from this excerpt. First, we see the significance placed on preservation of the language of respondents, not simply in terms of what was said but, as crucially, what was not said. Charlesworth views this process of preservation as vital to being able to understand the nature of working-class life. This is explicitly phenomenological since it seeks to identify *how* language is used by individuals to express their lives and draw attention to themes and practices that are shared by respondents. The second point is related to the challenges this poses for the scholar. As Charlesworth explains, it is impossible to perfectly represent the expressions and articulations of individuals as they describe their experiences. There is always a process of translation as these testimonies are used within academic writing. This draws attention to the power of the scholar to select the sections of the interview and subject these to their own scholarly analysis. This power inequality becomes more pronounced where the scholar is using theoretical approaches with which the interviewees may be unfamiliar. There is no simple means of avoiding these issues of power and they present a serious challenge for scholarship that is using qualitative methodologies. As we will see in Chapter 6, feminist scholars have been particularly attuned to questions of power within the academy and have advocated new participatory methodologies that seek to allow respondents to help frame research questions, and contribute to the processes of analysis and writing.

This phenomenological focus on the relationship between the individual and the material world opened a series of new pathways for human geographers. Rather than attempting to map spatial distribution, what it means to be human became a focus of geographical inquiry. This is often described as a shift towards exploring 'human **subjectivity**', literally referring to what it means to be a human subject. Specifically, geographers were interested in how individuals invest meaning in certain sites and space, using the term 'place'. Drawing on phenomenological ideas, geographers began to focus on the social construction of 'place' rather than previous preoccupations with mathematical models of space. A key proponent of this school was Edward Relph (1976: 43) who was instrumental in highlighting the creation of place through the thought and action of human agents:

> The basic meaning of place, its essence, does not therefore come from locations, nor from the trivial functions that place serves, nor from the community that occupies it, nor from superficial or mundane experiences . . . The essence of place lies in the largely unconscious intentionality that defines places as profound centres of human experience.

Let's give an example of this idea of 'place' by returning to the work of Simon Charlesworth. In order to illustrate the decline of Rotherham's traditional industries and consequent rise in unemployment and economic deprivation, Charlesworth provides two accounts. The first is a quantitative overview of the declining job opportunities, the differential rates of unemployment between men and women and the uneven nature of the declining job opportunities, with manual labour and semi-skilled professions disproportionately affected. This account is supported through the use of statistics, such as em ployment figures and local income figures. However, in line with his phenomenological perspective, Charlesworth goes on to provide testimonies from Rotherham residents as to the implications of this decline of the character and condition of the town. He is presenting how these processes shape perceptions of place. For example, he cites a 25-year-old man who sees the changing economic environment having profound consequences on the behaviour of residents:

> People are depressed an' nut 'ouses [mental hospitals] are fillin' up, thi's [there is] mo'ore an mo'ore people wot just can't tek [take] it any mo'ore, thi's too much strain fo' 'em. Rother'am's like a gaol wiy'aht [without] any walls. The'r [they're] stuck 'ere but can't see the walls. Wot it's like is, people can't see wot it is that's causin' ahr thi feel, it's like the'r ill an' doctor can't tell em what's up and ca(n)'t du 'owt fo' 'em. (Charlesworth, 2000: 53)

Here Charlesworth is carefully transliterating the oral testimony of the Rotherham resident into his account of urban social change in order to convey the feeling of hopelessness that is felt by segments of the town's inhabitants. Rather than simply providing an account of the statistics of economic decline, this phenomenology evokes the feelings of despair and isolation felt within the town. It demonstrates that the 'sense of place' in Rotherham has transformed since the decline of traditional industry, from a place of opportunity to feelings that it is a 'gaol', a site of incarceration.

So by drawing on Husserl's philosophical approaches, phenomenologists have attempted to study human consciousness and through this attempt to discern shared cognitive categories that underpin experiences. From this discussion we can draw out three traits of phenomenological studies:

1 *Subjectivity*: phenomenologists are concerned with how humans experience their environment in unique ways. Phenomenology is interested in human subjectivity.

2 *Construction of knowledge*: phenomenologists focus on how individuals give meaning to their objects of consciousness.

3 *Search for shared categories*: the goal of such study is the identification of elements in individual consciousness which control the allocation of meaning, what Husserl referred to as 'essences'.

Existentialism

Existentialism is the second, and closely related, philosophy drawn upon by humanistic geographers. While phenomenology is concerned with the human understandings of phenomena, existentialism is oriented more specifically towards understanding the nature of human existence. Seamon and Sowers (2009: 666) define existential geography as a set of approaches 'that insist human experience, awareness, and meaning must be incorporated into any study of people's relationships with space, place, and environment'.

The origins of existentialism are often traced to the work of Friedrich Nietzsche and Søren Kierkegaard, although neither thinker used the term itself. This highlights the care that must be taken when discussing the history of existential thought: many scholars whose research has subsequently been labelled as existentialist did not see their work in such terms. Jean-Paul Sartre (1905–1980) is credited with bringing existential philosophy to prominence in the French intellectual scene with the publication of *Being and Nothingness* (1948). For Sartre, existentialism comprises a resolute focus on human existence and in particular the unique capacity for humans to reflect on their own actions and, therefore, inscribe their actions with meaning. In elevating the significance of conscious action, Sartre was dismissive of those who did not accept the responsibility for their actions, and instead allowed factors other than their own consciousness to determine choice. Such an individual, according to Sartre, is living an inauthentic experience, living in bad faith (Entrikin, 1976: 622). As this celebration of human agency would suggest, Sartre wrote *Being and Nothingness* through an acute disaffection with Marxist thought, an intellectual tradition which he felt suffered from over-determinism and a related lack of appreciation of the human condition.

Therefore like phenomenology, existentialism can be viewed as a reaction against idealist or Marxist thought which abstracted thought from action, knowledge from emotion and humans from their existential situation (see Entrikin, 1976: 622). To this extent existentialism shares with phenomenology its anthropocentric nature: it is a philosophy that seeks to ground our understandings of the world in what it means to be human. And, like phenomenologists, existentialists were not concerned with developing an overly academic theory that lacked grounding in the lives of individuals. As Nicolas Entrikin (1976) explains, the main difference between the existentialist perspective and the phenomenology of Husserl is that Husserl is concerned with the *a priori* foundations of knowledge (or *essences*), while existentialists concern themselves with the nature of human existence. Therefore, as Johnston and Sidaway (2004) explain, whereas phenomenology assumes the primacy of essence – the allocation of meanings results from the existence of consciousness – for existentialists the fundamental root of existence is 'being'.

Existential philosophy has been extremely influential within geography. Indeed, geographers have argued that existentialism's focus on existence and human consciousness has required a more nuanced understanding of spatiality. This assertion centres on the claim that at the heart of human consciousness is the distancing between the self and the surrounding environment. This 'primal setting at a distance' is explained by Marwyn Samuels (1978: 27):

> Spatiality is more than a necessary condition of human consciousness. It is the beginning of human consciousness. Spatiality is meaningful here precisely because it constitutes a minimum definition of man as the only historic life form to emerge with a capacity to detachment. For this reason, too, the human situation is defined in existential terms as one predicated on distance. *Estrangement* – 'the primal setting at a distance' – is understood to be the human situation par excellence. Space is 'existential' precisely because it reveals the condition of alienation; i.e., the existential condition of man's distancing.

Therefore, existential thought brings our attention to the separation within human consciousness between man and the environment. This is a fundamental point as it acts as the origin for the existential argument that humans are 'alienated' from the world of things, set apart to such an extent that humans are locked into a perpetual struggle to make meaningful their surroundings. As we will see in the next chapter, it is on this issue of **alienation** that existentialist thought intersects with Marxist **ideology**.

In order to elucidate this concept of alienation, scholars have turned to the philosophical framework of Martin Heidegger, who was an assistant of Edmund Husserl at the University of Freiburg. The significance of man's relationship with the world and its central importance to existential thought is illustrated in Heidegger's concept of *Dasein*. Dasein refers to man's distinct mode of being, as always 'being there', 'being in position' or more prosaically 'being-in-the-world'. For Heidegger, being-in-the-world is not the same as an object's physical location, but refers instead to being in the distinctly human world consisting of intentional and emotional ties of men and women toward other humans and objects around them (Entrikin, 1976: 622). This draws our attention to the forging of relationships, how humans relate to other beings and objects in a given environment and thereby attempt to negotiate their alienation. In doing so Heidegger's work urges us to confront some of the fundamental philosophical questions concerning the nature of space and place in human existence.

From this discussion we can identify three traits of existential thought:

1. *Existence*: existentialists are, as the name suggests, concerned with understanding the structures of human existence.
2. *Consciousness*: rather than constructing an ideology or grand theory, existentialists reflect on humans as possessors of free will and study the nature of human meaning-making.
3. *Alienation*: meaning-making stems from humans' experience of alienation or complete separation from their surroundings. An *existential dread* stems from the experience that we are completely different from what surrounds us. Consequently humans strive to understand the world by inventing classificatory systems in order to make our environment meaningful.

We would make two points by way of summary to this discussion of phenomenology and existentialism. First, humanistic geographers have usually combined aspects of phenomenology and existentialism rather than adopting a single, pure philosophical perspective. Indeed, some of the prominent recent work in this field has not adopted a single perspective, but rather examined how the work of single scholars can contribute to geographical thought (see Elden, 2004, 2006). This highlights a key aspect of our exploration of humanistic geography, namely that we need to think of this form of inquiry *in practice* rather than simply in theoretical terms. For it is the application of these ideas to real-world settings that serves to highlight the strengths and weaknesses of humanistic approaches to the geographical discipline. Second, we should understand the adoption of phenomenological and existential modes of thought as a form of criticism, specifically as a means through which the certainty of spatial science was questioned by a section of the geographical discipline. In doing so, scholars often focused more on the philosophical underpinnings to humanism (namely the rejection of positivism) than the extent to which it offered a coherent and concise body of knowledge.

Humanistic geography in focus: the work of Yi-Fu Tuan

The scholarship of Yi-Fu Tuan provides a wide range of examples of innovative humanistic geography (see Box 3.1). Tuan's work is philosophically ambitious; it seeks to address some of the primary questions facing human beings, such as the nature of existence, place and humans' attachment to their environment. In doing so Tuan draws on elements of both existentialism and phenomenology, and reflecting this his research is underpinned by two central beliefs. First, Tuan's work demonstrates a belief in the separation of

Box 3.1

Yi-Fu Tuan (1930–)

It is very difficult to provide a concise summary of the career of Yi-Fu Tuan. His scholarship drew on a range of disciplinary perspectives, including philosophy, anthropology, biology, psychology and sociology. But his work has gained most popularity and significance within geography: the discipline which he taught in US universities for over 40 years. Tuan was born in China, but studied his bachelor's and master's degrees in Oxford, UK and his PhD at the University of California, Berkeley. Tuan's time at Berkeley was crucial to the development of his geographical imagination. While there he studied under Carl Sauer, the extremely influential cultural geographer who was responsible for drawing attention to the significance of human perception while strongly criticising the **grand narrative** and implicit racism of environmental determinism (see Chapter 1). This spirit of inquiry, and its phenomenological underpinning, was developed by Yi-Fu Tuan across ten books including *Topophilia* (1974) and *Escapism* (1998). More than advancing phenomenology in geography, Tuan's work has developed understandings of place, focusing on human subjectivity and drawing attention to the significance of questions of inhabitation and care. But, in addition, Tuan has managed to reflect on some of the primary questions that drive humanity – what is a good life? How can we build a shared understanding of a subjective notion such as morality? In many ways, Tuan's style of writing reflects the humanity of his scholarship – his texts are engaging and humble, and this is particularly reflected in his autobiography *Who Am I? An Autobiography of Emotion, Mind and Spirit* (1999). While Tuan's work has received criticism, in particular for underplaying the importance of questions of gender or class identity in structuring human geographies (see 'The Challenge to Humanity') his influence to the emergence of humanistic geography as a central aspect of geographical inquiry is unquestioned.

Further reading

Tuan, Y.-F. (1974) *Topophilia: A study of environmental perception, attitudes and values*, Englewood Cliffs, NJ: Prentice Hall.

Tuan, Y.-F. (1999) *Who Am I? An autobiography of emotion, mind and spirit*, Madison, WI: University of Wisconsin Press.

humans from the animal world on account of humans' ability to attach meaning to their environment:

> A case that illuminates human peculiarity is the importance that people attach to the biological events of birth and death. Animals make no fuss over them. To pragmatic animals locations have values because they satisfy current life needs. A chimpanzee does not wax sentimental over his past, over his birthplace, nor does he anticipate the future and dread his own mortality. Shrines dedicated to birth and death are uniquely human places.
>
> (Tuan, 1976: 269)

Second, Tuan believes that geographers could build on prevailing scientific knowledge rather than simply reject or surpass it. In doing so he appears to support the *extension* school of thought set out earlier in the chapter. Tuan argues that understanding the constraints facing humans requires an appreciation of the role of scientific factors on human behaviour, such as the role of biology, physics or, perhaps most importantly, psychology:

> Humanistic geography's contribution to science lies in disclosing material of which the scientist, confined within his own conceptual frame, may not be aware. The material includes the nature and range of human experience and thought, the quality and intensity of an emotion, the ambivalence and ambiguity of values and attitudes, the nature and power of the symbol, and the character of human events, intentions and aspirations (Tuan, 1976: 274)

These two underpinning beliefs fed into a range of monographs and papers, although *Topophilia: A Study of Environmental Perception, Attitudes, and Values* (1974) stands out as a key text in the field of humanistic geography. This ground-breaking book centred on an empirical consideration of 'topophilia', defined as the coupling of sentiment with place. In doing so, Tuan drew on both phenomenological and existential philosophies (see above) to evoke a more holistic sense of human experiences of the environment. This relationship was considered through a range of geographical and historical examples, from British holiday pursuits to Chinese landscape painting (and a considerable amount in between). This ambitious empirical canvas was held together by Tuan's particular perspective: the motivations, values and emotional attachments of humans with their environment. He sets out this perspective in the opening words of the introduction (1974: 1):

> The themes taken up here – perception, attitude and value – prepare us, first of all, to understand ourselves. Without self-understanding we cannot hope for enduring solutions to environmental problems, which are fundamentally human problems. And human problems, whether they be economic, political or social, hinge on the psychological pole of motivation, on the values and attitudes that direct energies to goals.

Here we see Tuan setting out the first two principles of a phenomenological approach: an interest in subjectivity ('understanding ourselves') and the construction of knowledge ('values and attitudes'). Tuan alludes to the third principle of phenomenology, the search for *essences*, on the following page (1974: 2):

> What can be the common ground between a detailed analysis of the shopping behaviour of housewives in Ames, Iwa and a grand survey of the Christian doctrine of nature? Or between the study of colour symbolism as a universal trait and the history of landscape painting? A possible reply is that somehow they all bear on the way human beings respond to their physical setting – their perception of it and the value they put on it.

This search for essences, then, centres on the shared biological attributes of the human species: 'all human beings share common perceptions, a common world, by virtue of possessing similar organs' (Tuan, 1974: 5). The organs that Tuan has in mind are the receptive senses: sight, touch, hearing and smell. Tuan contends that these senses set humans apart from other animals and facilitate the development of particular affective connections with places. Consider Tuan's example of the medieval cathedral versus the modern skyscraper (1974: 11):

> The medieval cathedral fascinates the modern tourist for various reasons, but one that has received little comment is this: the cathedral offers him an environment that stimulates the simultaneous use of three or four sense receptors. It has sometimes been said that the steel-and-glass skyscraper is the modern equivalent of the medieval cathedral. Actually, apart from the vertical bias the two buildings have very little in common. They do not illustrate the same principles of construction, they are not put to the same use, and their symbolic meanings are entirely different.

For Tuan, then, it is the sensory connectivity that sets these places apart (the dark, cool, stone-built cathedral and the bright, noisy, steel-and-glass skyscraper) that is of greater importance than their physical similarities (the vertical bias of the two constructions). Developing this framework, Tuan attempts to compare and contrast the subjective responses of differing cultures and historical epochs to their sensory environment. While his narrative is underscored by a search for essence – 'all humans share common perspectives and attitudes' (1974: 246) – he simultaneously emphasises the influence of culture and the physical environment on individual perception. For example, in the case of two groups of Botswana Bushmen he identifies the influence of geographical environment on the development of sensory perception (1974: 246):

> We can say that the development of visual acuity is related to the ecological quality of the environment. Thus the Gikwe Bushmen learn to identify individual plants in the dry season whereas the Kung Bushmen, living in a better endowed setting need only learn the location of plant aggregates.

As we shall see later in the chapter, this balancing act between individual subjectivity and the influence of wider stimuli has been the subject of intense debate within humanistic geography.

Tuan's scholarship has had a profound impact on geographical thought. We can witness this impact on three levels. On the first, Tuan's existential humanistic approach has proved influential for geographers examining social and cultural practices. The methodologies developed through his published work are vital in exploring what is meant by place and how humans construct attachment to particular places. Through this approach we can see a particular recovery of Tuan's focus on sensory perception in recent research on the role of emotion and **affect** in shaping human geographies, in particular around concepts of the home, dwelling and refuge. On a second level, Tuan's research drew attention to the personal nature of academic writing – Tuan's questioning of positivism included a rejection of the idea of a scientific gaze that did not include a consideration of the role of the scientist. As stated above, Tuan's approach involved an enduring quest to 'understand oneself', to appreciate what values and perceptions shape your relationship with the environment. We may observe this goal of self-understanding in more recent methodological reflections on the importance of reflexivity in social research, or the critical evaluation of one's own actions. Tuan was keen to set such **reflexivity** apart from a notion of 'bias'. 'Bias occurs', Tuan argues, 'when we are ignorant of our philosophical presuppositions or when we insist that a perspective is an all-inclusive system' (1976: 275). This focus on reflexivity constitutes a crucial part of recent geographical research into identity politics and new qualitative reflections on research position. Finally, on a third level, Tuan's humanism involved orientating primary philosophical questions into geography – perhaps most clearly in his question 'what makes a good place?' Tuan's scholarship therefore helped to widen the philosophical ambition of human geography.

The challenge to humanism

As we have seen, the adoption of humanist approaches into geographical thought served as a critique of earlier positivist approaches. New theoretical perspectives focused on the importance of human consciousness and developed new methodological approaches directed towards capturing how humans engaged with the world around them. But humanistic geography was not without critics: scholars who felt that it privileged human free will over societal structures and that it ignored the perspectives of women. We will attend to each of these criticisms in turn.

Structure–agency

One of the key criticisms of humanistic geography has focused on its portrayal of human life as individualistic and voluntaristic, whereby humans are considered to act on account of their own free will. Over the 1970s, concerns were raised whether humanistic methodologies and philosophies over-emphasised human agency. Such criticisms focus on whether constraints on human action, imposed by economic, political and cultural structures, are adequately incorporated into the analysis of humanistic geographers. For example, take two individuals living in London, one a top-earning international footballer playing for Chelsea and the other a newly arrived asylum seeker working in the restaurant industry. These two individuals have starkly different opportunities and constraints on their lives, through their material position, cultural resources and freedom of movement. The economy, society and the state all serve to constrain the free will of a population; human action is not simply the consequence of human preference. While humanistic geographers may be well-placed to assess the different human experiences of these two individuals, critics have argued that humanistic approaches fail to comprehend the different structural and material factors that shape life choices. As the geographer Derek Gregory (1981: 16) states, 'the materiality of social life *is* weakly developed in modern humanism, and as a result it inevitably encounters severe difficulties in comprehending "objective societal forces"'.

The relationship between human agency and constraining societal factors shaped the structure–agency debates that began in the wake of the rise to prominence of humanistic and Marxist approaches across the social sciences in the 1970s. On the 'structure' side scholars felt that human life is determined by structural constraints; perhaps the best known example is Marxist thought, where economic relations (such as class) are thought to structure human history and development (see Chapter 4). In its most extreme form, human agency is considered of little consequence; rather individual practices are determined by larger structural process and relationships. As we have seen, the 'agency' side foregrounds the existential freedom of humans to act and think according to their own free will. As Nicolas Entrikin and John Tepple (2006: 31) suggest, this work has focused on the role of 'experiences, attitudes and beliefs, as well as moral and aesthetic judgement, in making decisions that shape their environment'. In doing so, this work is considered to have overlooked the bounded nature of human life or, at least, failed to consider such boundaries a specific problem worthy of scholarly consideration. The structure–agency debate, held over a number of disciplines (such as political science, sociology and geography) has involved a prolonged attempt to establish a compromise position between these two poles of societal structures or human agency as the central factor in the development of human life.

Although not exclusively connected to humanistic geography (as we will see in subsequent chapters, the structure–agency debate features in many aspects of geographical thought) we can observe three distinct ways in which scholars have attempted to transcend the structure–agency debate. First, there has been an attempt to highlight the intersubjective nature of much humanistic geography – that this is an area of geographical thought that often attempts to establish shared conditions between human subjects. Although humanistic geographers critiqued the existence of an absolute truth waiting to be uncovered, their methodologies often involved rigorous analysis of case study material in order to establish trends or practices that may be observed in other contexts. Indeed, as we have seen, one of the philosophical principles of phenomenological thought is the existence of essences within human consciousness that denote shared notions of human consciousness. The geographer Susan Smith identifies the roots of humanistic methodological practice in the urban ecology of Robert Parks in Chicago in the 1930s. Citing Ralph Turner (1967: xxv), Smith highlights that '[t]here is a very special sense in which Park sees the city as a microcosm in which are exposed and magnified, as under a microscope, the processes taking place in larger society' (Smith, 1984: 358). Therefore, an intensive study on a small scale can provide insights that are applicable to other locations, contexts and scales. Here we witness again the notion of humanistic geography as an extension of the themes and methodologies developed within the paradigm of logical positivism, rather than a rejection of the claims of scientific inquiry.

The second means of transcending the structure–agency debate involved the adoption of new theoretical approaches into humanistic geography. Specifically, scholars have looked to the work of sociologists Anthony Giddens and Pierre Bourdieu to provide a more grounded understanding of human *practice*. The first example of such new theoretical insights is the adoption of structuration theory, developed by Giddens, by geographers in the early 1980s. Structuration theory emphasised the ways in which actions and practices interacted with structural constraints to both transform and reproduce social structures. Derek Gregory (1981: 8–9), a key figure in the adoption of structuration theory by geographers, explains its basic tenets:

> in the reproduction of social life (through systems of interaction) actors routinely draw upon interpretive schemes, resources and norms which are made available by existing structures of signification, domination and legitimation, and that in doing so they are thus immediately and necessarily reconstitute those structures: in short, 'the structural properties of social systems are both the medium and the outcome of the practices that constitute those systems'.

Structuration theory emphasised that social structures are socially produced, and are therefore continually re-made through the actions of individual human subjects. The empirical verification of structuration theory involved examining habitual practices – such as residential location decisions, the routes individuals take through urban areas or how individuals constructed their personal taste – and through such work highlighted that habits and routines reinforced existing social relations (see Entrikin and Tepple, 2006). Structure and agency do not, then, occupy two distinct positions but are rather mutually constituted through human practice. Another theoretical pathway to transcending the structure–agency debate has been the adoption of the concepts and vocabulary of Pierre Bourdieu, in particular his concept of 'capital'. For Bourdieu this economic metaphor highlights 'the struggle (or competition) for scarce goods for which the universe is the site' (Bourdieu, 1987: 3). Through aspects of behaviour, comportment, educational qualifications and social contacts individuals acquire different amounts of social and cultural capital. He explains (1987: 3–4):

> [The] . . . fundamental social powers are, according to my empirical observations, firstly *economic capital*, in its various kinds; secondly *cultural capital* . . ., again in its various kinds and thirdly two forms of capital that are very strongly correlated, *social capital*, which consists of resources based on connections and group membership, and *symbolic capital*, which is the form the different forms of capital take once they are perceived and recognised as legitimate.

Therefore, just as economic capital (money) may be accumulated, Bourdieu argues that individuals can accumulate social and cultural resources. One of Bourdieu's key insights was that each form of capital could be converted into other forms (see Painter, 2000), though with varying degrees of difficulty. Bourdieu conceived symbolic capital as the most influential as 'it is the power granted to those who have obtained sufficient recognition to be in a position to impose recognition' (Bourdieu, 1989: 23). Crucially, a key aspect of this power is control over other forms of capital, managing their conversion and exchange rates. Thus, Bourdieu constructs a theory of human practice that assesses the constraints and compromises placed on both social structures and human agency.

The third and final means through which humanistic geographers transcend the structure–agency debate is to question the validity of this binary as an organising narrative of human life. Rather than attempting to question the structural nature of determining structures (Giddens) or demonstrate how individuals strategise through conditions of scarcity of social and cultural resources (Bourdieu), a third pathway simply questions the possibility of knowing whether human action is the consequence of a hidden structure or a demonstration of free will. As Susan Smith (1984: 364) observes:

> Whether, in the final instance, action is free or determined has no practical consequences for the present, for neither can be proved, and the ambiguity of commonsense experience submits to the sensation of either or both.

Rather than focusing on whether human action is free or determined, Smith recovers an established humanist refrain in arguing that geographers should highlight the dynamics of social processes. In so doing she argues for a philosophical framework guided by **pragmatism**, a movement emerging in the late 19th and early 20th century aimed at grounding philosophical debates in everyday life. Smith (1984: 355) explains:

> From a pragmatic viewpoint, thought and conduct are continuous. Ideas are the means of creative activity, and experience refers not to the impression of a pregiven structure of reality, but to the process through which a world is realized.

Pragmatism therefore centres attention on human action – 'on what can be achieved by intelligent intervention in an imperfect and unequal world' (Smith, 1984: 369). This approach seeks to avoid generalising human action on the basis of grand ideologies (such as Marxism) and rather focus attention on the experienced reality of moral choice. This has been an important intervention on two levels. First, pragmatism serves as a reminder of what is potentially unknown or unknowable – there are ultimate limits to our ability to assess the intentions and guiding logic of human action. Consequently, the tenets of pragmatics suggest that human geography is grounded in practice rather than attempting to develop further theoretical abstraction. Second, pragmatism draws attention to the *uses* of geographic knowledge, focusing attention on the gap between academic practice and the practicalities of everyday life. Pragmatists view philosophy only as an instrument for the enhancement of people's lives through its practical application, rather than an intellectual pursuit in its own right.

Feminist geography

While we provide an in-depth analysis of the contribution of feminist perspectives to geographical thought in Chapter 6, it is worth drawing out some points in direct connection to humanistic geographies. The relationship between feminist geography and humanistic approaches such as existentialism and phenomenology is complex. As we have seen, humanistic approaches had developed geographical interest in the nature of human experience and the psychological and emotional processes through which place is understood. Such a focus on individual agency was closely allied to emerging feminist geographical concerns with the experience of gender and the processes through which women were marginalised spatially and socially in Western societies. The links between humanistic and feminist approaches are illustrated in their shared embrace of qualitative methodologies such as interviews, focus groups and participant observation. Both feminist and humanistic geographers saw the need to explore how meanings of space and place are constructed and articulated by individuals and groups.

But, despite these shared philosophical and methodological perspectives, feminist geographers were critical of certain humanistic scholarship. Their critiques take two forms: challenging humanistic concepts of the human subject, and criticising the breadth of topics covered by humanistic scholars. In terms of the human subject, there was a concern among feminist scholars that humanistic approaches can reproduce the idea of a single, uniform human subject unmarked by gender difference. In this way some humanistic work was described as 'essentialist': that it presupposed an underlying human essence that could be uncovered when all other factors such as culture and economics are stripped away (Seamon, 2006; Seamon and Sowers, 2009). Feminist scholars have argued that such an understanding of the human subject can underplay gender inequality. Rather than 'stripping away' cultural and political issues, feminist geographers have been more concerned with bringing such factors into debates concerning the marginalisation and oppression of women.

The second aspect of criticism by feminist scholars was the areas of study chosen by humanist geographers. For Janice Monk and Susan Hanson (1982: 14), human geographers had posed research questions that applied to both men and women but were analysed 'in terms of male experiences only'. This approach produced a sexist bias in geographical research that focused in particular on predominantly male workplaces, leisure activities and lifestyles. This leads, Monk and Hanson argue, to 'gender blind theory' that fails to grasp the differences in gender roles and the implications these have for the structure and practices of social life. This leads the authors (1982: 16) to argue that 'women are generally invisible in geographic research, reflecting the concentration on male activity and on public spaces and landscapes'. Feminist scholarship sought to correct this bias by examining the practices of women, in workplaces, public spheres and the home.

Conclusions

By way of conclusion it is important to stress that humanistic scholars sought to respond to these criticisms and identify misconceptions in the practice of critique. Geographer David Seamon, who wrote the influential text *A Geography of the Lifeworld* (1979), has provided a robust response to charges, over later years, that his phenomenological approach is based on the search for human 'essences', suggesting that critics have overlooked 'the basic phenomenological recognition that there are different dimensions of human experience and existence that *all must be incorporated* in a thorough understanding of human and societal phenomena' (Seamon, 2006). Seamon goes on to explain that these 'dimensions' include:

> (a) one's unique personal situation – e.g., one's gender, physical and intellectual endowments, degree of ableness, personal likes and dislikes; (b) one's unique historical, social, and cultural situation – e.g., the era and geographical locale in which one lives, his or her economic and political circumstances, his or her educational, religious, and societal background; the technological, communications, and media infrastructure that contribute to the person's or group's particular lifeworld; and (c) one's situation as a typical human being immersed in a typical human world.

In a sense, Seamon's response illustrates the complexity of humanistic scholarship: it is work that attempts to provide a universal philosophy of human experience, but in doing so illuminating the variety of processes, resources, emotions and contexts that shape such an experience. This ambitious style of scholarship has left a considerable legacy on the discipline of geography, not least through work on feeling, emotion and affect that has risen to prominence in recent years (see Chapter 12).

As we set out in the introduction, over the course of this book we are seeking to avoid easy narratives of the chronology of geographical thought. Geographical ideas do not neatly follow on another as a clear sequence of distinct theories. Rather they are folded into each other, new modes of thought critiquing certain aspects of previous structures of belief while building on and extending others. This is clearly the case in the example of the emergence of 'humanistic geography'. Although humanistic approaches were appropriated by geographers in the1960s and early 1970s, in many ways the discipline was simply engaging with modes of thought and action from earlier epochs and other scholarly traditions, in particular humanist philosophy. This widened the sphere of what was appropriate for geographical study, orientating scholars towards the nature of the human condition. We can trace the legacy of this turn towards human subjectivity in later work in feminist geography and non-representational theory.

Summary

- Humanistic geography explores meanings, values and experiences as they are formulated within human practice.
- We should understand the emergence of humanistic geography in context: as a response to positivist approaches in geography that were feared to have de-humanised the discipline.
- Humanistic geography draws on the philosophies of phenomenology and existentialism, two allied bodies of thought that promote the significance of understanding human experience.
- We need to be clear that humanistic geography did not do away with scientific approaches – many human geographers have used a combination of positivist and humanistic approaches. In much humanistic work we can observe the *extension* of the models of positivism through the inclusion of new qualitative methodologies.
- Criticisms of humanistic geography suggested that it over-emphasised human free will without taking account of the structures that govern human action.

Further reading

Charlesworth, S. (2000) *A Phenomenology of Working-class Experience*, Cambridge: Cambridge University Press.

Entrikin, J.N. (1976) Contemporary Humanism in Geography. *Annals of the Association of American Geographers*, 66: 615–632.

Relph, E. (1976) *Place and Placelessness*, London: Pion.

Samuels, M. (1978) Existentialism and Human Geography, in D. Ley and M. Samuels (eds), *Humanistic Geography Prospects and Problems*, London: Croom Helm, pp. 22–40.

Seamon, D. and Sowers, J. (2009) Existentialism/existential Geography, in R. Kitchen and N. Thrift (eds), *The International Encyclopaedia of Human Geography*, vol. 3, Oxford: Elsevier, pp. 666–667.

Smith, S. (1984) Practicing Humanistic Geography. *Annals of the Association of American Geographers*, 74: 353–374.

Tuan, Y.-F. (1974) *Topophilia: A Study of Environmental Perception, Attitudes and Values*, Englewood Cliffs, NJ: Prentice Hall.

Marxist radical geographies

Learning objectives

In this chapter we will:

- Explore who Karl Marx is and why his ideas on the accumulation of wealth continue to be relevant.
- Identify seven key Marxist ideas, namely: historical materialism, the economic base, the superstructure, ruling ideas, class struggle, class consciousness and commodity fetishism.
- Document the rise of radical geography and investigate how these ideas were later to translate into a committed project of Marxist geography.
- Consider how a Marxist geography can enable us to critically evaluate both the urban landscape and the natural environment.
- Reflect upon established critiques of Marxist geography and its continuing salience in the modern world.

Introduction

Prising open a box-fresh pair of sports trainers offers a real moment of anticipation. Carefully unwrapping the crisp white sheets of tissue paper inside, before delicately removing further balled-up tissues held within each one, we can at last slip into our much coveted footwear. Yet what makes one pair of running shoes (like the ones shown on p. 72) so much more expensive than another? Why do multinational

Sports shoes as fashion commodities
Source: Alamy images/David Pearson

corporations such as Nike outsource 100 per cent of the labour? How is it that trainers may use rubber from Indonesia, leather from India and be stitched in the Philippines? And what types of hidden labour and forms of exploitation are hidden within this 'depthless' commodity?

In exploring Marxist theories and radical ideas, this chapter seeks to provide a framework through which to interpret these relations. By the end of the chapter you should begin to develop concepts through which to analyse the uneven socio-spatial relations between capital, labour and commodities. The chapter begins by introducing you to a leading thinker of the modern age, Karl Marx. We then focus upon seven key Marxist ideas, which form a guide for Marxist criticism. Having outlined what Marxism is and the revolutionary insights it generates, the remainder of the chapter concentrates upon a variety of geographical engagements. We begin with exploring the emergence of 'radical geography' in the late 1960s, with its explicit focus on 'changing the world'. We then demonstrate a 'turn' to Marxism and Marxist ideas among certain geographers in the 1970s and 1980s. This is further evidenced in a following section on Marxist geography exploring the socio-spatial relations of class. Here, we reveal how human geographers take Marxist ideas into the city and produce new spatial insights on social class that speak back to the idea of Marxism. We further maintain that the potential of Marxist enquiry is not restricted to urban geographers as we turn to consider the '**political ecology**' of Marxism. This entails an environmental Marxist approach in which the decimation of physical landscapes and the unequal deployment of natural resources feature. Finally, we reflect upon the demise of early Marxist geography before casting our gaze towards future horizons and possibilities for its invigoration. Rather than view Marxism as a relic of the past, throughout we seek to demonstrate the power of this legacy and its relevance within new global times.

Karl Marx

Karl Marx lived and wrote in extraordinary times. He witnessed the emergence of the Industrial Revolution in the late-19th century and the introduction of manufacturing and production on a grand scale. It is a period associated with the rise of modern capitalism as we know it – the system of monetary exchange for labour.

It is no little exaggeration to declare that Marx is one of the most influential thinkers and writers of the modern age. The impact of his ideas can be traced not only in his writings and upon numerous disciplines but they have left an indelible imprint upon the political arena. His biography and the changing world in which he resided are captured in Box 4.1. For Marx modern **capitalism** may be a deeply alienating process that polarises people across the fault-line of class, but it is not without hope. One of the contradictions of capitalism is that it carries with it the seeds of its own destruction. Under such exploitative conditions Marx believed capitalism to be 'in crisis', making resistance and eventually class revolution inevitable.

Box 4.1

Karl Marx (1818–1883)

Karl Marx was born in 1818 in the town of Trier, Germany, close to the French border. He studied humanities and law at Bonn University before undertaking law and philosophy in Berlin whereupon he joined the Young Hegelians, a club dedicated to the philosopher Hegel. After completing his doctorate in 1841 Marx moved to Cologne, and married Johanna ('Jenny') von Westphalen, a Berlin professor, a year later. The couple eventually had three children but spent most of their lives in poverty, financially supported by Marx's compatriot Frederich Engels. Marx would spend the majority of his life in England, having been expelled from Paris, Brussels and Germany en route. Although a member of the Communist League and a fervent writer and later editor of radical journals, Marx was more of a philosopher than an activist.

He was also a historical observer and social commentator, documenting changing times. The London that Marx observed was an awe-inspiring global city where raw materials were funnelled in from around the world, processed in factories and spun out as finished commodities across the globe. He saw first-hand the markedly different lifestyles of aristocrats, industrialists, white-collar professionals, manual workers, the urban poor and the homeless. He witnessed the different geographies of the city, from those who resided in the splendour of large town houses, to those impoverished in slums and on the street. How could a country with so much wealth, he wondered, be so starkly stratified?

Marx carried forth this burning question into his writing. He was a prolific writer, publishing many books and several analytical volumes blending philosophy, economics and social history. While his most important work is probably *Das Kapital* (1867), a searing critique of the organisation of capitalist relations that would eventually be published in three volumes, he is probably best known for *The Communist Manifesto* (1848), written with his long-term collaborator Engels. This latter edition was published at a time when revolutionary fervour spread throughout parts of Europe, and hence captures a certain zeitgeist of the period. Although Marx published a voluminous amount of work, throughout his lifetime he remained little known. He died in squalor in London, 1883; his grave and monument can be found at Highgate Cemetery. Today, many of Marx's ideas are seen as indispensible, offering a valuable critique of capitalism as the organising principle in modern society.

Key Marxist ideas

Marxism has furnished us with numerous concepts and a voluminous amount of technical language. To provide analytic clarity this section will concentrate on seven key Marxist ideas. The Marxist concepts under discussion will include: **historical materialism**; the **economic base**; the **superstructure; ruling ideas; class struggle** including labour theories of value; class consciousness; and **commodity fetishism.** We argue throughout that these essential Marxist concepts can be applied to everyday social relations and are a useful analytical device for understanding the world in which we live. Indeed, a good way of testing your knowledge of these concepts is to try to apply these ideas to modern-day examples of social, cultural, economic, political and ecological transformation.

Historical materialism

Marx developed what he called a 'materialist conception of history' from which to understand the social world. He argued that in order to understand a person's social circumstances we need to examine his or her place in the economic base over time. This important conception, generally termed *historical materialism*, suggests that an individual's thoughts, ideas, attitudes, perceptions and disposition cannot be divorced from the forces of production but instead are a consequence of their economic being. As Marx surmises, 'It is not the consciousness of men that determines their existence, but their social existence that determines their consciousness' (Preface to *A Contribution to the Critique of Political Economy*, 1950). Historical materialism is premised upon an understanding that in order to survive human beings must produce and reproduce the material requirements of life. This implicates them within specific *relations of production*, arrangements that we now view as the class system. Marx's material conception of social relations is suggestive of the powerful bearing the economy has upon our social circumstances. He postulates that:

> Men make their own history, but they do not make it just as they please; they do not make it under circumstances chosen by themselves, but under circumstances directly encountered, given and transmitted from the past.
>
> (Marx, 1968: 97)

This materialist conception of history reveals that class does not only come to shape who we are now, but who we may yet become in the future. Historical materialism is then central to providing alternative possibilities and scenarios for economic geography. For this reason Marxist scholars and activists are concerned to challenge, critique and even overthrow capitalist relations as described below.

The economic base

Marx is concerned with the changing organisation of human societies. He demands us to confront the question of who benefits from the prevailing social structure, how and why? In observing the relationship between people over time, Marx noted how those who controlled the *economic base* have the greatest opportunity to amass and reproduce their wealth. In feudal times the economic base would have been ownership of land which could be used for growing crops or rearing livestock to make a living. As knowledge of keeping livestock and increasing crop yields becomes established, the power of landed owners increases. Over time more pastures may be acquired and labourers employed to harvest the land, thus increasing yields and the power of landowners. As land is passed on through generations this means that ownership of the economic base can have a multiplier effect in generating wealth. In industrial times the economic base has been centred upon capital and the ability to generate money. The economic base is then the driving force of society, which means that those with the most money – and knowledge of that money – are best placed to reproduce their wealth over time. For Marx this explains how human beings may reproduce wealth across generations and why capitalism is a system that serves the interests of those who are already rich and powerful. This means that those with little stake in the economic base to begin with are increasingly less likely to escape their conditions of existence.

A modern example of the reproduction of wealth can be seen with the recent property boom. As many Western cities have been undergoing state-sponsored urban regeneration, property investors have sought to buy up disused buildings, warehouses and large properties, securing large amounts of debt in the process. With an injection of capital this has enabled them to build a series of state-of-the-art apartments that maximise the density of residential living per square foot. Where there is a high level of demand, buyers are encouraged to purchase property prior to construction by placing a deposit. These deposits earn interest and can be put towards further building work and refurbishments. When the apartments are complete and exchange takes place, the total profit made at point of sale is usually far in excess of the original capitalist outlay and labour cost. Developers would claim that the 'surplus profit' achieved counters the risks they have taken and stimulates further growth in the economy.

However, it could be argued that the properties are designed around profit not people. This can be seen where many first-time buyers may be excluded from the housing market, traditional practices of community come under attack, and everyday spaces are transformed into spectacles of consumption. This suggests that the social costs of housing will tend to be obliterated by the cold logic of capital. In Marxist terms those who own or can purchase the **means of production**, that is the technologies necessary to transform raw material into commodities, are able to command a workforce from which to produce profit. We can reconsider this spatial inequality by returning to our opening example of the production of Nike trainers. In this example, when wages rise, trade union activity intensifies or environmental regulations are imposed, capital is seen to be 'footloose' and able to relocate elsewhere to enable the production of sports shoes to occur in other places. Where capital is free to move across borders, unskilled factory workers are invariably less mobile.

The superstructure

The logic of capitalism dictates that those with the most money wield the most power. As Marx would have it, those who own the means of production have the capacity to structure social life. On closer inspection it appears that these people are able to shape society in such a way that the prevailing social structures, laws, values and belief systems support their group interests. But how exactly is such a project achieved?

According to Marx there is an intrinsic relationship between the economic base and the superstructure. The superstructure comprises those state institutions that are formative of modern social values. This could include the field of education, the family, judicial law, the church and, more lately, global media. As individuals we each participate in the superstructure and are each subject to the influence of these social formations. These institutions are not mutually exclusive but intersect with one another to produce shared knowledge and ideas about society. As such they can be considered purveyors of *ideology* that carry with them beliefs about how to be and how to act. An ideology is a set of beliefs that envision a whole 'way of life'. For example, capitalist ideology may prioritise individualism, cosmopolitan mobility, entrepreneurialism and the display of numerous consumer status symbols. Implicitly those who risk challenging these orthodoxies are likely to be seen as failing. Ideology is communicated through the machinery of the state, the institutional superstructures and individuals within. It is at its most successful when it appears the social norm. Over time certain ideologies become dominant and are shared across much of the superstructure, exerting an abiding influence on our lives. Our attitudes to drugs, crime, work, sexuality, property or homelessness are formed in dialogue with these social structures. The superstructure is then an integral arena in which ideologies are produced, contested and negotiated. Although dominant ideologies can be resisted, challenging them creates a risk of being ostracised or stigmatised. This is an indication of their power.

Ruling ideas

As we have seen, ideologies represent a particular, if incomplete and contested, vision, from which to organise human society. For Marx and Engels, ideology functions as a set of 'ruling ideas' that marshal the interests of dominant elites. The most successful ideologies are those that appear as 'commonsense'. The ideology of what constitutes 'the family' or attitudes to work form part of this wider group consensus. In Box 4.2, Marx and Engels document the relationship between ideology and power to distil how the 'ruling ideas' are in fact those that are forged by dominant social classes.

Box 4.2
'Ruling ideas' – Marx and Engels

'The ideas of the ruling class are in every epoch the ruling ideas: i.e. the class which is the ruling material force in society is at the same time its ruling intellectual force. The class which has the means of material production at its disposal, has control at the same time over the means of mental production, so that thereby, generally speaking, the ideas of those that lack the means of mental production are subject to it. The ruling ideas are nothing more than the ideal expression of the dominant material relationships, the dominant material relationships grasped as ideas; hence of the relationships which make the one class the ruling one, therefore, the ideas of its dominance. The individuals composing the ruling class possess among other things consciousness, and therefore think. In so far, therefore, as they rule as a class and determine the extent and compass of an epoch, it is self-evident that they do this in its whole range, hence among other things rule also as thinkers, as producers of ideas, and regulate the production and distribution of the ideas of their age: thus their ideas are the ruling ideas of the epoch.'

(Marx and Engels, 1968: 64–65)

The relationship between base and superstructure is shown below.

IDEOLOGY

↑

Superstructure
(e.g. education, law, media, religion, family, etc.)

Relations of production
(e.g. bourgeoisie, proletariate, etc.)

Mode of production
(e.g. manufacturing, machinery, new technologies, etc.)

↑

ECONOMIC BASE
(e.g. feudalism, capitalism, etc.)

Looking at the passage in Box 4.2 you might be struck by the intense relationship between the ownership of power and the social production of ideas. For Marxist theorists there is an intrinsic relationship between political economy and culture. Although the extent to which the base 'determines' what goes on in the superstructure is hotly debated, most Marxists would argue that the wealthy have the greatest ability to shape the social structures of society. As Marx and Engels declare, the ideas of the ruling class are indeed the ruling ideas. In order to understand this relationship between the economic base and the superstructure we might turn to the diagram in Box 4.2. Marx's own ideas are perhaps more subtle than what we present here and he does appear to recognise that the relationship between base and superstructure is dialectical, conditional, dynamic and shifting. Since the 'cultural turn' (see Chapter 5), geographers have sought to rethink the role of culture beyond what is sometimes seen as a mechanistic, Marxist reductionist analysis which renders culture purely to the subservience of the economic. Rather than seeing culture as a mere 'epiphenomenon' of the base – a by-product of material relations – contemporary geographers have explored how culture is embedded within economic relations, how it may carry its own symbolic capital and how it can become commodified, as the example of Liverpool City of Culture in Chapter 5 distils. Further on in this chapter we will examine Zukin's (1988) study of 'loft-living' in New York to consider how a culture of 'Bohemian chic' is materialised in space. Culture, it seems, is intrinsic to economies, place and identity.

Class struggle

Class struggle is a prerequisite of Marxism. Marxists have long concerned themselves with the potential for class conflict to bleed into class revolution. Marx himself envisaged that the numerous points of friction within capitalism would lead to periodic challenge, conflict and ultimately outright class revolution. Such a revolution, it is imagined, would enable workers to free themselves from the bonds of capital. In doing so, Marxists believe that a more egalitarian society can yet evolve, where the interests of capitalists and workers are no longer pitted against one another. By replacing profit as the ultimate goal in favour of social well-being, Marx felt that changes in the organisation of society can lead to greater autonomy and equality.

Writing in the early industrial period, which began around 1840, Marx saw the division of labour enacted between capitalist proprietors and factory workers as formative of specific relations of production. These new relations of production formed across class lines and divided into two main camps: capital and labour. Here, Marx identified an emergent **bourgeoisie** which comprised a newly minted, enterprising middle-class, and a mass **proletariat** made up from casual workers and labourers who came to form a stratified and diverse working-class. Although the proletariat significantly outnumber the bourgeoisie, in terms of money, status and power they remain relatively disenfranchised. Members of the bourgeoisie could include factory owners, business men, property landlords and capitalists whose accumulated wealth could be used to expand their infrastructure and extend their profit margins. The proletariat, on the other hand, include those who only have their bodies to sell in exchange for monetary reward.

Marx and Engels were particularly drawn to the 19th-century factory as a modern site of mass production. This is not surprising as factories employ vast number of workers, are

Table 4.1 The fault-line of class: opposing interests leading to class division and potential conflict

Desires of a bourgeois factory owner	Desires of a proletariat factory worker
• Constant productivity	• Regular paid breaks
• Open-ended working hours	• Restricted working hours
• Freedom from responsibility of care to workforce	• Owners accountable for health, safety and well-being of workforce
• Low fixed rate pay scales	• Higher incremental pay scales
• Non-representation of workers' rights	• Representation of workers' rights through shop stewards and national trade unions
• Ability to make decisions about the workforce unimpeded	• Workplace decisions made in negotiation with unions and representatives
• Recognise the benefits of abundant surplus labour to keep costs down	• Recognise the benefits of loyal, reliable workforce to ensure productivity
• Unpaid leave	• Annual paid holidays
• Wages immune to profit	• Wages increase in line with profit

hierarchically organised and are involved in production on a grand scale. In their observations, Marx was acutely aware of the markedly different life-worlds occupied by factory owners and their workers. In the 19th century many factory workers were subject to brutal conditions. This included corporal punishment, child labour and inhumane working hours that could result in people losing limbs on machinery, as inexperience and physical exhaustion set in. It seemed that within the particular 'relations of production', the bourgeoisie and proletariat held sharply opposing and potentially conflicting interests based around different needs and desires. For Marx there is nothing natural about this; instead it is the outcome of a particular ideological regime that values entrepreneurs, wealth accumulation, mass distribution and profit. This social division of labour can be represented in the divided interests that exist between workers and capitalists. Although they may share some interests, their needs and desires are often polarised and may come to be split along lines of social class, characterised in Table 4.1.

Although many of the more oppressive forms of 19th-century factory life successfully have been contested – for example child labour, the inclusion of safety standards and regular holidays – the capitalist system itself has largely gone unchallenged. In the 19th century it was frequently stated that the bourgeoisie and the proletariat occupied 'two worlds'. Many of the novels by Charles Dickens, of which *Oliver Twist* would be emblematic, offer stark insights into these different social class groupings and how they appear to live, work, eat and rest in separate universes. In the modern global economy, the relationship between those who own the means of production and the workers they employ may be equally distant and remote.

The Marxist idea of *class alienation* suggests that many modern workers are undertaking arduous and repetitive labour for little reward and that this has a *dehumanising* effect upon their existence. Manual workers on a Fordist car assembly-line in Germany, or mange-tout crop pickers in a field in Kenya, may each feel alienated from the goods they produce. Labour theory related to the function or use value of a commodity, its exchange value in the marketplace and the surplus value it generates through profit, have meant that workers may feel distant from the corporation's work – be it BMW or Tesco. Such 'uneven' geographies are seen where a Kenyan crop-picker in a field is unlikely to ever meet anyone from the Western supermarket chain that places strict orders on the preparation, quality and size of the yield. This alienation is amplified when Western consumers enter into relations of capitalist exchange. This may entail the purchase of a neatly packaged, plastic-coated, handful of mange-tout from the supermarket shelf with little thought of the hidden labour belying their production. Many geographers investigate these commodity chains and the inequalities that transpire at different stages of production, distribution, retail and consumption (see Hughes, 2007). For many workers class alienation may lead them to feel they are living in a precarious, uncertain and risk-laden world. For Marxist geographers such as Neil Smith (1990) capitalism inherently leads to uneven development across time and space. A strict division of labour then splices the interests of the flip-flop wearing peasant, toiling under the heat of an African sun, with that of the white-collared corporate boss who sits in an air-conditioned office on the other side of the globe contemplating future profit margins with a spreadsheet. What we see here is a modern-day *Oliver Twist*, comprising two different worlds, two social classes.

Class consciousness

For Marx a starting point for revolutionary change is for workers to unite through a collective **working-class consciousness.** To do so workers have to see their plight as a shared one based upon mutual interests. These interests may be in opposition to those of the bourgeois employers, who are nevertheless reliant upon their labour. As Marx famously asserted, 'Workers of the world unite, you have nothing to lose but your chains.' This statement captures Marx's revolutionary zeal as well as his understanding that much 19th-century industrial labour was exploitative. In his observations of factory life Marx did not fail to notice the exhausting hours, dangerous conditions, severe punishments, lack of autonomy and meagre pay meted out to workers. Throughout the 19th century and beyond, philanthropic and socialist activists have sought to ensure more humane conditions for workers, for example by campaigning to raise the working age in a challenge to child labour, or through the introduction of the 10-hour bill designed to legislate the time that could be spent engaged in factory work. While we may take some of these reforms for granted today, they are the result of much protest, struggle and resistance. Recent 'No Sweat' campaigns and exposés on child labour echo the capitalist history of bodily exploitation.

The legacy of Marxism can be seen in industrial changes to the factory shop-floor. Labour is also organised through trade unions that enable local and national issues to be aired relating to the workforce. Workers who choose to sign up to unions will regularly pay a subscription for the support and backing of other comrades, which can be important in instances where redundancies are made, or decisions about strike action are taken. Today one of the ways in which labour is commonly represented in the factory system may be through the use of a shop steward or another intermediary acting on behalf of the workers. As a representative of the workforce the shop steward has the ability to speak with employers, raise the concerns of workers on particular issues such as pay and conditions, and negotiate settlements through dialogue with workers and employers. Despite these advancements Marxists would argue that factory life and other modern work relations are still highly exploitative and produce unequal class relations.

Since the working-classes numerically outnumber the bourgeoisie we may wonder why workers and the unemployed continue to remain oppressed. At a global scale we may pause to reflect upon why it is that developing nations with an abundance of resources continue to live under the shadow of Western capitalism. That people may not 'see' the conditions of their existence as exploitative, does not render this system any less oppressive. Marxist adherents have gone on to term this '**false consciousness**', where members of a subordinate class such as labourers, peasants or serfs are unable to see past the ideological fictions that enshroud them. Capitalist ideology, as a 'master narrative', is then systematically designed to conceal or obscure the realities of their subordination. Capitalism then creates distortions, mirages and blind-spots in the consciousness of workers that leads to acceptance of their conditions and place in society. Any conscious appeal to social justice is systematically erased.

The distorting quality of capital is evident in the 'American Dream', a utopian ideal that anyone can be successful with the right attitude and application regardless of their socio-economic circumstances. This theme is endlessly replayed in the plotlines of Hollywood film, the 'rags to riches' narratives of celebrities and the underlying premise of game shows such as *Who Wants to be a Millionaire*, *X-Factor* or *The Apprentice*. Each of these visions present us with a tantalising glimpse that everyone – yes you too – can make it; a spectacle that appears as ridiculous as it is compelling (see Bauman, 1988). In his age Marx regarded religion as 'the opiate of the masses', a spiritual distraction which purported that those who prayed and endured the injustices of the material world would be eventually salvaged in the afterlife: the meek shall inherit the earth. For contemporary Marxists, popular media productions such as the shows described can also be seen as opiates, inducing 'hallucinations' that distract from the task of seeing the oppressive conditions of one's own making. In celebrating capitalism, individual stardom and fortune, the suggestion is that false consciousness never strays far from the surface of everyday existence.

Commodity fetishism

During the Cold War, prior to the collapse of Communism, one of the most desirable commodities to bring into the Soviet Union and parts of Eastern Europe was a humble pair of denim Levi jeans. As a commodity Levi jeans can be said to have a 'use value' – a measure of their utility in human society – in that they cover our nakedness, keep us warm and function as a practical and enduring item of clothing. It is for these reasons that Levi jeans were designed at the height of the California goldrush of 1853: for prospectors, cowboys, cattle handlers and farm ranchers across the US. However, under capitalism the value of goods lies not in their usefulness to society but in their ability to be exchanged for other things, usually money. If we consider Levi jeans more closely they also contain a 'labour value', indicated by the amount of time, skill and expense spent on manufacturing a particular commodity. Finally, a commodity such as Levi jeans would also carry an 'exchange value' defined through the market which combines use and labour value but also includes a profit margin, or 'surplus value'. Marx exposed how a capitalist surplus can only be achieved if the labour value contributed by workers at the point of production is greater than the labour value of the actual wage they receive. In other words, exploitation of labour is a prerequisite for profit.

Marx was further fascinated by the way in which capitalism shrouds life in a veneer of mystique, ideology and fetishism. Here, fetishism refers to the acclaimed practice of attributing special powers to mundane material objects. This can provide goods with a 'phantom objectivity' that manifests in 'perverted form' (Watts, 2005: 541). The curious demand for Levi jeans in parts of the Eastern bloc can neatly be explained then, through the concept of 'commodity fetishism'. After all what makes one pair of jeans or trainers better than another? In the context of the Cold War, Levi jeans epitomised all that was glamorous about the West – freedom, fashion, sex and prosperity. Levis could easily be exchanged for money, labour, food or other desirable goods in the black market economy. It is perhaps tempting to think of material goods such as jeans as lifeless products, 'dead matter'. But as Michael Watts (2005: 534) reminds us, 'commodities have lives, or *biographies*. They are made, born or fabricated; they are fashioned and differentiated in a variety of ways; they are sold, retailed, advertised and ultimately consumed or "realized" (and perhaps even recycled!)'. It is these biographies, or commodity stories, that geographers can use to enliven seemingly inanimate objects by showing the invisible hand of human labour.

Indeed, one way in which geographers have sought to develop critical accounts of commodities is to 'follow the thing' (Cook, 2004), by examining how material objects move across time and space to form a 'commodity chain'. Whether that commodity is exotic fruit such as papaya (Cook, 2004), Jamaican hot pepper sauce (Cook and Harrison, 2003) or a pair of Nike trainers, if we follow the links that make up the chain – from production through to consumption – a unique and complex biography is likely to unfold. 'Following the thing' involves dissecting the geographical entanglements and spaces of flow that stretch across networks of production, circulation, regulation and consumption (see, for example, Castree, 2004; Cook and Harrison, 2003; Cook, 2004; Hughes and Reimer, 2004). If you were to open your kitchen cupboard and take out a commodity such as a box of tea, jar of pickle or tin of sardines you may soon begin to narrate at least part of its story from the fine print attached. In order to follow the thing beyond its point of retail you could place these goods within a 'commodity circuit' (Watts, 2005: 534) and bring to light their hidden geographies and untold histories by pursing the commodity chains across time and space. This circuit might also include diverse actors such as farm growers, crop-pickers, packagers, handlers, sales representatives, exporters, importers, drivers, distributers, retailers, marketers, purchasers and consumers.

In capitalist relations certain goods become endowed with myth and fetishism that sets them apart from other less desirable items and adds value to their stock. Commodities such as Coca-Cola and Levi jeans are said to have achieved 'classic status', intimating that their value is enduring. Consider once more the Nike trainers as emblematic of this fetishism, where a 'back story' or narrative adds zest to its mystique. In Greek mythology Nike is the winged goddess of victory, intimating that the product has mythical origins. The 'swoosh' symbol seems to convey speed, coolness and athleticism, effectively captured in the company's slogan, 'Just Do It'. The product is also frequently advertised by US black athletes locating it as 'hip', 'edgy' and a focus for a particular type of colonial fetishism rooted in the Western imagination. The power of **signs**, symbols and motifs to speak to our unconscious dreams and desires should not be underestimated. The way in which everyday material goods may be symbolically encoded in the social relations that arise between people and objects implies a powerful imaginative appeal. It is precisely this relationship between neoliberal capitalism, branding and signs that the populist commentator Naomi Klein rages against in her best-selling book *No Logo* (2000) in which we can find the embedding of social relations into the economy.

But while geographers have rightfully argued for a return to the materiality of commodities (Cook and Harrison, 2003; Jackson, 2000; Gregson, 2007; Miller, 2008) commodity fetishism is also about the immaterial – what is unseen. For geographers such as David Harvey (1990), whose work we address in further detail in this chapter, we need to get beneath the veil of commoditisation by rejecting the surface appearance of things and the inherent properties of the 'commodity fetish'; this involves tracing back the spatially uneven relations that enable capitalist commodities to materialise. Penetrating the 'mystical veils' of commoditisation that selectively seek to obscure or reveal particular meanings is a way of unravelling or deconstructing the hidden messages contained. In the symbolic construction of commodities, the sweat of human labour is all too readily absented as goods become disposable and are emptied out of meaning.

Contemporary society is frequently understood as 'hollowed out' and replete with a scattering of hallucinatory images, affects and artifice that give way to a type of '**hyper-reality**' (Baudrillard, 1994). Here, the symbolic value of items may far exceed their use value as the social act of gift-giving is given further meaning through the mystery of gift-wrapping, presentation, advertising and logos that imbue mundane objects with 'magical' properties. The magical appeal of everyday objects, and how this operates through global capital, is part of the 'economies of signs and space' (Lash and Urry, 1999) that pervade and proliferate late-**modernity**. For Marxists our tendency to be seduced by the latest gadget, designer wear or food fashion is precisely the spectacle of capital we continue to be transfixed by, even as it imprisons us. Bringing together the materiality of commodities with the immaterial desires and emotional feelings that surround them is then an important means for understanding our complex and often conflicting relations with commodities. The smoke and mirrors of capitalist ideology suggest all that is solid, soon melts into air.

Radical geography

In the immediate years succeeding World War II, human geography was primarily a spatial science concerned with relationship between space and environment (see Chapter 2). Under the grip of the Quantitative Revolution geographers sought to understand the social world through statistical analysis and mapping techniques. However, by the mid-1960s a new political consciousness was in the air: as the songwriter Bob Dylan nasally crooned, 'The Times They Are a Changing'. But what was the nature of this change, why did it matter and what effect did it have upon Anglo-American geography?

The changing times of the 1960s are often caricatured in colourful documentaries celebrating 'free love', 'spliffed-out' Hippies and screaming fans enraptured by Beatlemania. However, the cultural revolution in sex, drugs and rock 'n' roll occurred across a backcloth of political unrest. The student revolt that occurred in Paris in May 1968 appeared symptomatic of the demands of a new generation eager to shake up the existing social order. On the other side of the Atlantic the police brutality, which accompanied the Civil Rights Movements in the US, instantaneously spilled into the living rooms of the public through the new medium of television. Those campaigning for equal rights between African-Americans and their white counterparts began to reach a global audience. The shooting of Martin Luther King and war in Vietnam further highlighted the relationship between race and power as a new mood of social change pricked popular consciousness.

US university campuses were primary sites for peace protests, civil right sit-ins and student activism across a range of political issues. Geographers were not immune to these transformations and a concerted radicalism emerged around imperialism, environmentalism, gender inequality and development. More liberal quantitative analysts began to argue for a 'socially relevant' geography which was now claimed as necessary if the discipline is to remain topical and true in its commitment to world issues. Thus, the American Professor James Blaut (1970) viewed the Vietnam War as an extension of Western imperialism, which he plainly characterised as 'white exploitation of the non-white world'. But Blaut did not stop there; his subsequent work on ethno-development is an example of the productive cross-over between Marxist analysis and the post-colonial perspectives we consider in Chapter 11 (for detailed reflections of Blaut's contribution see special issue of *Antipode*, 2005: 7). The formation of the geography journal *Antipode* in 1969, based at Clark University in Massachusetts, captures the spirit of the time. Subtitled *A Radical Journal of Geography*, *Antipode* became an important arena for popular and academic discussion of poverty, spatial ideology, discrimination, human rights and development fittingly communicated in posters, essays, reviews, poems and cartoons.

The desire to instigate a 'geography that matters' is further evident in the work of William Bunge, a leading quantitative analyst, and perhaps unlikely radical. But Bunge (1969) was impatient with the stuffy, apathetic world of US academia and launched scathing attacks on what were decried as 'armchair geographers': those who sat back, mapped and described the social world instead of working to make a difference. To achieve a more meaningful geography Bunge augmented his quantitative analysis with theoretical approaches from which he hoped to envision a new geographical imagination. Bunge's new-found radicalism would result in him being refused academic tenure by Wayne State University in 1967, but it did not quell his acerbic ardour. A year later Bunge formed the 'Society for Human Exploration'. It is worth pausing on the nomenclature of this venture, for it perhaps captures something of the contradictory manner in which Bunge veered between a radical geography of social change and something much more conventional in practice.

Bunge set out to retrieve some of the tenants of early geographical enquiry, seduced by the thrill and allure of 'geography fabulous' from a previous era (see Chapter 1). Where the colonial narrative of past geographical history was fixed upon wild jungles and distant tribes, Bunge's lens was focused upon the equally exotic and seemingly 'unknown' quarters of inner-city Detroit, populated by black residents and the urban poor. He would later inspire similar 'expeditions' in Toronto, Sydney and London. Few geographers can fail to be moved by Bunge's admirable, if slightly grandiose ambition, to embark upon an expedition in which local residents become the planners and agents of social change in their communities. 'The purpose of the Expedition is to help the human species most directly', he opined. 'It is not a "nice" geography or a *status quo* geography', declared an ebullient Bunge. What Bunge essentially argued for is a participatory geography in which, 'Local people are to be incorporated as students and professors' (1969: 35). Reading Bunge's prose today it is certainly charged with vitality, yet there is a slight unease that his quest for expedition carries with it at least some of the imperial hallmarks of early excursions into the 19th-century city which were simultaneously designed to repulse and titillate the Victorian British bourgeoisie (Jackson, 1995). Bunge was unapologetic in advocating 'a geography that tends to shock' (1969: 35) as he attempted to include a spectrum of human urban experience, later focusing upon the marginalised plight of inner-city children (Bunge, 1973).

But is such an approach as novel as William Bunge would have us believe? And is he the doyenne of a new urban radical geography? A broader gaze across the disciplinary horizons might find that the Chicago School of Sociology and Anthropology had long before turned its attention to geographies closer to home with accounts of race, immigration and poverty in American urban society. Most famously, William Foote Whyte's expedition into the slum district he calls 'Cornerville', first published in 1943, develops an intimate portrait of 'Street Corner Society' in this Italian-populated neighbourhood. Rather than reflecting upon the chaos of slum life, Whyte delivers an eye-opening account of urban living in which neighbourhood gangs form highly organised relationships with the police, political groups and rackets as a response to prevailing economic conditions. By living with an Italian family for 18 months, learning the language and participating in the leisure activities of the younger generation, few would deny that Whyte, and other illustrious ethnographers from the Chicago School, had already begun to excavate an 'upclose' urban anthropology that was epistemologically and methodologically as sophisticated as anything radical geographers proposed. The fact that Bunge's mantra of 'social relevance' was rapidly embraced by the geography profession – becoming a centrepiece of future academic discussion – suggests he was less the pioneering radical he may have liked to imagine, and more a voyager of the broader political and cultural currents that geography was now caught within.

In critically evaluating radical geography it appears that much work might fall under the banner of 'radicalism'. Radical geography appears radical in intent but liberal in practice. It operated in ways not dissimilar to much mainstream geographical practice at the time. As Richard Peet (1998: 75), an early advocate, professes 'radical geography was radical in topic and politics but not in theory or method of analysis'. It remained tied to positivist approaches and statistical explanation. For Cloke *et al.* (1991: 34) the radical geographer like other spatial scientists is, 'still a technical expert, but now a more socially aware one'. In hindsight much radical geography appears idealistic and occasionally naïve. Although we would not depart from the criticisms made above, radical geography did air important questions about the purpose and nature of the discipline by recognising geography as a political practice. Recent discussions concerning the value of **participatory methods** in research and the need to empower respondents in the process also speak back to early radicalism (see Cahill *et al.*, 2007), as do contemporary concerns to impart a 'geography that matters' (see *Geography Compass*, 2, 2008). It is also worth underlining that while radical geography may have enamoured a number of geographers, many remained unmoved. However, if radical geography has achieved anything it surely lies in the manner in which it helped create – however unintentionally – the conditions, climate and environment in which a new and genuinely radical project could emerge: Marxist geography.

The 'turn' to Marxism

The roots of Marxist geography can be found in the protest culture of the sixties and the impulse to make the discipline socially relevant. The transformation from an altruistic, radical geography to an explicitly Marxist revolutionary agenda, which came to jettison many of the positivist principles human geographers held dear, signals a genuine paradigmatic breach. The seeds of Marxist geography then sprout from the rich and fertile soil cultivated by radical geographers, activists and critical theorists.

David Harvey, a leading quantitative geographer of the period, epitomises this transformation. His debut book *Explanation in Geography*, first published in 1969, is firmly rooted in the scientific revolution, perhaps exemplifying what Peet (2001: 75) declares as geographers' 'desperate attempt at becoming white-coated "scientists"'. Just four years later the British geographer Harvey would turn his back on the scientific tradition with the publication of *Social Justice and the City* (1973). Challenged by his experience of living in Baltimore, Harvey began to adopt a Marxist approach to urban theory, land use, wealth accumulation and market rent. In these readings the city can be seen as a mode of production and a site for the enactment of spatial inequality, as the example of the housing market reveals. While the high point of Marxist geography is undoubtedly the 15 years preceding publication of *Social Justice and the City*, it would be naïve to presume that more traditional geographical enquiry was stymied. For many geographers it was, as Cloke *et al.* wryly comment, 'very much a case of business (literally) as usual' (1991: 40).

Even so, the smouldering embers of radicalism could no longer be ignored when fanned in the conjoining of Marxism and geography. It appeared that a number of academics burning with a desire for social change at last had a coherent theory, vocabulary and political mode of expression that was sophisticated enough to capture the complexity of the changing social relations. For Harvey the embrace with Marxism is no idle dalliance but a committed love affair, as documented in such Marxist titles as *Limits of Capital* (1982), *The Urban Experience* (1989), *The Condition of Postmodernity* (1989), *The New Imperialism* (2003) and *A Brief History of Neo-Liberalism* (2007). A particularly useful insight into Harvey's thinking can be seen in *The Condition of Postmodernity* in which he provides a strenuous critique of postmodern theory developed through a focus on material relations and political economy. In other words, Harvey regarded the later postmodern turn – discussed in Chapter 9 – as a distraction from the capitalist struggle.

If the collection of essays in *Social Justice and the City* (1973) mark the beginning of new Marxist geography, for this to have transpired into a widespread disciplinary 'tradition' would require other adherents, disciples and practitioners. Key influences included luminaries at the Frankfurt School in Germany whose Marxist critiques of urban life and popular culture were enormously influential in developing approaches to everyday life based on 'critical theory' rather than numerical analysis. Urban theorists such as Manuel Castells, who would later write about information technology and Sharon Zukin, who would chronicle the changing culture of city space, each produced accounts in which power was the primary unit of analysis. Along with David Harvey other geographical exponents of Marxism would come to include Neil Smith writing on the political economy of the nature–culture binary; Doreen Massey on the economic geography of labour markets; and Don Mitchell who focused his attention on cultural landscapes and workers, before later turning his attention to the enactment of public space and the exclusion of the homeless. Each of these studies, and many more besides, came to form the bedrock for a distinct brand of critical Marxist geography.

Marxist geography and spatial constructions of class

If there is one area that geographers have contributed most in Marxist debate it concerns how *spatial practices* produce and reproduce social inequality at different scales. Where

Marx concerned himself primarily with a historical materialist analysis of economic relations over time, radical geographers have considered how power and inequality are expressed through space. David Harvey (1989b) would describe his approach as formative of a 'historical-geographical materialism' that took Marxist ideas and applied them to the spatial environment. In doing so Harvey produced an analysis that was not simply 'additive' to more familiar Marxist lines of enquiry, but genuinely takes Marxist theorisation in new directions through considerations of its spatial dimension and expression over time.

A rigorous example of the new 'historical-geographical materialism' is executed by Neil Smith (1990) in his important edition *Uneven Development: Nature, Capital and the Production of Space*. The term uneven development refers to the way in which development is an unequal process that occurs over time and space. We might see this unevenness taking place at different scales, for example in the widening inequalities that exist when it comes to education, health, child morbidity, income and life expectancy in comparisons between the global north and global south. These differences can also be mapped at a local scale where stark difference may exist between neighbourhoods in the same city. Throughout the book Smith uncouples the binary that appears to construct space and society as two distinct phenomena. Using Marxist analysis Smith reveals how the forces of production and the logic of capital are inherently spatially organised. Smith captures this sense of estrangement and its accompanying forms of alienation:

> The point is that uneven development is the hallmark of capitalism . . . The uneven development of capitalism is structural rather than statistical . . . uneven development is the systematic geographical expression of the contradictions inherent in the very constitution and structure of capital, (Smith, 1990: xiii)

Unevenness is not, then, incidental to development, but it is endemic to it. This leads Smith to assert how, 'uneven geographical development at different spatial scales is a necessity of the logic of capitalism'. To demonstrate this we will go on to consider Zukin's (1988) observations of the transformation of SoHo in New York City from a rundown industrial area to one of the wealthiest enclaves in Manhattan. What is evident is that the dialogue between Marxism and geography is not a one-way street but a coming together, formative of new thinking on social class. As the American urban geographer Ed Soja poignantly reflects, 'Marxism itself had to be critically restructured to incorporate a salient and central spatial dimension' (1989: 335). Rather than seeing space as a vacuum-sealed container in which other social processes occur, geographers instead emphasise how space can also be exposed as a social structure in the ordering of societal relations.

In his book *The Urban Question: A Marxist Approach* (1977), the urban sociologist Manuel Castells draws upon the insights generated by the French Marxist structuralist philosopher Louis Althusser. Although Castells would later develop more post-structuralist accounts, here he argued that cities function for the reproduction of capital and labour power. This perspective challenged older work on urbanism conducted by members of the Chicago School of Sociology through its refusal to abstract urban culture from the material world. Both Castells and Harvey were critical of the Chicago School for favouring culturalist rather than economic explanations for urban growth. Castells would go on to write how the capitalist city is made up of complex social structures that appear to determine how we live and act, with little apparent room for agency or intervention. However, under the rubric of a new Marxist geography it is also possible to transcend this determinism. An exposition of these ideas can be found in Henri Lefebvre's meticulous account *The Production of Space* (1991), first published in France in 1974. Lefebvre, a Marxist-inspired philosopher of the sociology of 'everyday life', has been described as developing 'the finest product of the Marxian conception of space' (Peet, 1998: 101), although until his work was translated in 1991 he was largely ignored. *The Production of Space* is a multilayered analysis with a central focus concerning how 'The class struggle is inscribed in space' (1991: 55).

For Lefebvre the architecture of the city is both an expression and an engineer of capital. Here we get a glimpse into how a focus on geography can advance Marxism:

> Space as a whole enters into the modernized mode of capitalist production: it is utilized to produce surplus value. The ground, the underground, the air, and even light enter into both the productive forces and their products. The urban fabric, with its multiple networks of communication and exchange, is part of the means of production. The city and its various installations (ports, tube trains, etc.) are part of capital. (1979: 287)

Lefebvre's spatial account of Marxism certainly reveals how capital is produced through space. He demonstrates how power penetrates space, seeping into the pores of everyday life almost unnoticed in the habitual practices and routines of living. Lefebvre produces a more open-ended and multiple sense of spatial practice than the 'over-determining' capitalist relations Castells writes of. What is interesting about his approach is that it remains resolutely Marxist yet also intensely humanistic, focusing upon the rhythms and habits of daily life. This is not the 'inflexible', cold logic of Marxist structuralism, but as Ed Soja remarks, 'For Lefebvre, the lived spaces were passionate, "hot", teeming with sensual intimacies' (1996: 30). The combustible approach of Lefebvre may owe something to his own biography as a Marxist firebrand. For Lefebvre would experience class struggle first-hand as he became involved in the 1968 student protests which began in Nanterre University where he was appointed as Professor. His Marxist analysis of space, class and capital is then an animated account in which the city is very much 'hot-blooded' and alive.

However, Marxist geography does not have to be limited to the urban environment. The relationship between rural landscapes, labour and the economy are carefully illustrated in Don Mitchell's (2003) story of migrant strawberry pickers in Southern California. He reminds us how seemingly natural landscapes are cultivated through the logic of economics. When it comes to strawberry growing, farmers whose fields lie in environmentally favourable areas can command the highest fees for producing the best fruit or highest yields. Mitchell further reflects how what appears a 'natural' commodity such as the strawberry says nothing of the labour that makes it. It is, therefore, 'dead labour, that is, labour ossified, concretized, materialised into a definite thing with a definable shape and structure' (2003: 237). Mitchell's point is that the blemish-free berries we purchase, wash, slice and consume hide an extraordinary amount of 'living labour'. Consider the strawberry as a material construct engineered through social processes – its molecular and cellular structure is the culmination of countless years of laboratory experimentation made to make it less susceptible to pests, disease and the rigors of storage and transport, yet still appealing to the eye of the consumer when the product appears on the supermarket shelf. But this is only one aspect of the labour process. Crops are often picked by migrant workers who may sleep in ramshackle conditions and spend the day bent double beneath the glare of the California sun as they inhale pesticides and dust, pick up allergies and risk severed hands and fingers in the cultivation of the 'devil's fruit'. For Marxist geographers these are the invisible fingerprints of exploitation, the 'dead labour' we wash from our produce to serve up on a hot summer's day.

In surveying geographical research of the time, it is interesting to note the widespread incorporation of Marxist terminology into debates on space. For example, David Harvey would ruminate on the 'spatial fix' of labour compared with the hyper-mobility of capital to exploit these relatively static markets as and when it saw fit. Neil Smith saw 'uneven development' as the necessity of a rapacious capitalism that needed to be directly challenged. The economic

Box 4.3
Loft living: culture and capital in urban change

To illustrate the practice of Marxist geography and its capacity to shed light upon spatial form we can turn to a study undertaken by Sharon Zukin in her book *Loft Living* (1988). In her observations of New York residency undertaken in the SoHo district of the metropolis, Zukin explores how a number of New York artists were drawn to studio lofts for the cheap rent, floor space and light they offered. These artists were willing to put up with dirt, noise and the inconvenience of not living near local amenities. Zukin's account documents the changing economic and cultural space of SoHo as manufacturing gives way to a new aesthetic in urban living.

Zukin, herself a loft resident in the Greenwich Village area of the city, observes a startling transformation as capital and space intermesh to dramatic effect. She recalls how by 'around 1970, as the bare, polished wood floors, exposed red brick walls, and cast iron facades of these "artist's quarters" gained increasing public notice, the economic and aesthetic virtues of "loft living" were transformed into bourgeois chic' (p. 2). At this point Zukin found that a number of middle-class elites began to move into loft spaces and pay architects and designers to conduct expensive renovations. Where the artists had used these spaces for living and working the lofts for upper-middle-class people were used solely for residency. The attraction of loft living was enhanced by the idea that the middle-class could achieve a Bohemian lifestyle as they rubbed shoulders with artists, craft-workers, dressmakers and other creative people.

In his preface to *Loft Living* David Harvey proclaims the study 'an exemplary piece of work in the venerable tradition of historical materialism' (p. x). The reproduction of capital over time can be seen where property developers and investors begin to buy-up loft spaces. At this point lofts are transformed from places of production to a spectacle for cultural consumption. Zukin remarks how light

Loft living as an industrial aesthetic
Source: Alamy Images/Red Cover

manufacturing is annihilated in this transformation when lofts can no longer be used for machine shops, printing plants, dress factories or die-cutting operations. For Zukin, 'The residential conversion of manufacturing lofts confirms and symbolises the death of an urban manufacturing centre' (p. x). Zukin's research offers an interesting exemplar of the practice of Marxist geography. It reveals how capital accumulation can transform space in which lofts can become part of a wider terrain of conflict between different social actors. This may include such diverse social class groups as independent artists, middle-class tenants, small manufacturers, real estate developers, wealthy upper-class elites, local politicians or corporate financers. The value of this study is seen where processes of gentrification similar to the one Zukin describes in the SoHo district of New York are now commonplace throughout most Western industrial cities. In this way geography can be seen to be at the heart of the struggle for place, class and identity.

geographer Doreen Massey developed a notion of the '**spatial division of labour**' to emphasise the particularity of place and its role in the process of capital restructuring. In this Marxist revisionist account Massey went on to reveal how uneven capitalist development is situated in the landscape of employment, economy and class. These brief examples serve to demonstrate how the technical aspects of Marxism have become part of the grammar through which human geography is now inscribed. To get a firmer idea of Marxist geography in practice we can turn to a study by Sharon Zukin (1988, first published in 1982), exploring urban transformation (see Box 4.3). Through the application of Marxist economic geography we can gain insight into how urban areas change over time. As you read through the passage you may wish to consider how an explicit focus on Marxist geography can help explain social change.

The political ecology of Marxism

The application of Marxist ideas to urban life has been a centre-point of geographical analysis. It is evident in the approaches of the work discussed above by David Harvey, Henri Lefebvre, Neil Smith, Doreen Massey, Ed Soja and Sharon Zukin. However, a number of geographers has also deployed Marxism to interpret the natural environment. In doing so they challenge the easy split made between 'nature' and 'economy' by demonstrating how these relations cut into one another like teeth on a saw. Political ecology refers to the relationships humans, animals and plants have with the non-living environment and the ways in which this is structured by the politics of global capital. This can be seen in debates on water resources, timber, mineral extraction, food scarcity, oil or the preservation of wildlife where the market intersects with the environment. These examples suggest that natural resources can readily be transformed into commodities that are sold or exchanged in the global marketplace. It is even possible for an 'absence' to materialise into a 'presence' through the alchemy of capital. For instance, there have been recent political discussions concerning the possibilities of nation states trading in 'carbon off-setting' to avoid environmental and economic incursions. The political ecology of Marxism is evoked in Neil Smith's potent metaphor of 'El Nino capitalism' (Smith, 1998). Just as El Nino shapes the global climate, a crisis in capital can wreck havoc upon the financial landscape that has similar worldwide repercussions. References to a 'global meltdown' and 'global recession' following the collapse of world markets in October 2008 are indicative of this 'El Nino' effect.

An emphatic example of the political ecology of Marxism can be seen with the impact and aftermath of Hurricane Katrina which swept through New Orleans in 2005. It was largely the poor who died, whose houses were destroyed, who were left behind or were displaced and forced to take refuge in airless sports stadia. Yet the devastating effects of the hurricane can make it appear an arbitrary act of misfortune. Instead, a Marxist political ecology could certainly provide a compelling account of the relationship between environmental damage, material deprivation and survival. At the same time social class was not the only determinate influencing an individual's life chances. Endless news pictures revealed a further inescapable truth: the majority of victims abandoned on rooftops or left to die were black. How do we account for this and what function might race and class as vectors of power play in organising these seemingly random social relations? Any consideration of these questions must engage with the post-colonial legacy of empire, the forces of capital and the geographical displacement of people from parts of Africa and the Caribbean by way of the slave trade (see Chapter 11). Although the period of slavery may have passed, many Southern African-Americans continue to remain dispossessed. These communities were brought to Mississippi against their will to work as slaves in cotton fields in the Deep South.

The history of the Deep South and the history of race are then inseparable. Where New Orleans was once known for being a former slave-owning 'cotton town', today its black roots are celebrated in blues, jazz and exotic cuisine such as Jamaican jerk chicken or spicy jumbali. These more recent examples of post-colonial commodification reveal how the past tears holes in the fabric present. Nowhere is this historical materialism more apparent than in the case of Hurricane Katrina. New Orleans contains some of the most deprived neighbourhoods in the US with subsidised housing concentrated in impoverished, predominantly black areas. It was these places that were worst hit by the floods. The concentration of African-Americans in these neighbourhoods was enhanced by poverty and longstanding forms of residential and schooling segregation. Nearly 80 per cent of the residents living in the flooded areas of the City of New Orleans were non-white. Moreover, a third of the African-American population were entirely reliant upon public transport; their mobility was fatally impeded when the storms came. Stanton (2000) previously notes how many urban parts of New Orleans were regarded by whites as a dead-zone. He records how, 'Canal Street, the teeming main downtown New Orleans, was described as "dead" by white residents when I arrived in the city. The energetic "Third World" and African-American commercial presence there was not registered as a realm of the living' (p. 129). It may come as no surprise then that the African-American population are disposable, for they had long been cast as the 'living dead'.

The aftermath of Katrina has been no kinder to the region's black residents. The shocking pictures of black bodies floating past river banks – a haunting reminder of past Mississippi race murders – and survivors crying out for food, water and medical supplies is a salutary reminder of those most affected. As President George W. Bush surveyed the scene from above in a helicopter that did not touch down, the voices of those demanding state intervention were, often literally, drowned out. The radical musician Gill Scott Heron poetically reminds us of the power of ideology in moments of crisis – 'The revolution will not be televised.' The racialised power of capital is seen where in New Orleans it is homeowners that have been placed as 'high priority' for state support, while those in the rental sector, which includes numbers of minority ethnic families, remain dispossessed. Without any healthcare or insurance to recoup financial losses, their exposure to risk is magnified through historically entrenched patterns of racial segregation. This reveals not only the stratified nature of US society but how the processes of race and class inequality are reproduced across time and space. The geography of impoverishment and discrimination is not accidental but structural. This is, then, no act of God or haphazard matter of circumstance. As Cindi Katz (2008: 17) has recently discussed, the historical geography of structural inequality in New Orleans meant 'Katrina revealed the contours of visceral racism', once hidden in the headlights of neoliberalism. Forced to live and rent houses in high-risk areas when the levis broke, that is the raised banks used to prevent flooding, black communities were especially vulnerable to environmental danger, trapped in the 'eye of the storm'. They remain the inheritors of 'risk' and the victims of disaster. Such environmental disasters inform us of the value of a Marxist geographical inquiry to an understanding of political ecology. At the same time they also impart how other social relations – in this case constructions of race and place – exceed and complicate the Marxist class model. With this in mind, we will explore some of the limits of Marxist geography and how it is being revised in light of contemporary criticism.

The limits of Marxism

The end of the Cold War and the demise of socialism signalled a hammer-blow for the purpose and value of Marxist ideology. The break-up of the former Soviet Union and its ironclad-grip on allied Eastern bloc 'satellite states' was emphatically symbolised with the tearing down of the Berlin War in 1989. Where a geo-political divide had once separated the world into its constituent parts of capitalist or communist states, the triumph of global capital has appeared all but complete. Today the growth of capitalism in former communist states has also been accompanied by the rise of new markets in the developing world and has seen the economic rise of countries such as India and China. Inevitably the relevance of Marxist geography as a radical project has to come under political and theoretical scrutiny.

However, the challenge to Marxist geography was not simply external but had long been taking place within the heart of the discipline and more generally across the arts, humanities and social sciences. Prominent here is the turn to post-modern theory as a sophisticated means to capture the complex and fragmented world, through what was deemed a less blinkered geographical eye (see Chapter 9). For post-modernists Marxism was revealed, perhaps unfairly, as an unwieldy, totalising theory that stuck too rigidly to the logic of capital for its answers. It was seen as a 'grand narrative', a big story for explaining what was said to be more diverse and complicated global relations. The tendency towards 'big thinking' of this kind is that there is always a danger of overlooking or explaining away the unique nature of place, difference and dissonance when it is encountered. Traces of the 'grand narrative' approach can even be seen in the work of Harvey, a sophisticated Marxist exponent of geographical practice.

> Phenomena as diverse as urbanisation, uneven geographical development, inter-regional interdependence and competition, restructuring of the regional and international division of labour, the territoriality of community and state functions, imperialism and the geo-political struggles that flow there from all should stand to be elucidated and incorporated into the corpus of theory that Marx bequeathed us. (Harvey, 1985: 32–33)

The tendency to use Marxism as a large umbrella underneath which a multitude of processes must gather offers a *reductionist* view of society, condensing complex social analyses within an inflexible Marxist framework.

In response some Marxist revisionists would turn to elaborated accounts of the superstructure, such as those advocated in the neo-Marxist work of Louis Althusser, Antonio Gramsci or Pierre Bourdieu. These important lines of thinking offer a persuasive means for understanding the relationship between political economy and culture by recognising how power relations are struggled over across competing arenas. Inspired by the Gramscian approach adapted by the Centre for Contemporary Cultural Studies (CCCS) at the University of Birmingham, which yielded ground-breaking research on working-class youth culture (CCCS, 1981), the geographer Peter Jackson discusses the value of this encultured approach in his landmark introduction to cultural geography, *Maps of Meaning* (see Jackson, 1995: 47–65). Despite a splintering of different Marxist practices, critics posited Marxism to be overly *functional* with a tendency to subsume the individual within the logic of capitalism. The determining effect of ideology made it appear that 'there was little scope for the individual to have an impact in changing society' (Cloke *et al.*, 1991: 45).

While some geographers such as Neil Smith and David Harvey remained wedded to a strong Marxist account of political economy, others such as Doreen Massey, once a formative Marxist geographer, grew increasingly disillusioned. She would later enter into a series of heated exchanges with Harvey, in particular about his neglect of feminist theory and gender within the covers of his popular edition, *The Condition of Postmodernity* (1989a). With its focus upon the public world of work, Marxist economic geography in particular can at times appear a '*masculinist*' pursuit that overlooks the private world of the domestic sphere. Women's unpaid labour in the home, childcare and gendered divisions of labour in the new economy offer important arenas for incisive Marxist-feminist interventions. These elaborations are now a familiar part of the post-structuralist Left and can be seen in recent attempts to rethink the economy in more open ways that recognise difference while paying attention to forms of inequality (Gibson-Graham, 2008).

As a post-structuralist impulse developed from structuralist and Marxist literary criticism in the late-1980s, scholars began to emphasise the multiple and fragmented subjectivities of individuals. They asserted that human beings are not just class-determined subjects but their lives intersect with gender, **ethnicity**, sexuality and other competing identifications. For post-structuralist writers, and at least some feminist and anti-racist activists, the privileging of class above other forms of oppression – such as racism or sexism – means that these dynamics tend to be sidelined to the periphery of the superstructure. As we saw in the example of Hurricane Katrina, a purely class-based analysis cannot adequately explain why black people were the primary victims of flooding, without taking into account the social and spatial privilege of **whiteness**. The bringing together of race and class is then a vital cornerstone for interpreting the human and environmental disaster in Southern Mississippi. It reminds us of the compelling explanatory power of Marxist geographical analysis and a revisionist need to be open to 'differences of nationality, gender, sexuality, geographical location and so on' (Castree, 1999: 153).

Future horizons

Although the high point of Marxist geography may have been surpassed it would be disingenuous to view radicalism as a momentary hiatus in the flow of mainstream geography. After all, radicalism by its very nature could never be *the* dominant paradigm in geographical thinking. In this closing section we briefly uncover strands of radicalism within the discipline today and speculate upon the future horizon for Marxist geographies. Of particular interest here has been the rise of **New Social Movements** that may imaginatively use Marxism and radicalism to challenge the authority of global capital.

A cursory examination of topics on most geography undergraduate curricula reveals engagement with inequality to be a popular theme. This is apparent in human and physical geography, as reflected in critiques of the 'uneven' nature of global capital or in studies concerning the abuse of natural resources and its impact upon the environment. What is perhaps harder to find is any application of these ideas through forms of direct action, user engagement and protest. However, there has been a renewed emphasis upon 'participatory geographies' in a quest to develop more ethical ways of conducting research (see, for example, a special issue of *Environment and Planning A*, 39, 2007 and a review by Pain (2004) on social geography and participatory research). The formation of the Participatory Geography Working Group in 2005 is suggestive of the vitality of this field. These approaches offer a methodological means of working in collaboration with participants and user groups, but can also involve an epistemological challenge to subvert the flow of dominant power relations. In particular Kesby (2007) explores the contribution geographers can make to participatory approaches and the potential for the creation of new spaces of interaction between researchers and respondents. Participatory research then, bears at least many of the hallmarks of early radical geography in terms of method and approach.

Marxist geography as a singular enterprise may no longer be flavour of the month, but is giving way to diffuse practices and broader notions of 'critical geography'. For example the 'Crit-Geog' forum offers a creative, if occasionally feisty arena for academics, researchers and postgraduates to engage in the cut and thrust of political debate in cyberspace. Geography has also witnessed a marriage of different perspectives, such as in Marxist-feminist writing. The aim in such revisionist approaches is to provide a critique of political economy while being open to identity, difference and diversity. At the 2006 Association of American Geographers (AAG) conference in Chicago a debating stream on the decimation of the Twin Towers on September 11, 2001, prised open a thought-provoking dialogue between left-wing activists and post-structuralist political geographers. Although such rifts in the spaces of academia are rare, opportunities for inclusion remain. While the journal *Antipode* may have lost some of its initial revolutionary fervour it continues to showcase Marxist thinking, as a recent symposium on Marx's volume *Grundrisse* testifies (see 2008, 40(5)). In surveying recent journal editions you are likely to find a number of articles in the broad field of radical geography offering biting critiques of neoliberalism, water privatisation, homelessness or development. Occasionally there are interviews with political activists and polemic calls for social justice. There is also space for personal debates on the value of public intellectualism and the problems, paradoxes and possibilities for conjoining political activism with academic teaching and writing (see, for example, Pickerill, 2008). Radicalism has the potential to materialise when geographers, students and activists collectively organise and engage in protest movements. In recent years a fertile terrain for these interactions has included anti-war demonstrations, environmental movements and anti-globalisation protests against the World Trade Organisation and the G8 (see photo). As Blunt and Wills (2000) lucidly demonstrate, these encounters form examples of new geographies of resistance.

New geographies of resistance
Source: Reuters/Erik de Castro

In 2008, a new left-wing journal entitled *Human Geography* was launched under the stewardship of Richard Peet and Derek Gregory. The editors are motivated by two main concerns. First, to retain control over the value produced by academic labour, since most journals are owned by multinational conglomerates and based on profit making. In recent years this has had an effect upon the social production of knowledge. The editors aim to charge 'modest' subscription fees with profits put towards sponsoring radical enquiry. Second, the editors are driven by a feeling that radical geography and critical politics 'are being dissipated in philosophical-theoretical niceties and empirical evasions' (http://lists.aktivix.org/pipermail/knowledgelab/2008-February/000729.html). They claim that young geographers are forced to bury radical motivations, while core issues concerning the Iraq War, global finance and environmental crises are marginalised. If Peet and Gregory's assessment of the discipline today is correct, then Marxist inquiry still has a place in the production of geographical knowledge and practice as illustrated in recent student protests against the impending increase in fees in British universities.

It is widely accepted that Marxist radicalism is disintegrating in the West as a number of nation states undergo post-socialist transitions. The break-up of the former Soviet Union, in particular, has led to some premature declarations concerning the end of socialism. However, if we cast our gaze beyond Europe towards Latin America, we find a marked revival of socialist politics and activism. In Bolivia Evo Morales, a former cocoa farmer and llama herder has become the nation's first indigenous president. In Venezuela, President Hugo Chávez represents a far-left way of thinking, famously giving five million barrels of heating oil to the country's 25,000 poor. Luiz Inacio Lulla Silva (commonly known as 'Lula') began life as an impoverished shoe-shine boy and peanut seller in Brazil. Later, working in a metal plant, he became a union leader and eventually head of the Workers' Party – which is now in power. A further example of radicalism can be found in Chile where Michelle Bachelet, a former political prisoner whose father was murdered under the Pinochet regime, has emerged as the nation's first woman president. What these brief political cameos demonstrate is the successful manner in which indigenous movements and socialist politics can come together to inspire New Social Movements based on collective action. Here we discover how a genuinely radical geographical imagination can rethink and transfigure the nation state.

Conclusions

This chapter has outlined how a select number of geographers sought to prosecute radical ideas through the discipline. This involved engaging with social, economic, political and environmental issues directly. Radicalism had its moment particularly in the late 1960s and early 1970s, where it later conjoined with Marxist forms of inquiry in the development of an explicit Marxist geography; this was to see the incorporation of Marxist ideas, themes and terminology into the discipline. At the same time this type of geographical practice spoke back to Marxism by demonstrating how inequalities generated by global capital are inscribed in space. The high point of Marxist geography was from the 1970s until the mid-1980s when critiques from neo-Marxist, feminist and post-structuralist geographers started to recast the geographical imagination. The rise of post-modernist forms of thinking (see Chapter 9) explicitly challenged what were seen to be the more 'functional', 'deterministic' and 'reductionist' aspects of Marxism, and the unmoving geographical focus upon political economy. It had also become apparent that the revolutionary vision of Marxist geography had not been realised. At an international scale, the end of the Cold War and the fall of the Berlin Wall in 1989 symbolised the end of a socialist era and the victory of global capital.

While Marxism and radicalism may no longer be at the forefront of the discipline, they do exert as an enduring presence. The well-intentioned idea of working with communities familiar to early radical geography is currently being replicated in debates concerning 'participatory geographies'. Similarly critical theorisations of globalisation and an interest in New Social Movements, environmentalism, uneven development and socio-economic exclusion suggest that Marxism has prepared the way for new critical geographies in the contemporary moment. The recent financial collapse of global markets, dwindling oil supplies, uneven development and world debt suggest that a rigorous Marxist geographical project may still have much to offer in turbulent times.

Summary

- Marxism offers a radical critique of political economy and capitalist ideology, providing a paradigm for revolutionary change.
- Radical geographers are concerned with changing the world through the production of 'geographies that matter'.
- Radical geography developed theoretically and methodologically when it engaged with Marxism.
- Marxist geography explores the spatial expression of capitalism and its uneven effects at different intersecting scales.
- Insights developed from Marxist and radical geography can be seen in work on New Social Movements, political ecology and methodological debates on participatory methods.

Further reading

Blunt, A. and Wills, J. (2000) *Dissident Geographies: An Introduction to Radical Ideas and Practice*, Harlow: Prentice Hall.

Castree, N. (1999) Geography and the Renewal of Marxian Political Economy. *Transcations of the Institute of British Geography*, 24(2): 137–158.

Cloke, P., Philo, C. and Sadler, D. (1991) *Approaching Human Geography: An Introduction to Contemporary Theoretical Debates*, London: Sage.

Cook, I. (2004) Follow the Thing: Papaya. *Antipode*, 36(4): 642–664.

Harvey, D. (1973) *Social Justice and the City, Geographies of Justice and Social Transformation*, Athens, GA: University of Georgia Press.

Harvey, D. (1989) *The Condition of Postmodernity: An Enquiry into the Origins of Cultural Change*, Oxford: Blackwell.

Klein, N. (2000) *No Logo*, London: Harper Collins.

Mitchell, D. (2003) Dead Labour and the Political Economy of Landscape – California Living, California Dying, in K. Anderson, M. Domosh, S. Pole and N. Theft (eds), *Handbook of Cultural Geography*, London: Sage, pp. 223–248.

Human geography and the cultural turn

Learning objectives

In this chapter we will:

- Discuss what is meant by culture.
- Identify early traditions of American cultural geography.
- Outline the defining features of British cultural studies.
- Document the use of the 'new cultural geography' and its eventual acceptance in the discipline.
- Demonstrate how a turn to culture has re-shaped economic, political and social thinking in the discipline.

Introduction

Throughout 2008 the city of Liverpool (see photo) celebrated its status as the European City of Culture. To be paraded upon a European stage – a feat only previously achieved by its football teams – is no mean feat given its recent history includes high levels of unemployment, urban conflict, crime and social deprivation. The City of Culture bid has resulted in urban regeneration, international tourism, and public and private enterprises fostering the arts, media and knowledge economy.

What Liverpool's bid shows is that, first, becoming the 'Capital of Culture' is as much a cultural statement as an economic one. Where Marxist geographers had tended to relegate culture to the 'superstructure' – those social institutions that exist outside of the economy but are nevertheless subject to it – the cultural turn has offered new ways of interpreting these relations. Culture is no longer that part which is secondary to or left

Liverpool's shining imperialism
Source: Alamy Images/Alan Novelli

over from the economy; rather 'each is in a leaky relationship with the other' (Barnes, 2001b: 558). Second, as part of the City of Culture bid Liverpool promoted itself using the slogan, 'The World in One Place' to signify its 'brand' and unique past as a port city with successive histories of immigration. This informs us of the way in which culture now operates as a product or 'commodity'. However, a third more interesting point arose concerning the relationship between culture, power and the struggle for representation. For some residents the slogan represented a forward-looking modern multiculturalism in which the 'local' and the 'global' intersect towards the construction of a 'global sense of place' (Massey, 1996). At the same time, for many of Liverpool's black British population the strap-line was seen to gloss over, obfuscate and conceal Liverpool's imperial history as a city built upon wealth generated from the slave trade and the sweat of enforced human labour. Although these contested geographies are ongoing, partial and incomplete in their rendering, the example reveals how lines of geography hide lines of power (Barnes, 2001b). This type of 'place branding' erases the conflict, inequality and social disharmony that are also a part of the city's less legitimate history. What we see enacted is a spatial struggle over the politics of place. Appropriately the anthropologist Clifford Geertz (1993) depicts culture as a system of knotted-together signs, whose meanings are dependent upon **positionality** and subject to the social act of interpretation and representation. Through this powerful act of *representation*, people and places are given new meaning, but it is not necessarily one of their own making.

The cultural turn that stretched across the humanities and social sciences from the late-1980s has shaken-up how we might think about the world. With its focus on representation it has challenged the idea of a single objective truth. This has placed an emphasis on the partial, incomplete and contested nature of meaning. Those writing from within the cultural turn emphasise that reality is not 'out there' waiting to be discovered but is put together in words, images, figures, graphs and tables. It is through these devices that a version of the outside world is constructed. In this sense 'reality', or at least what we once thought of as reality, is now being understood as a *representation*.

The insight that 'reality' folds into representation is a critical moment in social science thinking. As Ola Söderström declares, **semiotics** – the study of signs, symbols and other representational codes – disrupts the older regime of truth once harnessed by human geographers and others:

> With modern semiotics a gap is opened between words and things, which will never be closed again. In the process, representation becomes in the human sciences a questionable and active fabrication instead of a source of certainty. (Söderström, 2005: 13)

The suggestion here is that the realist view of the world has been torn open to the extent that we can now see how this vision is constructed. What had once masqueraded as objectivity, now appears just another construction, what we might consider a 'dominant representation'. For Söderström the power of representation is ultimately to displace and implode 'reality' in this way. The realisation that all knowledge is constituted through the medium of representation – most commonly language – means that any idea of a truth that exists as exterior to culture is misleading. This suggests that the meanings generated through language, rather than reflecting an underlying reality, actually produce that 'reality'. This fiction is made to appear 'real' through the repetition of narratives that over time become part of an accepted **discourse**, a commonsense way of understanding and narrating the world. In this sense we can never grasp 'reality' but must access it by taking a detour through culture and its codes of representation. More significantly, the turn to culture is characterised by a reflexive awareness to the discursive construction of the world, the openness to multiple meanings and a critical attitude to foundational truths and 'objectivity'.

The cultural turn has seen ideas of culture move from the margins to the centre of various academic debates. This approach has influenced a number of social science disciplines. In human geography the engagement with culture is particularly interesting given the discipline's roots as a once pragmatic subject serving imperial expansion (Chapter 1) that later attempted to be recognised as a spatial science (Chapter 2). Under the auspices of 19th-century imperial geography, cultural traits were once attributed in a fixed and defining manner to local tribes, nations and whole continents. In many cases the simplistic designation of imagined cultural practices to particular groups formed a type of 'cultural racism' (see Chapter 8) based on notions of absolute difference and ascriptions of the 'exotic'. Following the collapse of the empire and later attempts to situate geography through quantitative methods (see Chapter 2), culture was largely ignored in the pursuit of a 'pure' spatial science. Moreover, culture appeared slippery and unquantifiable for a subject that now prided itself on the concrete value of quantification. For British geographers in the 1950s and 1960s, culture seemed largely irrelevant to the task of geographical inquiry. So, why was the turn to culture and representation vigorously pursued by geographers from the late-1980s, how did cultural studies transform core parts of the discipline, and what practices took place to enable 'cultural geography' to emerge as a sub-discipline in its own right? It is these questions that this chapter will address.

The chapter begins by discussing the meaning of culture before considering how the term was deployed in early cultural geography writings pioneered by Carl Sauer and students of the Berkeley School, California. We then turn to British cultural studies which offered a different understanding of culture, viewing it as an ideological process that is inherently political. These insights were to form the basis for what is described as the '**new cultural geography**' that emerged in the late 1980s. We document this rise and show how ideas of representation, power and meaning as practised in the 'new cultural geography' were transported from the margins to the centre of geographical thinking. Here, we reflect upon the explosion of cultural geography, its subsequent institutionalisation, and the way in which the cultural turn more broadly has influenced other areas of the discipline including political and economic geography. We conclude this chapter by considering some of the critiques that have accompanied the cultural turn. This includes concerns that the materiality of everyday life might be overlooked, that older social inequalities may be absented in a theoretical turn to text, and that the rush to representation has led to the erasure of more sensuous understandings of the world as interpreted through immediacy of bodily sensation, performance and practice. Despite these observations we remain in little doubt: culture matters.

The meaning of culture

Culture is all around us. It is intrinsic to human civilisation and part of the vocabulary through which we learn to decipher and articulate the world. But what exactly do we mean by this term?

Famously, the cultural studies writer and historian Raymond Williams suggests that culture is one of the most complicated words in the English language. In his book *Keywords* (1976) Williams draws attention to the original meaning of culture as a process derived from the cultivation of crops and animal rearing; what we now know today as 'agriculture'. Williams provides us with some useful definitions of culture that refer to: the arts and artistic practices; the symbolic features of a particular way of life; and the dynamic process of development. What is evident in Williams' discussion of the term is that rather than seeing culture as a static object or 'thing' he draws our attention to it as a *lived practice*, what he describes as a 'way of life'.

There are two common-held assumptions concerning what culture is, both of which should be approached carefully. A first is often to relate it to the idea of national cultures – the traditions, customs and practices familiar to a given population. In these readings of 'culture' it is often set apart as the preserve of non-Western societies, indigenous communities and people who reside in the global south. Such interpretations equate culture with the '*exotic*' and those who have 'apparently escaped the trajectory of "progress" in whose image the cosmopolitan West has been made' (Anderson, 1999: 1). As we will go on to establish in our critique of early cultural geography such representations have a great deal more to impart about the West than they do about those people and places cast as, 'colourful', 'unfamiliar' and a source of infinite 'curiosity'. A second popular understanding of culture is to regard it as a rarefied practice; perhaps associated with the 'high' arts, classical music, literature, opera or Shakespearean drama. We might regard this familiar 'high-brow' depiction as an elite or *dominant* version of culture. Just as the first trope of culture is written through the imaginative inscriptions of the West, the second is also a *situated* reading, this time specifically located through a nexus of social class relations. In this case being seen as 'lacking in culture' is one of the many ways in which class is encoded (Bourdieu, 1995). Yet as Raymond Williams (1989) attests, much culture is simply 'ordinary'. In this way cultural pursuits and activities might just as easily refer to watching television soap opera, reading 'chick lit' novels, using social networking sites such as Facebook or listening to pop music. The point that these activities may not readily be associated with 'culture' implies that they have a *subordinated* status that makes them appear superficial, fluffy and trivial. Nevertheless such pastimes and practices can be considered as popular expressions of culture.

For Williams culture is transmitted through the creation and use of symbols. This may include explicit signage such as that displayed by way of traffic lights, national flags or monetary currency – the complex sign systems we have learned to decode without a moment's hesitation. But the sign-wearing world is also enwrapped in more subtle markers that may barely register in our consciousness such as those deployed in architecture, advertising, fashion, interior design and the wider sphere of everyday life. In this way we might consider 'culture as the signifying system through which necessarily (though among other means) a social order is communicated, reproduced, experienced and explored' (Williams, 1976: 13). As this illustration makes clear, Williams is concerned with the *doing* of culture. We shall return to his ideas and those of other cultural studies scholars later in this chapter as they encourage us to take culture seriously and were to inspire 'new cultural geographers' in the late 1980s and 1990s to develop more critical accounts of cultural phenomenon. In order to understand the new architecture through which these writers were building more intricate, complex and detailed expositions of culture we need to begin by briefly examining the traditional scaffolding of early cultural geography.

Culture:

- Includes highbrow, popular and everyday practices.
- Is that system of shared meanings and understanding which constitute an identity, nation or 'community'.
- Can be understood as a 'way of life' so may include ordinary pastimes (e.g. going to the pub, listening to soul music, playing computer games).
- Is a point of ideological power struggle between competing group interests.
- Is not an 'object' or 'thing' but a fluid set of processes.
- Is constituted through practice and embodied in the material world where it can have an exchange value in the market.

Early traditions of cultural geography

The roots of geography entwine around the work of imperialism, development and nation building. In Chapter 1 we discussed the violence of these histories and their bearing upon geographical practice. A theme addressed in the section on empire is the prevalence of environmental determinism – the belief that certain environmental conditions produce what were deemed as patterns of 'natural' behaviour. While geographers have long departed from deterministic interpretations of the natural environment and their influence upon human characteristics, up until the 1920s these ideas held credence. Although environmental determinism was ultimately underscored by its inability to account for human variation across space, it also became the focus for a sustained critique in what was to become known as 'cultural geography'.

A pre-eminent figure in this emergent subfield was Carl Sauer, who is best known in the discipline for his pioneering work in cultural geography and cultural ecology. Sauer's influence upon geography, in particular upon American geography, cannot be underestimated. The voluminous PhD students and numerous academic acolytes he fostered rapidly reshaped and reinvented geographical practice (see Box 5.1). As environmental determinism began to reel from its own inadequacies, Sauer delivered what was effectively the knockout blow. Published in 1925, his essay, 'The Morphology of Landscape' (2008), punched through the failings of environmental determinism and in doing so turned the deterministic approach on its head.

For Sauer it was not the natural world that produced human culture, but it was culture that worked through and upon nature. In prioritising *culture* Sauer made a decisive breach with environmental determinism which had sought to match changing landscapes with changing human characteristics and behaviour. Instead Sauer understood landscape as 'both physical and cultural' (2008: 98), a material manifestation of the cultures that made it. Reading the changing formation of landscape, what he termed its 'morphology', could then offer valuable insight into human culture and its variation over time. In examining the interaction between humans and nature Sauer was concerned with the way people shaped rural landscapes through hedgerows, crop plantation, deforestation, bridges, barns and houses. Through these 'materials' Sauer sought to excavate a *material* understanding of human culture. As he saw it the study of the cultural landscape remained a largely 'untilled field' (2008: 103), that is until his students began to vigorously pursue Sauer's methods giving rise to a distinctly Berkeley School approach to the study of geography.

For Sauer and his followers, this method of peeling back the accumulated layers of cultural history embedded upon landscape offered a radical way of rethinking the connections between people and place, culture and nature, and geography and history. Sauer's contention was to reveal 'cultural history in its regional articulation'. He regarded landscape as the product of culture and believed in the value of areal description to distinguish the natural environment from the cultural impressions left by humankind. Writing against the backdrop of the 20th century Sauer observed the cultural impact that machinery, technology, electricity, explosives and transport were having upon a rapidly changing landscape. 'There is a strictly geographical way of thinking of culture', he went on to declare, 'namely, as the impress of the works of man [*sic.*] upon the area' (2008: 100). Sauer brought together understandings of soil, rocks, climate, water and vegetation with modernisation. His focus is then upon the materiality of culture including the aggregation of such artefacts as concrete buildings, steel structures, barns, fencing posts and a variety of housing and other materials. The indexing of such mundane cultural phenomena revealed the processes through which landscapes developed and the ways in which they were mutable to the human touch. By cataloguing change in this way Sauer, and the profusion of cultural geographers who followed in his footsteps over the next half century, endeavoured to archive the manner in which 'a new landscape is superimposed on remnants of an older one' (2008: 104).

Despite the rigour with which Sauer and other Berkeley School adherents provide a working cultural history of landscapes, there remain some shortcomings in their approach. As these have been the focus of previous accounts (Cosgrove and Jackson, 1987; Jackson, 1995; Mitchell, 2005) we will briefly address three concerns related to culture, focus and method.

Sauer assumed rather than explained what culture was; taking it to simply mean the actions of humans upon the natural environment. He explained how landscape may be imbued with 'a succession of cultures' (2008: 100) but uses the term in a loose and general manner, referring to a seemingly homogeneous 'culture group' (2008: 103). Although some cultural geographers such as Wilbur Zelinski (1973) in his influential text *The Cultural Geography of the United States* would theorise culture more specifically, there remained a tendency in the work of the Berkeley School to see culture as 'super-organic'; that is as an entity that was independent of people, existing in and of itself. In Sauer's writing culture is ascribed a mystical type of agency where the ability to mould the environment 'lies in the culture itself' (2008: 104). Treating culture as an active thing in this way remains problematic, as we shall further explain in our engagement with new cultural geography.

For these latter writers culture is a process in which people are actively engaged in and make meaning by way of its symbols and repertoires. Rather than seeing culture as an abstract force shaping people and landscapes, these scholars opine, 'the decisive agents in constructing social life and landscape are not inherited baggages of customs, but rather people' (Anderson and Gale, 1999: 4). In this respect people are not empty vessels who absorb past cultures uncritically. We do not approach 'culture' in a pure undiluted form, but are creative and critical disseminators who may in turn rearticulate these habits to produce new meanings. In contrast Sauer appeared uninterested in the 'inner workings' of culture and concerned only with its outcome – the imprints left upon landscape.

A second limitation with Sauer's approach to landscape was his parochial focus upon rural backwaters and small hamlets. In many ways he turned his back upon urban spaces leading to a type of nostalgic depiction of the rural idyll. As the British social geographer Peter Jackson espoused, 'Sauer betrayed an anti-modernist tendency that went hand-in-hand with a fundamentally conservative outlook' (1989: 15). In this respect Sauer's cultural geography largely omitted urban areas and tended to look to the past for its answers. Given the vitality of much recent cultural geography work on cities, the dismissal of urban inquiry seems short-sighted. Sauer's approach remains preoccupied with folk geography and, at least for Jackson, 'betrays a reactionary attitude towards social and cultural change' (1989: 16). Indeed, Don Mitchell (2005) observes how this narrow gaze encouraged an anti-modern tendency in future generations of American cultural geographers who too easily dismissed urban areas, global political events and macro-economic processes in many of their accounts.

At the level of method Berkeley School adherents were admonished for providing an endless archive of information pertaining to the changing landscape. Here the meticulous accounting of artefacts became something of a pursuit in itself. Peter Jackson (1989: 19) complained about cultural geographers' 'almost obsessional interest in the physical or material elements of culture rather than in its more obviously social dimensions'. Charting changes in the landscape is one thing, but failing to provide accounts that challenge the prevailing social order or engage with power is another. In this way it has been alleged that Sauer approaches the fieldwork site as a 'blank sheet', relying upon 'a stroll, the drawing of a sketch, the taking of a photograph and the pencilling of a few notes' (Cloke *et al.*, 2004: 2). For Cloke *et al.* this intuitive way of operating engenders a 'magical translation' (*ibid.*) of the physical landscape, a process that continues when Sauer returns to the office to develop his report. The criticism lies with rendering the landscape as an *objective reality* which can be neatly accounted and accurately described. Jackson (1989) also critically reflects on a method of cultural geography concerned with documenting log buildings, graveyards, barn styles and gasoline stations. In many of these studies culture tends to be reduced to material artefacts and fixed to particular objects or traits. The disembodied approach associated with Berkeley School cultural geography is slightly surprising given Sauer's own interest in phenomenology, folk traditions and cultural ecology but it remains a lingering residue. As we shall now discover it was developments in British cultural studies and a broader turn to culture that was to provide the platform upon which a new cultural geography could arise and re-invent itself. Those looking to short-circuit this inter-disciplinary influence and move to geographical debates may turn to the section entitled 'The new cultural geography' (p. 111).

Box 5.1

Carl Sauer (1889–1975)

Carl Ortwin Sauer is best known as the founding father of American cultural geography. Born in 1889 he grew up in a German Methodist farming community in Warrenton, Missouri. This background was formative of his longstanding interest in rural communities and pastoral ways of life. After teaching at the University of Michigan, Sauer moved to the University of California, Berkeley in 1923. Here, he was concerned with the impact of human civilisation upon the land. Sauer witnessed first-hand the way in which rapid industrialisation encroached upon the state of California, and he grew increasingly disenchanted with the cultural and ecological impact of this transformation. Sauer's landmark essay, 'The Morphology of Landscape' (first published in 1925), shaped the direction of geographical exploration for many years. A fierce critic of environmental determinism, he turned instead to the role that human culture played in transforming the landscape. In doing so he drew upon the anthropological critiques of Franz Boas, Alfred Kroeber and Robert Lowie and regularly turned to the arts and cultural history to develop his ideas. An advocate of international fieldwork approaches and detailed description of the Earth's surface, Sauer remained concerned with phenomenological understandings of landscape which he regarded as the 'unit concept of geography' (2008: 98).

Sauer's methodology considered the landscape as a record of human activity. A feature of his work was a concern to chronicle changes in the natural landscape to expose how these transformations reflected the requirements of successive human cultures as they imposed their will upon the land. Sauer would go on to supervise around 40 PhD students who came to epitomise a tradition of cultural geography that became synonymous with the 'Berkeley School'. A number of these students would in turn become leading figures in cultural geography, reproducing and elaborating upon Sauer's early model for reading landscape. In time his ideas on culture and his tendency towards romanticising the pastoral idyll would be critiqued, most notably by British 'new cultural geographers' who sought to reinvigorate the sub-discipline. Despite these shortcomings Carl Sauer remains a founding cultural geographer, a pioneer in landscape studies and a highly respected figure in cultural ecology and sustainable living. His influence upon American cultural geography is such that it is felt in the present.

New maps of meaning: British Cultural Studies

From the late 1960s onwards, British writers from the University of Birmingham's Centre for Contemporary Cultural Studies (CCCS) were to emerge as some of the most eloquent and influential exponents of culture. Their writings reflect a deep commitment to Marxist and neo-Marxist traditions, the roots of which can be found in the work of Karl Marx, Frederich Engels, Louis Althusser and most evidently in Antonio Gramsci's theories of power and **hegemony**. The CCCS, or 'Birmingham School' as it was to become known, formed the heartbeat of what materialised as British Cultural Studies. Established in 1964 under the directorship of Richard Hoggart, before the baton was passed on to Stuart Hall and then Richard Johnson, the Centre had various incarnations. It initially functioned as a post-graduate research centre, publishing landmark books and a series of incisive stencilled papers, before later opening the doors for undergraduate programmes. It underwent various levels of restructuring that saw it positioned with media studies and then sociology, before it was brutally axed by university management in 2002 – despite its global reputation for research. In this section we shall briefly consider four areas of the CCCS approach which have been particularly relevant in re-shaping geographical ideas. This includes the contributions made to working-class history, youth subcultural studies, feminist and post-colonial readings of the nation state, and later work on media and communications.

Working-class histories

At its inception, the CCCS developed a commitment to taking popular or working-class culture seriously. Buoyed by Raymond Williams' conviction that 'culture is ordinary', early writers on the British Left sought to implode the literary canons and high elitism that held culture at a remove from ordinary folk. In an evocative phrase that at once captures its ephemeral and material aspects, Williams (1985) described working-class culture as a 'structure of feeling'. By this he means the tacit understandings, meanings and values that are rarely spoken but deeply felt in many traditional working-class communities built around the shipyard, pit, mill or factory. For Williams these structures, such as they are, are historically informed but lived out in everyday life in ways that generate emotions, affects and particular values with regard to respect, parenting, money, work, family, leisure and so on. In this way they are not dissimilar from the class 'habitus' Pierre Bourdieu (1995) describes: a way of thinking and being that arises out of one's place in the economic order. Although such '**structures of feeling**' may carry with them a masculine whiff of nostalgia that romanticises physical labour they offer a sensory means for understanding the psycho-social aspects of class experience. These dimensions can be traced in work in British **post-industrial** cities such as Rotherham (Charlesworth, 2000), Newcastle (Nayak, 2003a, 2003b), Sheffield and Manchester (Taylor *et al.*, 1996) even after the decline and near-disappearance of heavy industry. This work offers a way of understanding the emotional content of class relations and the resilience to change in 'new times'.

Alongside the heart-warming narratives of Williams, Hoggart's book, *The Uses of Literacy* (1957), became a founding text for cultural studies. Where some people saw a contradiction in the radical socialist politics of Williams, with his focus upon predominantly elite literary sources, Hoggart more consciously based his account upon familiar past times and a range of mass literature including magazines, tabloids, cheap paperbacks and

popular music. His work also afforded scope to celebrate aspects of working-class communal life such as dance groups, British seaside holidays, pub life, dominoes, choral societies or pigeon fancying. The point was to recognise the skills, histories and traditions of the 'masses' as an important, if much maligned culture of Britain. *The Uses of Literacy* is spliced into two parts aimed at documenting the changes in working-class culture in the past few decades before the rise of mass consumerism in the 1950s. Like Sauer before him, Hoggart values older ways of life and past traditions that are rapidly eroding under wider social fragmentation and consumption. Having documented the practices of 'an "older" order' in part one, Hoggart turns an unappreciative eye, 'yielding place to the new', in the book's second section. Here, he focuses upon the growing market of youth culture, modern entertainment, American-style milk-bars, glossy advertisements, sex-and-violence novels, science fiction, pop music and what he euphemistically terms 'spicy magazines'. For Hoggart much of what materialises in these new art forms is a formulaic compound: 'flat paper, crude print, vivid glossy cover' (1957: 251). Although these cultural products are occasionally resisted by more discerning consumers, for Hoggart they amount to little more than 'sex in shiny packets' (1957: 246), entailing a hollowing out of the rich experiences, oral traditions and past times of working-class communities.

Instead, for Hoggart, such flimsy modes of cultural consumption were helping to produce a modern world 'which is largely a phantasmagoria of passing shows and vicarious stimulations' (1957: 246). For example, Hoggart is scathing of the faux American styles adopted by a new generation of British working-class youth, the Juke-Box Boys, who frequent milk-bars in order to play records on the nickelodeon. 'Compared even with the pub around the corner', he protests, 'this is all a particularly thin and pallid form of dissipation, a sort of spiritual dry-rot amid the odour of boiled milk' (1957: 248). One can only shudder to think what Hoggart would make of today's youth fascination with hanging out at American burger bars dressed in baseball caps and the latest global sportswear. While Hoggart may see some forms of working-class expression as more 'authentic' and legitimate than others, his over-arching concern is to give value to popular forms that would largely have been overlooked or despised by middle- and upper-class groups. Although he is critical of mass consumerism, advertising and Americana with its 'shiny barbarism' (1957: 193), Hoggart has a keen eye for folk traditions, popular culture and social change. He recognised that culture can only be understood within the contexts within which it was being practised and that there is much to value in 'ordinary' working-class life.

This concern with the social reproduction and transformation of class relations became a feature of early historical contributions in cultural studies. Raymond Williams, Richard Hoggart, Stuart Hall and Richard Johnson revealed how family, work, education and leisure were an integral part of the struggle and reproduction of 'class cultures'. This recognition stretched to encompass the 'making' of histories as complex ideological activity, subjectively given meaning to in selective acts of remembrance and forgetting (Johnson *et al.*, 1982). In other words, dominant groups such as the English middle-classes could define and deem worthy what they saw as 'culture' (literary classics, modernist art, classical music) and ignore or denigrate the rest. By recuperating forgotten aspects of working-class tradition these writers challenged the very notion of culture and revealed it to be formative in the struggle for ideas. A biographical approach to education is resonant in this work. Williams reflects on being raised in a working-class household on the Welsh borders before taking up a Cambridge scholarship. Hoggart draws upon his boyhood experience as a working-class child who went to grammar school and the different, sometimes conflicting worlds this opened up to him. Hall also writes of his experience as a post-colonial migrant coming from Jamaica to take up a place at Oxford, where he felt detached and alienated. Hall would later discuss the hybrid or mixed attachments he felt to both Britain and the Caribbean, as he took up academic posts in Birmingham and eventually at the Open University. What the educational biographies of Williams, Hoggart and Hall have in common is a sense of estrangement and displacement. This perspective enabled a critical cultural studies approach to education evident in the CCCS publication of *Unpopular Education* (1981) which showcases the important relationship between social class, history and schooling.

Youth subcultures

Throughout its 38-year lifespan, the CCCS produced a surfeit of work on popular culture. The vigorous engagement with the popular was to develop and extend the early 'critical theory' approach adopted by associates from the Institute for Social Research at the University of Frankfurt. More popularly termed the 'Frankfurt School', Theodore Adorno, Walter Benjamin, Herbert Marcuse and Max Horkheimer were leading lights of German Marxist critiques on 'mass culture'. Utilising Marxist notions of 'false consciousness' (see Chapter 4) the Frankfurt School viewed popular culture as a type of 'bubble-gum for the masses', ideologically imposed from above for ordinary people to ruminate upon in an uncritical, de-politicised manner. In contrast, writers from the CCCS drew upon Gramsci's theories of hegemony, concerning the struggle and enactment of power, to reconceptualise popular culture as a shifting terrain in a 'war of manoeuvre'; marked as much by coercion as consent. For CCCS scholars hegemonic consensus was not an ideological given – the culmination of dominant ruling ideas holding sway over subordinated classes – but achieved by an on-going process of resistance, negotiation and incorporation. In this dynamic reworking, popular culture was no longer like pink candy-floss imposed from above to mask ideology but was *itself an arena for class conflict and struggle* between dominant and opposing groups. This radical shift challenged the top-down perception that working-class people were simply passive recipients who absorbed culture uncritically as if injected by a hypodermic syringe. An example of the complex hegemonic struggle happening within culture itself can be seen in vivid illustrations of youth **subcultures**. CCCS writers explored a series of youth subcultures including Mods, Hippies, Teeny Boppers, Skinheads, Punks, Rastafarians and Bikers. They remarked how subcultural practices are ultimately 'rituals of resistance' enacted by working-class youth in response to the break-up of traditional communities and an unbridled post-war consumerism that was creating a sharply visible, unequal distribution of wealth. If we take the example of the Punk, working-class resistance can be traced in the symbolic transgression of style, music, lyrics, values, attitudes, dress, posture and beliefs (Hebdige, 1979).

Phil Cohen's (1972) seminal work on subcultural conflict and working-class community, which was to evolve into a detailed cultural geography of social relations in London's East End (see Cohen and Robins, 1978), was in many respects a template for future CCCS work on youth subcultures. Cohen argued that post-war British youth subcultures engage in an 'imaginary' relationship with older working-class traditions and past times. Here, the exhibition of a subcultural identity is a means of expressing and 'magically resolving' the crisis of class relations – at least at the level of the symbolic – through territorial practices and stylistic gestures. The symbolic aspects of class resistance are then obliquely signified and circuitously carried in youth subcultures. The focus on subculture, social class and resistance is emblematic of a number of essays found in the CCCS youth studies collection *Resistance through Rituals* (Hall and Jefferson, 1975). The majority of these Marxist-inspired essays apply a complex form of textual semiotics (meaningful **deconstructions** of words, images or signs) to interpret the cultural meanings thought to lie 'beneath' youth styles and activities. Key ethnographic studies of young people were to include Paul Willis's (1978) vibrant analysis of Hippy and Biker subcultures as well as his critically acclaimed book *Learning to Labour* (1977) exploring how working-class young men ended up in working-class jobs on the factory shop-floor, thereby reproducing their class position.

The influence of this latter work can be seen in feminist geography through Linda McDowell's (1997) research on 'redundant masculinities' which argues that in a post-industrial service economy young men are more likely to be 'learning to serve' than 'learning to labour'. The CCCS studies on youth culture would also offer a material backcloth for the 'post-subcultures' approach undertaken on the 'geographies of youth' (Skelton and Valentine, 1998) where the focus was more emphatically upon clubbing and consumption rather than the school-work transitions and production – a sign perhaps that the cultural turn needed to return to the 'social', an issue we shall address towards the end of this chapter. Feminist interventions into youth studies would incorporate research on young women's transitions from school to work (Griffin, 1985) and Angela McRobbie's (1981, 1991) discussions of girls' magazines and the gendered cultural worlds of music, fashion, romance and domesticity. We will discuss McRobbie's work further in this chapter, although the roots of feminist cultural studies can be traced in the CCCS edition *Women Take Issue* (1978) exploring aspects of women's patriarchal subordination, a theme developed in our discussion of feminist geographies in Chapter 6.

Race, ethnicity and nationalism

Along with a focus on class and later feminist politics, the CCCS made a number of important interventions into the study of race and ethnicity. These excavations were initiated in the early work by Stuart Hall and his colleagues, culminating in the book *Policing the Crisis* (1978). This volume produced a meticulous account of how the term 'mugging' was imported from the New York Bronx to British inner-city areas, where it was used to explain almost any crime involving black British youth. The press reporting of 'mugging', with its stark racial undertones, implied that there was something distinct and more abhorrent about black criminal activity that required a new vocabulary. The way in which the nation state was organised around notions of whiteness, Britishness and racial superiority as much as social class is powerfully conveyed in a later CCCS thesis, *The Empire Strikes Back* (1982). The book, written at the height of the Thatcherite New Right, documents the way in which the nation state was then being conceived through a *laissez-faire* free market economy and a 'popular authoritarianism' bolstered by heavy policing, strict penalties and increasing demands for law and order. The edition considers how an exclusionary sense of Britishness comes to be defined, as against immigrant communities who are cast as 'guests', 'outsiders' and a source of threat or 'moral panic' to the nation state.

The centrality of whiteness in the production of racialised Others is eloquently rehearsed in Dick Hebdige's popular account *Subculture: The Meaning of Style* (1979), which also offered valuable insights into ethnicity and nationalism in post-war Britain. Following the inner-city uprisings in British multi-ethnic areas in 1981 and again in 1985, race pushed its way to the heart of the Birmingham School's cultural analysis. Paul Gilroy's book *There Ain't No Black in the Union Jack* (1987), which took its title from a National Front slogan of the time, offered a paradigm shift in British race relations. Gilroy convincingly argued that blackness and Britishness should no longer be seen as mutually exclusive identities but formed part of a new multicultural Britishness already in the making. These ideas were given empirical weight, as richly developed in Simon Jones's (1988) neighbourhood ethnography of multi-ethnic youth relations in inner-city Birmingham and Les Back's (1996) subsequent work in parts of South London. Each of the above CCCS accounts considers the material and symbolic interplay of black and white stylistic forms and their potential for creating new 'cultures' from which to re-imagine Englishness beyond a vituperative imperialism. The suggestion here is that national culture – even at its most chauvinistic – can be reworked, transfigured and transformed through the practices of everyday life.

Popular culture and media theory

Having tentatively sketched out some of the extensive CCCS portfolio of research on working-class history, youth studies, feminist and post-colonial readings of the nation state, a final area of work we shall explore concerns media theory and communications. Some of this critical approach to media representations has already been signalled in the discussion of *Policing the Crisis* (Hall *et al.*, 1978). A cornerstone for critical accounts of the media can be found in Stuart Hall's work on representation, beginning with an early essay, 'Encoding/Decoding'. Here, Hall (1980) challenges the simplistic idea that audiences receive and understand mediated messages in the ways intended. The ideological meanings 'encoded' in newspapers, advertisements, films, music, magazines and soap-operas can be 'decoded' by audiences in multiple ways according to their situation and context. Hall suggests at least three positions exist from which an audience may 'make sense' of a media text such as a television programme. He identifies 'dominant hegemonic', 'negotiated' and 'oppositional' readings as part of this framework of decoding.

In the *dominant-hegemonic* position 'we might say that the viewer is operating *inside the dominant code*' remarks Hall (1980: 136), accepting the scripts and values disseminated. In the *negotiated* reading the viewer may be critical of aspects of what is represented (characters, plotlines, realism, etc.) but is likely to accept the general ethos and values of the text, for example the idea of the American Dream or the ideology of heterosexual romance as purveyed in countless Hollywood blockbusters. In the *oppositional* reading, the dominant framework is directly contested to provide a counter-narrative that disrupts the seamless flow of media information, narrative and signs. This can be seen in the critical readings that later arose in response to the genre of 'Nam films' which glorified the heroic invasion of Vietnam by American troops to form part of a 'mythic' imaginary. In time these oppositional readings have given way to more ambivalent filmic portrayals that make for less comfortable viewing. The existence of these dominant, negotiated and oppositional audience interpretations was empirically supported in Morley's (1980) study of the BBC regional current affairs programme *Nationwide*, demonstrating the critical awareness of respondents.

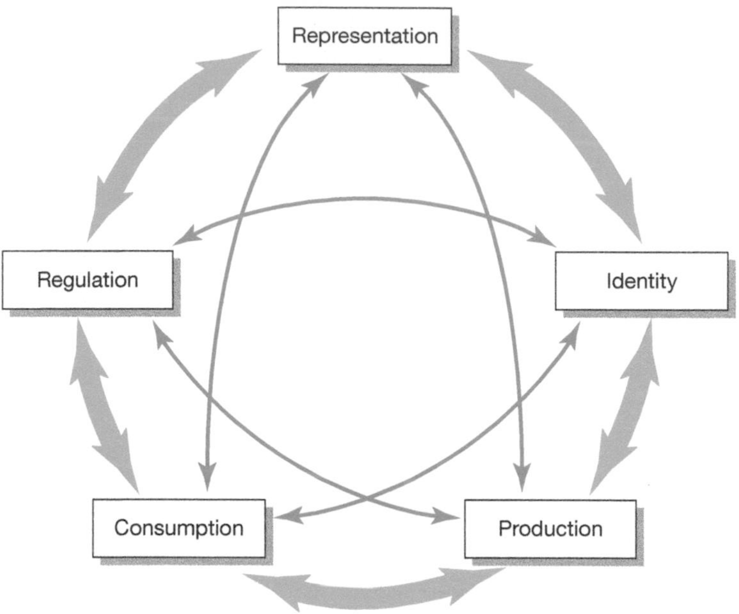

Figure 5.1 The circuit of culture
Source: du Gay, 1997

As a more rigorous cultural studies analysis of the media developed, McRobbie's (1978, 1981, 1991) early readings of 1970s girls' magazines such as *Jackie* would later be critiqued by new media theorists and cultural studies writers. By the 1980s the CCCS was developing complex, interactive understandings of the media which challenged the Frankfurt School tendency to regard the public as 'cultural dupes', passive recipients enslaved by dominant media messages. In a similar way, McRobbie viewed *Jackie* from a feminist perspective as reproducing a 'culture of femininity' that cohered around the ideology of romance. The search for a 'fella', the privileging of true love and the emphasis upon repetitive beauty routines were understood as an induction into the future world of marriage and the routines of domestic labour. In opening out these relations, members of the Birmingham School argued for the need to place media representations within a 'circuit of culture' (Johnson, 1986) to explore the complex interrelationships between media production, consumption, regulation and identity (see Figure 5.1).

These ideas were taken up by media theorists such as Martin Barker (1989) at the University of the West of England. In a close engagement with comics, magazines and fandom, Barker challenges the feminist assumption of *Jackie* as 'bad for girls'. His analysis suggests that a history of production enables us to understand how magazines are produced within a context of technical constraints and compromises that change over time. Factors relating to the technology of magazine production such as machinery, financial resources, artistic input and marketing have a material effect upon the commodity produced and further complicate notions of ideological 'reproduction'. During her time at the Birmingham School, the renowned cyberfeminist Sadie Plant extended these relations to the virtual reality world of cyberspace comprising 'a dispersed, distributed emergence composed of links between women, women and computers, computers and communication links, connections and connectionist nets' (1996: 182).

A clear distillation of the circuit of culture is discussed in a study of the Sony Walkman (Hall *et al.*, 1997) and would also be applicable to Apple's iPod and iPhone. Here the design, production and marketing of these products are essential to their success. The place of Sony and now Apple as global firms is a vital component of the story of Walkmans and iPods. How these products are used, worn and operate as fashion items iconic to youth markets informs us of the role they play at the level of identity production and representation. In the sphere of consumption these items also connect up with other modes of consumerism, such as the use of Apple's iTunes and contracts with other multinational companies that enable consumption, downloading and communication to transpire. Of course, this does not mean that illegal downloads, file sharing and illicit recording cannot occur as a subversive form of the 'symbolic creativity' Paul Willis (1990) referred to, where young people may occasionally refigure market capitalism. The miniturisation of the original iPod to the mini-iPod and now the Nano, along with the marketing of bright colours, reflect the place of items such as iPhones and iPads as the highly desirable new 'fetish objects' of late-modernity. Finally, the regulation of new technologies has seen restrictions on the noise levels of iPods particularly in spaces of public transport, public health debates concerning auditory impairment to the wearer, new legislation about the use of mobile communications for car drivers and aeroplane passengers, and wider questions concerning the role of electronic media in the lives of children and young people. In these examples it is clear that any representational deconstructions of objects and text need to be firmly situated within the wider 'circuit of culture' outlined. We now turn our attention to what was to become known as the 'new cultural geography' which emerged in the late 1980s and consciously drew upon the insights developed in cultural studies, most notably the idea of culture as *political*.

The new cultural geography

By the mid-1970s geographers were distancing themselves from the traditions of Berkeley School cultural geography. For the most part the ideas of Sauer and his disciples appeared archaic and increasingly 'irrelevant' (Mitchell, 2000). Culture remained a neglected area in geographical studies, although in cultural studies, anthropology and parts of the humanities more complex interpretations of culture were already underway. The belated turn to culture in geography is signalled by Peter Jackson who even in the twilight of the 1980s bemoaned how 'many geographers do not seem to regard culture as at all germane to their work, ignoring it altogether or treating it as a trivial residue left unexplained in more powerful analyses of political economy' (1995: 25). The over-riding sentiment is that as geography moved from a spatial science concerned with statistical mapping and quantification (Chapter 2) to embrace radical and, eventually, Marxist perspectives (Chapter 4), it had become overly deterministic in its approach. The concentration on political economy meant that culture was rarely discussed and where it was invoked it was merely as a sideshow to illustrate the power of global capital. In other words, there was little of the complex forms of resistance, transgression and strategic consensus depicted in CCCS accounts of culture, power and hegemony.

Despite substantial deficiencies in the conceptualisation of culture proffered by Sauer and the Berkeley School, geographers were slow to offer a sustained critique of these approaches although some debate was underway (Jackson, 1980). In 1987 Denis Cosgrove and Peter

Jackson sought to map out an agenda for a new cultural geography that would depart from the direction undertaken by Sauer and the Berkeley School. In sum, their plea was for a 'revitalised cultural geography' (Cosgrove and Jackson, 1987: 99) that aimed to be:

- contemporary as well as historical
- social as well as spatial
- urban as well as rural
- concerned with domination as well as resistance
- concerned with 'reality' as well as 'representation'
- concerned with the partial, situated and social production of knowledge
- concerned with the centrality of culture in human life.

Cosgrove and Jackson (1987: 96) criticised Sauer's approach as 'rural and antiquarian' with an over-reliance on physical artefacts. They viewed Berkeley School cultural geography as adopting a unitary view of culture, arguing that 'cultures are politically contested' (p. 99) which in turn implies the existence of a plurality of cultures. James Duncan (1980), in particular, was deeply critical of the Berkeley School's 'super-organic' approach to culture that appeared to give agency to the abstract. Indeed a number of debates brewed concerning the past traditions and future trajectories of cultural geography that served to provide a context for situating these new directions.

The value of Cosgrove's and Jackson's (1987) review for Lily Kong (1997), writing a decade later, is that it acted as an effective 'stock-taking' exercise of the new attitude to cultural geography being adopted, thus heralding its 'renaissance'. As Kong explains, Cosgrove and Jackson's outline for the 'new cultural geography' was not plucked from the air but was grounded in the context of geographical discussion and cross-disciplinary debates. In advocating 'new directions for cultural geography' Cosgrove and Jackson's (1987) concise paper acted as an important and highly visible signpost at the crossroads of cultural geography. While for Kong the essay may have somewhat exaggerated the points of departure between a 'new' and 'old' cultural geography, it nevertheless encouraged an awareness of the renewed significance of culture in contemporary debates of the time. The crystallisation of this moment, as Don Mitchell (2000) poignantly surmises, sees a 'new' cultural geography being assembled against one that, in the process, became 'old'.

The influence of the CCCS upon geography was most clearly seen in the work of Peter Jackson's book (1995, first published in 1989), which takes its title from a familiar cultural studies term, *Maps of Meaning*. Jackson's useful text provides an important introduction to cultural geography and was to spearhead the 'new cultural geography'. A feature of this book is its commitment to developing a cultural geography that departs from Sauer and the Berkeley School and instead turns to the type of political understandings of culture fostered by the CCCS. In doing so Jackson sees culture as 'a contested term' (p. xi), rejecting the unitary view of culture as a purely artistic activity or intellectual product of the elite. He affirms the value of popular culture and its potential for challenging dominant values, emphasising how the new cultural geography can develop through the recognition of a 'plurality of cultures' (p. 1).

Following the cultural studies mantra that '*the cultural is political*' (1995: 2) *Maps of Meaning* is an effective bridge-building enterprise between cultural studies and human geography. As Jackson goes on to explain, it offers a two-way street for intellectual traffic between disciplines. He remarks how 'culture is not only socially constructed and geographically expressed [...] it must also be admitted that culture is spatially constituted' (p. 3). Through a critical analysis of gender, sexuality, racism and social class, Jackson paves the way for a new cultural geography concerned with how power, ideology, discourse and forms of representation are enacted within space. His contention is that, 'geography is conceived of not as a featureless landscape on which events simply unfold, but as a series of spatial structures which provide a dynamic context for the processes and practices that give shape and form to culture' (p. 48). In bringing together social and spatial processes, possibilities open up for a new, politically charged cultural geography. The dialogue between British cultural studies and human geography can be traced in the collection *British Cultural Studies* (Morley and Robins, 2001) which also showcases the work of geographers such as Phil Crang and Peter Jackson writing on consumption, Linda McDowell on gender and employment, Peter Taylor on the political geographies of the state and David Sibley on travellers, youth spaces and drugs.

Landscape as text

Where cultural geography operated as a byword for much human geography in the US, in Britain it remained a niche terrain. Most British human geographers concerned with issues of social inequality, such as housing distribution, population studies, ethnicity and migration, and health and welfare, tended to be trained in social geography. What little cultural geography there was remained more connected to the arts and humanities through its association with landscape. From the late 1980s a number of cultural geographers sought to envision the landscape as a type of cultural text replete with material and symbolic encodings (Cosgrove and Jackson, 1987; Cosgrove and Daniels, 1988). Thus, James Duncan (1990: 17) conceived of landscape as 'one of the central elements in a cultural system, a text', one that, 'acts as a signifying system through which a social system is communicated, reproduced, experienced and explored'. The turn to representation was in part inspired by insights from the arts, literary and cultural studies. This offered a radical challenge to our understandings of landscapes by revealing that far from being natural, these sites are socially constituted phenomena that could be physical or symbolic. Denis Cosgrove emphasises the role of song, film, poems, paintings and other textual modes as important devices for giving meaning to landscape. For Cosgrove (2008: 81) the cultural landscape can be approached as a text in which, 'The many-layered meanings of symbolic landscapes await geographical decoding.' The art of 'reading' the landscape became a familiar cultural geography activity. The value of this approach is that it allows us to see landscapes as manufactured, cultivated and carefully manicured sites that carry meaning (see Mitchell, 2005). Indeed, for Mitchell the landscape needs to be seen as a site upon which capitalist relations of labour are to be found. Rather than simply being a product of the natural environment we can consider the role of hereditary land rights, technological knowledge, access to labour power and finance as crucial components of this 'scene' (see Chapter 4) and formative of what Carl Sauer (2008) previously termed 'the morphology of landscape'.

Deconstructing the environment in this way allows us to figure the landscape through particular relations of power. For example, the feminist cultural geographer Gillian Rose (1993) has shown how landscape is not only imbued with the power of capital, as Mitchell observed, but is also a space upon which relations of gender are struggled over. In his important book *Landscape and Englishness*, David Matless (1998) demonstrates how rural landscapes are often regarded as sites where true 'Englishness' resides and tend to be represented through a familiar iconography of thatched cottages, winding hedgerows and red telephone boxes in quaint villages. This work reveals how such images are far from neutral representations but connected to a web of power relations that serve to knot together ideas of whiteness, class and a nostalgia for the past, leaving some people to feel 'out of place' in these spaces (see Chapter 8 on rural racism). The 'call to nationhood' is also found in Pyrs Gruffudd's (1994) exploration of rural Wales in the inter-war period. The British new cultural geography on landscape then paid close attention to relations of power and how class, gender and nationhood may be 'imagined' and symbolically given meaning to. An early collection entitled *Geography, the Media and Popular Culture*, edited by Burgess and Gold (1985), demonstrates how cinema, television and other mediascapes may constitute our sense of place and make strong connections with cultural studies. This is further illustrated in David Clarke's (1997) edited collection *The Cinematic City* and the edition *The Place of Music* (Matless *et al.*, 1998). The suggestion is that novels, music, film and other texts are place-making devices that influence our understanding of rural and urban landscapes. Titles such as *Writing Worlds: Discourse, Text and Metaphor in the Representation of Landscape* (Barnes and Duncan, 1992), or Cosgrove and Daniels's (1988) popular edition, *The Iconography of Landscape*, exemplify the impact of the cultural turn at this moment and the heightened attention given to discourse, symbols, rhymes and representation as ways of 'reading the landscape' (Anderson and Gale, 1999: 13) (see Box 5.2).

This humanities-oriented, textual approach to landscape as developed through humanistic geography had been evolving for some time. The limits of this approach are later explored in Chapter 12, where contemporary cultural geographers are challenging representational modes through sensual and affectual approaches. What is interesting for now, though, is how this specific strand of humanistic and historical cultural geography became integrated with fresh approaches on music, media relations, race and ethnicity, consumption and so on, and designated as the 'new cultural geography'. While the differences between work on the historical cultural geography of landscape and the politics of popular culture were palpable, together they were seen to offer a critical mass of scholarship that could at least suggest that the cultural was an increasingly important area of geographical investigation. This amalgamation is not insignificant. To begin with it enabled the new cultural geography to root itself in the discipline. Moreover, British cultural geography was a fragile enterprise that at the time could easily have been trampled upon by more established parts of the discipline and left broken and discarded. This porcelain fragility was captured when James Duncan (1993: 372) pithily remarked how for a time one could 'describe UK cultural geography in three words: Cosgrove, Daniels and Jackson'. For most American cultural geographers the 'new cultural geography' in its formative years was regarded as a quintessentially British pursuit. In many ways the new cultural geography, already underway at the time of Cosgrove and Jackson's synopsis of the subfield, really flowered from the 1990s onwards and by the mid-1990s it appeared a more secure terrain. A core aspect of this work was to move beyond the traditional political economy theorisations of society – a post-structuralist manoeuvre which drew the interests of at least some economic geographers – to consider a much wider palette of experiences. In the process, cultural geography was transformed from a marginal pursuit to a mainstream practice.

Box 5.2
Reading the landscape

A close example of reading the landscape can be found in a paper by Daniels and Rycroft (1993) which deployed the novels of Alan Sillitoe to map the modern city of Nottingham. Sillitoe is probably best known for his novel *Saturday Night and Sunday Morning* (1950) which was later transposed into a film classic and associated with a genre representative of rebellious, 'angry young men'. In reading the landscape the authors draw upon novels, maps, autobiographies, photographs, posters, poetry, journalism and travel writing, combining them with an interview with Alan Sillitoe. The bringing together of literary fiction and geographical texts in a 'field of textual genres' is an attempt to understand the complex overlaps and interconnections between the worldliness of literary texts and the geographical imagination. At the level of method the cultural turn can be seen to enable creative understandings of people and place to come to the fore.

Through these techniques Daniels and Rycroft (1993) consider how the novels expose conflicts in the modernisation of working-class neighbourhoods in and around Nottingham from the 1920s to the 1950s. During this period Nottingham was fashioned as a 'modern city', economically progressive and socially forward-looking. Sillitoe was fascinated by maps and the aerial photographs he came across when working for the Air Training Corps during the War. Daniels and Rycroft argue that the heroes in Sillitoe's novels are resolutely masculine, occasionally misogynistic and always possess a gritty exterior. In contrast they suggest that the modern city of Nottingham – which became known as 'Queen of the Midlands' – is a feminine space with a copious amount of female labourers working in the lace, hosiery and clothing industries. Unlike promotional material depicting modern Nottingham as bright and spacious, Sillitoe's novels paint the city as a shadowy, industrial and masculine world. These openings allow us to see how planners may promote Nottingham as enlightened, progressive and optimistic while Sillitoe chooses to portray civic life as violent, oppressive and exclusionary. What Daniels and Rycroft distil in their complex cultural reading of modernisation processes in Nottingham is the competing and conflicting geographies that make up the city, and how it is understood and experienced.

Source: Daniels and Rycroft, 1993

The cultural turn from the margins to the centre

Institutionalising cultural geography

As the previous section starts to reveal, the cultural turn was never a solitary exercise. Moreover, geography was only one of a number of social science disciplines that was influenced by this movement so it is important to see the turn to culture as a broader project. The turn to culture was then a coming together of different strands and activities that included cultural studies, structuralist and post-structuralist literary criticism, historical geography as well as insights on power and subjectivity derived from feminist methodologies. The cultural turn is then a 'borrowing' comprising of different thoughts, ideas and traditions. It is not a unified project but a constellation of moving ideas and practices. As our focus is with geographical research in this section, we shall examine how these seemingly different moments could become institutionalised and distinctively labelled as 'cultural geography'. This interest in culture emerges from humanistic geography and later comes to shape a number of areas in the discipline.

One of the clearest ways in which we can evidence the turn to culture in the discipline is through its institutional arrangements. In January 1988 the Institute of British Geographers' (IBG) Social Geography Study Group declared that it would now go under the banner of being the Social and Cultural Geography Study Group. This institutional accommodation towards cultural geography meant that funding support, special events and seminars could be held to foster this now recognised cutting-edge area of debate. The publication of papers from the conference proceedings of this newly formed group, compiled by Chris Philo and entitled *New Words, New Worlds: Reconceptualising Social and Cultural Geography* (1991), captures the prevailing zeitgeist. As the title suggests, words, representations and discourse were seen to hold the potential for opening up social and cultural geography to other worldly horizons. The marriage between the social and cultural would be further cemented over the next decade. This would culminate in the birth of a new journal in 2000 entitled *Social and Cultural Geography*, which reflected the vitality of this subfield at the turn of the millennium. The journal largely encompassed the 'new cultural geography' and so offered an alternative theoretical and empirical trajectory to established US journals in the area. Of course these latter journals were also shaped by developments in the subfield and mirror the various ways in which cultural geography is done. Indeed, the doing of cultural geography is now an established feature of numerous volumes in the field. This includes the publication of conference proceedings (Cook *et al.*, 2000; Philo, 1991), specialist journals published from the mid-1990s such as *Ecumene/Cultural Geographies*, as well as a suite of student texts (Crang, 1998; Mitchell, 2005; Anderson, 2010), dictionaries (Atkinson *et al.*, 2005), key readers (Oakes and Price, 2008) and edited collections aimed at an undergraduate market (Anderson and Gale, 1999; Atkinson *et al.*, 2005) often with a focus upon the methodological practice of cultural geography (Shurmer-Smith, 2002; Blunt *et al.*, 2003).

What all this demonstrates is that if cultural geography was once a marginal pursuit in British academia, then this is no longer the case. Arguably the most significant of these more recent publications has been the scholarly *Handbook of Cultural Geography* (Anderson *et al.*, 2003) which, although neatly sectioned out, stands at 580 pages: hardly a pocket-size handbook! Even so, what this edition signals is the breadth and depth now apparent in cultural geography at this point. The volume parades the growing international dialogues occurring in the area, even if largely restricted to the wealthier Westernised academic spaces of the US, Australia, Singapore, the UK and other parts of Europe. To get a richer flavour of the extent of these global intersections one can further turn to the journal of *Social and Cultural Geography* which regularly contains snap-shot portraits of cultural geography from developing countries, post-socialist nations and small islands. The array of cultural geography work canvassed is in itself worthy of a Berkeley School-inspired morphological description, mapping and accounting. What such a model might reveal is that cultural geography is widely disseminated across the globe, but more densely compacted in some places rather than others. What it may not record, though, is the variety of guises in which cultural geography is now to be found. On the international scene contemporary cultural geography is found to be particular to the issues arising in a specific part of the world. For example, in a review of social and cultural geographies in the Anglophone Caribbean region research on environmental disasters, Caribbean food, art and music have featured along with longer-standing debates on migration and **post-colonialism** (Dodman, 2007). Significantly, cultural geography is no longer a Western tradition – if indeed it ever was – but rooted in the histories, environments and cultural practices of people around the world, as evident in Rana P. Singh's (2009a, 2009b) accounts of Indian cultural geography, where ecology and religious beliefs hold sway. While cultural geography may be eclectic, diverse and sometimes marginalised, it is international in its scope and composed of different regional traditions.

Notwithstanding this international interest, we would be mistaken to think that cultural geography was simply incorporated into the echelons of the RGS as a matter of will. What counted as geography was still a terrain of struggle that cultural geographers had to negotiate. If Cosgrove, Daniels and Jackson were once the sum part of new cultural geography, it was the following generation of British geographers that would form the critical mass for change. This new generation included PhD students who had broken away from Sauerian traditions to explore how culture is linked to power, ideology and the broader field of representation. Geographers such as Alastair Bonnett, Mike and Phil Crang, Tim Cresswell, David Matless, Catherine Nash, Steve Pile, Gillian Rose and others were formative in prosecuting a new politicised agenda for cultural geography. This entailed explorations of the discursive representation of race, class and gender; penetrations into the 'economy of signs and spaces' (Lash and Urry, 1999) that mark production and consumption; and an interdisciplinary use of post-structuralist, feminist and psychoanalytic approaches to everyday life. These new scholars turned away from the positivist scientific approach of geographical inquiry and embraced one another's ideas; a type of collective 'group hug' against staid objectivity. In adopting social constructionist understandings of thinking they recognised knowledge as partial, selective and the product of historical forces. Where cultural geography had comprised little more than a cell of individuals by the end of the 1990s interest had, as Mitchell remarks, 'exploded' (2005: 63).

In doing so, what was once 'new' inevitably began to translate into the norm, perhaps suggesting that the contrast between 'new' and 'old' has been somewhat conflated (Kong, 1997). The dissemination and take up of the new cultural geography by those in the US, Australia and other parts of Europe is indicative of the global reach of British geography and the wider circulation of these ideas. Textual representation became a key site of critical analysis, as the cultural meanings conveyed through language, image, music, symbols and visual signs became as important as the text itself. Ethnographic research, qualitative interviews, film and media analysis, as well as historical and visual accounts, offered alternative trajectories for developing new methodologies in cultural geography. These interests are reflected in the edited collection *Doing Cultural Geography* (Shurmer-Smith, 2002) which contains essays on post-structuralism, **feminism**, post-colonialism as well the use of methods such as interviews, focus groups, textual analysis, field observation, secondary data analysis, archival work and case studies. What can be discerned here is a shift to re-define cultural geography through qualitative methodologies and new critical theoretical approaches.

Recasting political and economic geography through the cultural turn

Another way in which the influence of the cultural turn can be felt upon the discipline lies in the manner in which other geographical subfields have responded to developments in cultural geography. The rise of a type of popular **geo-politics** within political geography has seen discursive lines of inquiry made applicable to political cartoons, speeches, music, news, film and television, as discussed in Chapter 10. This turn to text, images and soundscapes offers an arresting moment for those political geographers who traditionally based their accounts upon the views and opinions of political experts and official channels of state information. As official state media are often controlled by government, an exploration of a popular geo-politics from 'below' can show how political ideas are received, parodied and resisted in public circles. The 'circuit of culture' (Figure 5.1) connecting the organs of the state that produce political messages (religious institutions, state media, national curricula, official textbooks) and the variety of audiences who receive these messages allows for a more complete understanding of struggle and contestation. With a focus upon popular culture these grounded understandings can reveal much about political discourse in action and the banal forms of nationalism performed in everyday life (Billig, 1995). At the same time we should be wary that the frenzied and often playful investigation of media texts does not see cultural representations render political 'realities' into oblivion.

As with aspects of political geography, the cultural turn has also transformed parts of economic geography and questioned the prioritisation given to the role of political economy. Much of traditional economic geography is largely descriptive in approach, compiling information about resources, markets and services in different regions. The turn to culture in geographical debates has led some economic geographers to critically rethink what is meant by 'economic'. What is now widely known as the '**new economic geography**' is a moving constellation of ideas circulating between geography, economics, sociology and cultural studies (Barnes, 2001b). The 'new economic geography' is characterised by an interest in:

- consumption as well as production
- social relationships and networks
- the cultural practices particular to specific firms and industries
- ideas of performance and embodied difference
- new theories emerging in other disciplines
- rethinking economy through culture.

Our opening example of Liverpool's successful City of Culture bid for 2008 suggests the value of a new economic geography towards bringing more nuanced understandings of culture, representations and identities into accounts of political economy. Here, we saw how a notion of 'multiculturalism' was transformed into a saleable 'commodity'. We suggested that much of what was once thought of as culture – film, arts, media, retail, urban architecture, tourism, music – is thoroughly embedded in and intractable from the economy. Cultural accounts of the economy include manufactured presentations of display in restaurant chains (Crang, 1994), post-colonial histories and geographies of 'exotic' commodities such as Caribbean hot pepper sauce (Cook and Harrison, 2003), investigations of second-hand markets such as car-boot sales and charity shops (Gregson *et al.*, 2000; Gregson and Crewe, 1997) and explorations of how the global marketing of alcohol shapes the ownership and architecture of the night-time economy (Chatterton and Hollands, 2003). What these studies achieve is that they transcend the narrow dualism between economy and culture, no longer giving primacy to the former at the expense of the latter. In each of these accounts culture is not a mere side-show, but a central part of how economic relations work and are given meaning to in specific places through embodied interactions.

The cultural turn within economic geography, as Trevor Barnes (2001b: 546) ventures, meant 'not only did the use of specific theories alter, but the very idea and practice of theorization changed'. His point is that it is insufficient for economic geographers to simply 'add in' culture into their accounts; rather, the very premises of economic theory have to be rethought from the ground up to forge the 'new economic geography'. As Gibson-Graham

(2005) has shown, a feminist post-structuralist reading of the economy, for example, allows us to shed light upon the informal practices that exist beyond the exchange of labour for capital. This may include voluntary work, domestic work, emotional labour, care work and other forms of unpaid labour. If we take the example of housework it is often women who are expected to undertake unpaid labour which has a subordinate status compared with paid work and the attributes accorded to being the 'breadwinner'. In this way, the masculinism of economic geography which has privileged the visible, public world of men's work needs to be reconfigured through more nuanced cultural understandings that appreciate the gendering of economic relations. Although financial exchange is often the focus of much work on the economy, the turn to culture has seen non-monetary transactions brought into the fold. This includes practices of gift giving, charitable donations, voluntary work, illegal transactions and the exchange of time or favours. The embodied manner in which cultural ideas about gender are 'put to work' in the economy is illustrated below through a synopsis of McDowell's (1997) work on banking and financial services (see Box 5.3).

Previously in this chapter we argued that the culture turn exceeds the discipline having influenced the social sciences and humanities more generally. Where some geographers have taken on these new ideas, other economic and political geographers remain unmoved, critically reflective or underwhelmed by the recent engagements with culture. Indeed, some geographers have expressed concern that 'the cultural turn has been *too* successful, has become *too* hegemonic, and has led to the realms of (say) economic and political geography making too many accommodations with a cultural orientation' (Philo, 2000: 28). As we shall go on to discuss in the following section, the overly textual approach associated with the cultural turn has been criticised for 'turning the world into the word' and draining it of its vitality; an issue we more fully address in Chapter 12 on non-representational practices. The charge here is that more grounded accounts of geographical phenomena are required

Box 5.3
Capital culture: gender at work in the city

Linda McDowell's work on gender relations in the London banking sector offers a useful illustration of the role that culture plays in business. In many ways McDowell's account can be seen as an example of 'new economic geography' in practice. To begin with *Capital Culture* (1997) is a book that derives from mainly qualitative methods, including surveys and interviews with merchant bankers and employees in three City banks. Second, McDowell makes use of the tools of feminist theory on performance, embodiment and identity. Her use of the ideas of the feminist philosopher Judith Butler (1990, 1993), combined with sociological understandings of the workplace, are indicative of a broader interdisciplinary approach. By deploying gender theory, McDowell identifies how the world of banking is steeped in a staunchly masculinist environment, making it difficult for women in particular to become accepted. Furthermore strong class codes predominate to the extent that middle- and upper-class forms of behaviour are the expectation and norm. This means that women and the few working-class people employed in the sector are likely to be excluded from wider social networks, after-work leisure activities and particular forms of knowledge and communication. Third, McDowell explores the culture of banking itself through an exploration of bodily presentation. She discovers that there are highly exacting, gender-specific forms of personal presentation that are expected in terms of dress as well as bodily comportment and persona. A key contribution of *Capital Culture* is that it demonstrates how financial transactions occur through social relations which are culturally embedded within a patriarchal gender order. As such it reveals that economic relations cannot be extracted out from cultural processes.

whether the issue concerns the political geographies of territorial conflict or the role and function of the modern economy. These concerns have most clearly been expressed through a need to 'rematerialise' culture and 'reclaim' the social.

Rematerialising culture, reclaiming the social

When examining the plethora of topics undertaken in the name of cultural geography, it is tempting to think that culture is everywhere and, perhaps, everything. Yet in many ways culture is nothing. It is after all only a term we use to explain particular practices we have come to understand as 'cultural'. This has not stopped geographers from unintentionally giving concrete meaning to the term. Where Anderson and Gale (1999) write of the cultural 'domain', Cosgrove and Jackson (1987) refer to it as a 'medium' and other geographers speak of cultural 'arenas', 'spheres' and 'fields'.

But, as Don Mitchell (1995) clarifies, a problem with such metaphors is that they risk *reifying* culture, setting it apart so it is treated as a 'thing' that can be captured, documented, aggregated and analysed through the method of cultural geography. Giving culture **ontological** weight – claiming it exists as a 'real' category or unit of analysis – makes authentic what is ultimately insubstantial. The lack of explanation from 'new cultural geographers' concerning what culture *is* has led Mitchell to declare that 'Cultural geography has remained incapable of theorizing its object' (1995: 104). Although new cultural geographers were more subtle than Sauerians in their approach, Mitchell argues that 'These ways of seeing "culture" do not avoid reification, rather they perpetuate it by smuggling right into the heart of geography what are still a quite mystified set of assumptions about how social practice proceeds' (pp. 103–104). Mitchell's complaint is that while new cultural geographers were keen to depart from the Berkeley School's 'super-organic' interpretation of culture, inadvertently the metaphors deployed reinstate culture as a proper object instead of seeking to understand its phantom status. He contends that 'For cultural geographers culture *exists*' (p. 103), going on to remark that, 'while there is no such thing as culture, the idea of culture becomes very real indeed' (p. 104).

In many ways there is much common ground between Mitchell and the work of new cultural geographers, as Peter Jackson (1996) was to make clear in a subsequent response. Where some regarded this exchange as 'unnecessarily testy' (Kong, 1997: 181) Mitchell himself admits that his position is derived through the insights delivered by 'new cultural geography' and cultural studies. We suggest that Mitchell's advance can be viewed as a refined surgical procedure undertaken upon the body of new cultural geography rather than a dagger thrust into its very heart. Mitchell's 'hollowing out' of the term culture is also a useful reminder of the way essentialist ideas can linger and carry over from the past to the present. His point is not to deny that culture is important – he has after all made it a focus of detailed expositions of landscape – but to convince us that when we deploy the term 'culture' we are engaged in a *politics of representation* and a corresponding ascription of power. In this way Mitchell contends that culture has no solid ground; to substantiate it in this fashion is to concede to the illusion that culture is authentic. Overall, Mitchell's (1995, 2000) insights offer theoretical elaboration and clarity on the uncritical and prolific use of culture generated in a number of geographical studies of the time. The 'hollowing out' of culture, to see it as an empty sign, is something that new cultural geographers in the rush to embrace 'the cultural' frequently underemphasise. In his response Jackson concedes, '"culture" cannot explain; it is the thing to be explained' (1996: 572).

However, while culture may be no more than an 'idea' – in the same way that the idea of race operates (see Chapter 8) – Jackson warns against ignoring the manner in which these categories materialise and give meaning to landscapes, bodies and spaces. 'Without such a concern for the material world', writes Jackson (1996: 573), 'we risk producing a thoroughly anaemic cultural geography where the struggles are over language and the politics of representation'. If 'ideas' of race or culture are purely discursive and imaginary then, perhaps all that is needed is new representations, signs and symbols to offset what has gone before. Such a suggestion is strikingly naïve. As we witnessed in the 'rebranding' of Liverpool's imperial past the displacement of one cultural legacy for another is never complete but a process open to contestation, friction and resistance to the uneven geographies of place-making. To place these struggles outside of their material histories and geographies, and render debates about race or culture merely to the realm of language, can have the profound effect of de-politicising them resulting in the type of 'anaemic cultural geography' Jackson warns against.

The risk here is that in operating on a purely discursive terrain, human geographers end up producing what Chris Philo (2000) regards as abstract 'dematerialised' and 'desocialised' geographies. It is worth reconsidering early cultural geography in the light of Philo's first assertion. Much of the work undertaken by geographers at the Berkeley School is emphatically materialist, documenting how humans manipulate the environment (e.g. build mines to excavate raw materials) or transform the landscape (e.g. through buildings, factories or transport links). The concern to map and observe such material objects as barns and fence-lines contrasts with what Philo (2000: 33) identifies as a 'dematerialized human geography'. The hallmarks of this later approach reveal a pre-occupation with:

> immaterial cultural processes, with the constitution of intersubjective meaning systems, with the play of identity politics through the less-than-tangible, often-fleeting spaces of texts, signs, symbols, psyches, desires, fears and imaginings. (Philo, 2000: 33)

What we can observe here is a shift from 'reality' to representation that has the effect of deconstructing the social and cultural into immateriality. Philo's second concern, that of 'desocialising' geography, connects with fears that the social divisions of race, class, gender, sexuality or (dis)ability are magically erased and replaced by representational interpretations of identity, signs and discourse. In other words, the social becomes disposable as it is evacuated and emptied out.

In response to such concerns within geography, there has been a renewed interest to 'reclaim' the social as well as to 'rematerialise' social and cultural geography more generally (Gregson, 2003; Philo, 2000; Jackson, 2000). Jackson (2000) in particular argues for the need to 'rematerialise' social and cultural geography in a period where symbolic representations hold sway. Bringing together new cultural theory with a material understanding can add depth to empirical and theoretical interpretations. Here, we might return to our example of the Apple iPod both as a material product formed through market economics and technological advancements, as well as a cultural item rich in symbolic meaning through its association with youth, fashion and quick-silver **modernity**. As we saw, the materiality of the commodity is essential to understanding how it is understood within the wider 'circuit of culture' in which it operates and the way in which its material construction makes advancements or places limits upon how it can be used. In returning to material objects, bodies and things in this way we might be able to explicate the more abstract cultural processes and performances that surround them.

The sense that the 'social' is being absented in the turn to a more discursive, representational cultural geography is signalled in various geographical discussions. Where cutting-edge cultural debates seemed to be advancing geographical thinking, the Marxist geographer Neil Smith has suggested that in comparison 'social geography has languished' (2000: 25). While such perceptions are debatable, and inevitably come to define the social and cultural as against one another, these sentiments have been widely expressed (Gregson, 2003; Philo, 2000; Pain and Bailey, 2004). Quite simply, for some geographers the act of rematerialising geography has a strong social appeal that 'means refocusing attention on the material realities and lived experiences of oppression and injustice' (Pain and Bailey, 2004: 320). The plea here is often for more empirically focused accounts that engage, and preferably help ameliorate, social inequality. While issues of social justice may be side-lined in some cultural accounts, this does not necessarily mean that what was once thought of as the 'social' has elapsed. Indeed, older social geography approaches to housing, welfare or forms of racial segregation have themselves altered in method, scale and approach.

In recent years a rather different set of 'social geography' accounts has emerged, as current work on children's geographies, sexuality and space, homelessness, disability and religious exclusion testify. At least some of these topics would have been unfamiliar areas for social geographers of the 1970s. However, if we were to take the example of work on fear of crime there is little doubt that mixed methods incorporating the differently patterned experiences of women and children have added analytical depth to broader understandings of neighbourhood crime based upon the British Crime Survey (Pain, 2000, 2001; Nayak, 2003c). Although more traditional forms of social geography and spatial mapping are widely practised, the cultural turn has also seen a rather different set of social geographies emerge. The approach in recent studies is often small-scale empirical work using predominantly qualitative methods (e.g. interviews, surveys or focus groups), with awareness to marginalised identities and forms of oppression. As the social geographer Nicky Gregson argues '"the social" has been reconfigured' (2003: 45), rather than evacuated. The suggestion is that social geography is alive and well, but has adapted a rather different set of guises. Even so, Gregson goes on to suggest that contemporary social geography has moved away from some of the wider structural issues once fervently addressed towards more personal experiences of marginalisation and oppression. In doing so, issues of embodiment, identity and emotional geographies have come to the fore, often drawing upon new theoretical insights generated in feminist and cultural geography. Although such work has been theoretically influential, it operates at a very different scale from the social geography policy research that has been of national significance in reports on immigration, housing, health and education. In this way, the reconfiguration of social geography has on the one hand enabled more nuanced accounts of difference to transpire, while on the other has seen a shift from large structural inequalities towards issues of identity. A more detailed explication of the 'turns', 'shifts' and 'reconfiguration' of social geography can be found in Vincent Del Casino's (2009) recent treatise of the subfield.

Looking back at the extensive proliferation of new cultural geography from the late 1980s, much of this work actually engages with material practices, ideology, power and politics (see, for example, Jackson, 1995; Mitchell, 2005). It is far from colourless or anaemic. Rather surprisingly it is the later turn in cultural geography away from the politics of representation towards non-representational theory and practices that has arguably seen more indulgent and ephemeral geographies transpire (see Chapter 12). At the same time this turn from culture to performance has certainly illuminated some of the limits of representation. Nigel Thrift (2000: 2) has argued that the 'cultural turn is dominated by tired constructionist themes' to the extent that the methods themselves fall short and turn inwards towards affirmations of 'empirical truth', 'evidence' and scientific notions of 'data'. He explains:

> Cultural geographers have, over time, allied themselves with a number of qualitative methods, and most notably in-depth interviews and ethnographic 'procedures'. But what is surprising is how narrow this range of skills still is, how wedded they still are to the notion of bringing back the 'data', and then re-presenting it (nicely packaged up as a few supposedly illustrative quotations). (2000: 3)

Thrift's point is that cultural geography has currently reached a point where it has foreground textual representation at the expense of presentation, performance and the senses. He accuses the cultural turn of 'letting theory outrun the data presented' (Thrift, 2000: 1) in what might be a consequence of the endless quest for meaning. It seems the initial fascination with discourse can itself leave us feeling that we are 'prisoners of language' (Barthes, 1972), unable to express our thoughts or research findings outside the representational codes which strait-jacket us. The limits of language with its inability to capture the sensory overflow that accompanies action and experience are addressed in Chapter 12 on non-representational theory and affect. Here, we pick up the threads of representation once more in order to explore a new moment in the unfinished project that is contemporary cultural geography.

Conclusions

Demonstrating the significance of culture to geographical analysis has been a focal point of this chapter. We began by explaining what culture is and how it has been approached. We then turned to the Sauerian tradition of cultural geography as undertaken by the Berkeley School and pointed to some of the limits of this method. These constraints became strikingly apparent when placed alongside much later work that derived from British cultural studies where there was an explicit focus upon culture as a process, an everyday terrain of struggle in which power is enacted. In discussing what became known as the 'new cultural geography', we saw how these dynamic ideas could be used to develop the more familiar humanities approach to landscape, art and music by evoking a much more politicised account of everyday life. These theoretical developments enabled cultural geography to develop from a fragile subfield to an area that eventually has become established and internationally recognised within the discipline. The contribution of the broader turn to culture in the social sciences, arts and humanities is extensive and has influenced theory and method. Finally, we pointed to some of the problems of 'cultural excess', where everything is seen as cultural. In response some geographers are calling for a return to material relations that do not write out the 'social'.

Summary

- The cultural turn in human geography has seen a shift from empirical truth to a focus on representation, discourse, images and language.
- Culture is a process that materialises in time and place so is of geographical and spatial significance.
- Cultural studies provided geography with new 'maps of meaning' through which to understand culture as a point of political struggle and contestation.
- The early tradition of American cultural geography was to eventually give way to the 'new cultural geography' in Britain in the late 1980s, which went on to become an established subfield within the discipline.
- As the cultural turn has become institutionalised, contemporary geographers have argued for the need to rematerialise culture and reclaim the social within social and cultural geography.

Further reading

Barnes, T. (2001) Retheorizing Economic Geography: From the quantitative revolution to the 'cultural turn'. *Annals of the Association of American Geographers*, 91(3): 546–565.

Crang, P. (1994) 'It's Showtime': On the workplace practices of display in a restaurant in south-east England. *Environment and Planning D: Society and Space*, 12: 675–704.

Jackson, P. (1997) Geography and the Cultural Turn. *Scottish Geographical Magazine*, 113(3): 186–188.

Mitchell, D. (1995) There's No Such Thing as Culture: Towards a reconceptualization of the idea of culture in geography. *Transactions for the Institute of British Geographers*, 20: 102–116.

Oakes, T.S. and Price, P.L. (2008) *The Cultural Geography Reader*, Abingdon: Routledge.

Philo, C. (2000) More Words, More Worlds: Reflections on the 'cultural turn'; and human geography, in I. Cook, D. Crouch, S. Naylor and J.R. Ryan (eds), *Cultural Turns/Geographic Turns*, Harlow: Pearson, pp. 26–53.

Part 2

Geographies of difference

Feminist geographies

Learning objectives

In this chapter we will:

- Identify the political roots of feminist geography in first and second wave feminist political activism.
- Explore how geographers have drawn on feminist thinking to challenge mainstream understandings of space, place and the discipline of geography.
- Focus on the implications of feminist approaches in human geography.
- Introduce some recent theoretical work in affiliated disciplines – in particular sociology, cultural studies and critical international relations – that have attempted to rethink the category of gender.

Introduction

In May 2009 Laura Whotton and her two sons, three-month-old Joshua and four-year-old Thomas, went to their local swimming pool (similar to that shown on the photo), the John Carroll Leisure Centre in Nottingham, UK. After a short swim Laura began to breastfeed Joshua at the side of the pool, while keeping an eye on Thomas who was still enjoying the water. On feeding Joshua, Laura was confronted by a male pool attendant: 'Are you breastfeeding?' he asked. 'Yes,' Laura responded. The attendant continued: 'You are in a public area; you can't breastfeed because there are children here.' Incensed by this attitude Laura took both her sons home and complained to Nottingham City Council and contacted

A public swimming pool
Source: Alamy Images/Philip Hympendahl

the national press. Laura remarked, 'It's the most natural thing in the world – and I was made to feel like I was doing something terrible.' Despite issuing an apology, the City Council responded to Laura's complaint by arguing that the official was enforcing health and safety rules that ban eating and drinking at the poolside. But Laura felt that the reaction had more to do with the possibility of seeing female flesh in public, and this was the impression given by the pool attendant's initial remarks. 'I wasn't embarrassed because I didn't have anything on show,' Laura responded, 'people in bikinis were showing more skin and breast than I was.'

This story, which featured in the British newspaper the *Daily Mail* on 19 May 2009, illustrates a complex set of geographical issues relating to gender and sexual identity. At first sight, this is a story about public space and expectations of appropriate and inappropriate behaviour. This is evident in the male attendant disciplining the feeding mother for transgressing particular expected norms of covering her body at the side of the pool (the suggestion that 'children might be watching'). But Laura's response points to a more complex picture. Women in the pool were wearing bikinis and were perhaps showing more flesh than the feeding mother, suggesting that it was not as simple a situation as offence caused by skin, but rather that this was the 'wrong' type of skin, that women wearing bikinis did not offend, although the feeding mother could upset the children. In so doing, these comments suggest a set of wider social expectations of gender roles and behaviour. Finally, the figure of the pool attendant and the subsequent interventions by the City Council illustrate the ways in which certain gendered activities are encouraged while others are legislated against. In understanding the gendered nature of social life we need to take seriously the ways in which government bodies and individuals use legislation to promote certain gender behaviour in public (wearing bikinis) and exclude others (mothering). In doing so a series of interlinked spaces of citizenship are presented by the act of breastfeeding: the body (in a bikini or nurturing a child), public space (the side of the pool) and the spaces of government regulation (the City Council or the UK state).

Of the many advances in the discipline of geography initiated by feminist scholars, perhaps one of the most valuable has been the ability to examine the structures and workings of everyday life to reveal hidden power relations. We start with this example to demonstrate the ways in which ideas about public/private, man/woman, appropriate/inappropriate circulate within everyday life, but in doing so they confirm and reproduce particular understandings of gender behaviour. In this chapter we will explore both this process and the ways in which geographers have sought to understand the social construction of gender and transgress its oppressive effects.

The first five chapters of this book have illustrated the changing nature of geography, as the intellectual trends within the discipline have been shaped by wider political movements and events. As we have seen, over the late 1960s and early 1970s geographers began to draw on the writings of Karl Marx in order to bring class analysis into the discipline. But this move into what has been termed *radical geography* was threefold: first geographers started to explicitly address questions of social and economic inequality and oppression. They therefore widened the field of topics that were deemed appropriate for geographical inquiry. Second, scholars were committed to social change: geography began to explicitly foreground political objectives of transforming society through better understanding of patterns of coercion and exclusion. The move to radical geography therefore explicitly politicised the discipline. Third, the writings of Marx on class oppression encouraged geographers to look at their own working contexts and practices and see a landscape of elitism and, more recently, the incursion of the interests of capital into the university system, at least in the UK, through the language of profit, sub-contraction and private–public partnerships. Radical geography was therefore interested in turning scrutiny onto the exclusions, inequalities and tacit oppression of the practice of conducting geographical inquiry.

We include this quick overview of radical geography in order to draw attention to the multiple ways in which wider social changes can re-shape the priorities, politics and practices of human geographers. This threefold move is particularly evident in the case of feminist geographers. Influenced by the wider political movements of Women's Liberation in the 1960s and 1970s, geographers began to explicitly address the exploitation of women in their research and teaching. The geographer Linda McDowell, a key figure in the incorporation of feminist ideas into geography, identifies four reasons for the absence of women or issues relating to women from geography in the 1970s. 'What tend, somewhat dismissively, to be termed "women's issues" were excluded from consideration for many years on one or several of four grounds – that they are trivial; that they are set at the wrong spatial scale, for example the domestic; that the methods used to examine these issues are not respectable (not science, inappropriate to geography); that the work is biased, subjective or, worse, political' (McDowell, 1992: 404).

But, just as with radical geography, we can see three separate moves within the adoption of feminist ideas into geography. First, we can see a widening of the topics of research to include women and issues relating to women's lives. Geographical study had tended to foreground the lives of men, in particular through surveying techniques that, although they professed gender-blindness, often relied upon data that were gathered through the 'head' of the household, invariably a man. Feminist geographers sought to widen the topics and techniques of geographical research in order to bring in the attitudes and experiences of women. Second, feminist geographers sought to advance a particular political agenda: the overturning of inequalities between men and women. Much of this research was therefore directed towards social transformation, and moved away from the more scientific language of research neutrality.

Third, feminist geographers, like other radical geographers, sought to transform the geographical discipline itself. As we will see later in the chapter, feminists argued that geography was a *masculinist* discipline; this meant that they felt that geography was a male-oriented discipline that marginalised women's contributions and served to reproduce wider social inequalities between men and women. Many feminist geographers sought to overturn this inequality and foreground women researchers, teachers and students.

The chapter is divided into five sections. In the first we explore the history of feminist ideas and how they have challenged existing social and political structures in the late 20th century. In the second section we examine the incorporation of feminist ideas into geography, paying particular attention to the Women and Geography Study Group of the Institute of British Geographers, whose activities and publications stimulated the adoption of feminist perspectives in the wider geographical discipline. The third sections examined the forms of politics that feminist geographers have drawn upon: making a distinction between radical and socialist feminism. In the fourth section we explore the practice of feminist geography, exploring the theoretical and methodological innovations that have been afforded through the incorporation of feminist perspectives and methods. In the final section we identify the key legacies of recent feminist scholarship within geography: the advancement in understanding of identity through a rethinking of the concept of gender. This challenging theoretical work draws on other theoretical innovations covered in this book, particularly post-structuralism, to draw attention to the importance of *practice* as to how we understand gender.

First and second wave feminism

In order to understand feminist geography we need to examine the wider political and intellectual context from which it emerged. Feminism is a movement that seeks to promote women's liberty by dismantling unequal power relations between men and women. In doing so, feminism has challenged the modes of thought and social practices that reproduce inequality between men and women. While such unequal relations may be a timeless feature of human organisation, we can discern two phases to the feminist movement: a 'first wave' from the mid-19th century to the early 20th century and a 'second wave' lasting from the early 1960s to the early 1980s. Some commentators identify a 'third wave' of feminism, from the early 1990s to the present, reflecting a break in 'second wave' thinking in recent years. We will expand on these recent theoretical developments in feminism later in the chapter.

First wave feminism

The geographers Alison Blunt and Jane Wills (2000) identify the inception of first wave feminist political organisation to the women's movement in the US of the mid-19th century. As Blunt and Wills outline, '[a]lthough an organised women's movement emerged earlier in the United States than in Britain, both developed in similar contexts, where women were barred from most institutions of higher education, did not have the vote, and lost their right to own property after marriage' (Blunt and Wills, 2000: 96). From this list of exclusions, it was the right to vote (or suffrage) that served as the focus for political organisation in the late 19th and early 20th centuries. The term *suffragettes* was coined by the popular press in the UK to refer to women who campaigned for the right to vote. This movement argued that the exclusion of women from parliamentary elections denied them of their right to citizenship. Women were therefore without a voice that could be heard through the conventional democratic channels. The suffragette movement saw women express their citizenship through alternative means to voting such as popular protests, mass demonstrations and peaceful resistance. There was, however, a split in the UK suffrage movement between a constitutional wing that sought political change through parliamentary systems, and a militant wing that sought change through direct action, confrontation and violence.

In historical prominence it is the more militant activities of the suffragette movement that have gained notoriety. In the UK, the campaign for women's suffrage was spearheaded by the Women's Social and Political Union (WSPU), led by mother and daughter Emmeline and Christabel Pankhurst. Founded in 1903, the WSPU used increasingly militant acts to gain votes for women, such as hunger strikes, arson and physical violence against political opposition. One of the defining acts of the WSPU occurred when Emily Davison, a committed WSPU activist, threw herself under the King's horse, Anmer, at the 1913 Epsom Derby (see photo), dying four days later of her injuries. Davison is buried in the Northumberland town of Morpeth, and her grave is inscribed with the WSPU slogan: 'Deeds not words.' During World War I the WSPU suspended its militant activities, embracing instead nationalistic causes and supporting the British army. The WSPU was finally dissolved in 1917 as Emmeline and Christabel Pankhurst formed the short-lived Women's Party.

The objectives of the WSPU were partially met in 1918 with the UK Representation of the People Act which allowed women over 30 to vote providing they were on (or married to someone who was on) the Local Government Register. The Act also enabled women to become MPs (Blunt and Wills, 2000: 99). It was not until ten years later, with the 1928

Emily Davison throwing herself under the King's horse, Epsom 1913

Source: Getty Images

Representation of the People Act, that women were granted the vote at the age of 21, the same age as men. Of course, the equal right to vote may grant formal political equality between the sexes, but this does not necessarily equate with substantive equality in terms of opportunity or rights. After the victories of first wave feminist political movements, these issues of lingering sexual discrimination were attended to in a second wave of feminist politics, emerging in the 1960s.

Second wave feminism

The political activities of the first wave of feminism were directed towards expanding the political rights of women: perhaps most explicitly seen in the case of women's enfranchisement. The second wave of feminism emerged in the early 1960s in the US and was less directed at the formal political rights of women, and more towards engrained forms of the discrimination existing in society and perpetuated through academic study. This is not to say that formal political rights were not part of this struggle: in the UK the second wave feminist movement won significant parliamentary battles, with the 1970 Equal Pay Act legislating that women and men should be paid the same for the same work, and the 1975 Sexual Discrimination Act which made it a criminal offence to discriminate in education, employment and housing (among other services) on the grounds of sex (see Blunt and Wills, 2000: 99).

But in many ways the ambition of second wave feminism was much broader than simply seeking parliamentary change. Second wave feminism sought to transform the structure of society and the way in which we understand sexual and gendered identities. Second wave feminists have described Western society as *patriarchal*, referring to the continued oppression of women by men. Feminists sought to bring this to light, to resist its discriminatory practices and transform society. This feminist movement was energised by the wider political movements emerging in the US and Western Europe centred on civil rights and class politics. Feminists sought to identify the ways in which mainstream society normalised social roles and identities that were oppressive to women. Box 6.1 illustrates one example of a second wave feminist political movement: the Wilds communes that were formed in the 1970s. While we can gather a remarkable array of political activities under the banner of second wave feminism, this example demonstrates the ways in which this movement differed from first wave feminism. Rather than attempting to alter constitutional frameworks that disadvantaged women, second wave feminists have used direct action to imagine a different arrangement of society that is more equitable between men and women.

The second wave of feminist scholarship and activism challenged the very mechanisms and assumptions through which knowledge is produced. It did this in three ways: first it questioned the biological basis of gender differences, drawing attention to the social construction of gender. Second, feminist research highlighted the relational nature of gendered identity: that femininity was only possible due to the existence of masculinity. Third, feminist scholars used these twin insights to question the production of knowledge, in particular drawing into question the rational, detached, Enlightenment thinking that had dominated scientific and social scientific research over the previous three centuries. We will consider each of these advances in turn.

Box 6.1
The Wild communes

In the early 1970s a group of women came together in London, with the aim of creating a different way of living and rearing children to the norm of the nuclear family. This movement argued that the conventional family arrangement in Western society represented a key symbol of patriarchal oppression. As an alternative, they established a commune in the north London borough of Islington, where conventional ideas of 'men's' and 'women's' work were disregarded in a spirit of collaborative child rearing. One of the distinctive aspects of this movement was the rejection of conventional surnames, which the group suggested were evidence of the patriarchal nature of society. Instead, all the children born in the commune were given the surname 'Wild'. The idea of the Wild communes, as they became known, spread across the country and, from the initial experiment in Islington, communes were formed in Oxford, Leeds and Sheffield. Al Garthwaite, a member of the Leeds commune, describes her motivations for joining the movement: 'A lot of us felt very constrained growing up in the 1950s, which was a very dismal and conventional decade.' These comments reflect the wider desires of the Women's Liberation movement to challenge the ways in which patriarchal oppression were normalised in the post-war period in Western countries as reflecting 'convention'. The Wild communes argued that conventions were just the accumulation of repeated oppressive social practices, and as such could potentially be changed by imagining an alternative system of society. Al Galbraith describes the collaborative system of child rearing practised in the communes: 'We all looked after the children. They didn't have a primary relationship with their mother. Non-biological parents were as equally important as biological parents.' This way the traditional roles of 'mother' and 'father' were rejected and primary care was provided by a number of women in the commune. While there were men allowed in the commune, this was on the condition that they assuaged their 'male guilt' through housework.

The Wild commune experiment lasted until the mid 1980s, but at this time the communes began to split up. Partly this was a consequence of wider intolerance to the movement in British society: the communes were often the target of attacks or their participants verbally abused. Al Garthwaite's daughter, Shelley Wild, describes the misconceptions people had towards the commune members: 'Because we were called Wild, people assumed we were feral and rampant, barefoot in the street, with no adult supervision. In fact, if anything it was the opposite – we were given so much attention. Wild was simply to do with freedom, to be who we wanted to be' (Williams, 2009).

First, one of the key insights of early second wave feminism was the exploration of gender, as opposed to sex. Where sex refers to biologically determined differences between men and women, predominantly based around differences in sexual and reproductive organs, gender refers to the social construction of feminine and masculine traits. As the feminist geographer Isabel Dyck (1990: 460) explains '[g]ender identities are understood to be actively constructed through practices varying over time and space'. This observation is often traced back to the work of groundbreaking feminist philosopher Simone de Beauvoir, and in particular her 1953 book *The Second Sex*. In this text de Beauvoir uses the ideas of existential philosophy to argue that 'one is not born, but rather becomes, a woman' (1953: 267; see Chapter 3 of this book for a discussion on existentialism). This revolutionary approach looked to the ways in which social practices and forms of knowledge produced women as marginalised and oppressed members of society. This approach to gender has been labelled as *social constructivist*, since it turns attention to the social context within which gender identities are produced and understood.

Second, feminist scholars drew on social constructivist approaches to illustrate the relational nature of gendered identity. At the heart of this approach is the insight that concepts of femininity are created *in relation to* concepts of masculinity. This relational idea of identity formation relies on the binary distinction between 'Self' and 'Other'. Specifically, this position suggests that we all formulate our identity in relation to an imagined Other, against which we can specify what we are not. As the geographer Gillian Rose (1993) has identified, this process of identity formation is not power-neutral: the concept of masculinity is socially constructed as superior to femininity. In so doing, femininity serves as the Other to masculinity, a necessary counterpoint to ensure the existence of coherent masculine identity. As Rose suggests (1993: 5), masculinity depends on femininity for its existence. This process of gender making is not simply one of social exclusion and oppression, but it is related to the wider processes of knowledge production in society.

The third conceptual advance of second wave feminism applied these social constructivist ideas to the production of academic knowledge. As we have seen, feminist approaches sought to expose the often hidden mechanisms through which the dominance of men over women was reproduced in society. One of the principal mechanisms brought under critical scrutiny was the assumptions that underpinned scholarly knowledge. Most specifically, feminist scholars argued that scientific and social scientific scholarship had attempted to produce rational knowledge based on the detachment of the observer from the observed. This form of detachment is often traced back to the philosopher René Descartes, who famously pronounced in the 17th-century *cogito ergo sum* ('I think, therefore I am'). This phrase has been taken to illustrate a primary separation in Enlightenment thinking between the subject and the object of research. Feminist scholars consequently drew this separation into question. Indeed, scholars such as Gillian Rose have argued that the very concept of a detachment between the subject and object of research is a masculinist fantasy: an invention that relies on 'a knower who believes he can separate himself from his body, emotions, values, past and so on, so that he and his thought are autonomous, context-free and objective' (Rose, 1993: 7).

In order to counter masculinist objectivity, feminists called for greater appreciation of the *situated* nature of knowledge production. Put simply, feminist thinkers wanted to highlight that all knowledge emerges from a particular perspective and, in the main, this perspective is masculine. As Jarvis *et al.* (2009) identify, the majority of scholarly literature is unwittingly *androcentric*, meaning that it is written from a male point of view. In response, feminist scholars have argued that there is no view from nowhere, but rather specific accounts of society that privilege particular points of view. By understanding the situated nature of knowledge we begin to understand why particular ideas are dominant, and particular segments of the population marginalised. The feminist thinker Donna Haraway has been at the forefront of critiquing masculinist objectivity, suggesting that this approach performs a 'god trick' of seeing everything from nowhere, a universal gaze that fails to adequately acknowledge its status as an individual perspective rather than universal narrative. By way of an alternative, Haraway suggests cultivating feminist objectivity, which is concerned with 'limited location and situated knowledge, not about transcendence and splitting the subject and the object' (Haraway, 1997: 59). Feminist perspectives therefore resist universalising, stressing instead the partial and particular nature of knowledge claims, grounded as they are in the context in which they are produced.

Political perspectives of feminism

One of the key messages of second wave feminist thought is that there is not a single style of feminism, but rather the term acts as an umbrella for a number of (potentially divergent) political and intellectual positions. The UK-based Women and Geography Study Group (WGSG), which is part of the Royal Geographical Society with the Institute of British Geographers and we will discuss later in the chapter, has identified two distinct feminist approaches: radical feminism and socialist feminism. Prior to examining the adoption of feminist thought into the geographical discipline it is important to understand these two approaches.

Radical feminism

Radical feminists view the inequality between men and women as the primary, and fundamental, form of oppression in all societies. The WGSG (1984: 26) explains the radical feminist position:

> The primary cause of conflict in society is located in the struggle between men and women over the social relations involved in biological reproduction. Thus there are two fundamental classes in society, consisting of women and men, rather than classes such as feudal lords and serfs or capitalists and workers.

Like the wider feminist movement, radical feminists view human society as patriarchal. Radical feminists have pointed to the spatial effects of patriarchal society, where women have been relegated to a 'private' sphere where their child-rearing, sexual and domestic labour is undervalued and ignored. The WGSG (1984: 29) explains how this shapes the political action of radical feminists:

> Because patriarchal attitudes and practices so clearly restrict the nature of the activities that women can undertake, and also the places where they can carry them out, radical feminists are united in seeing men as the enemy in the struggle for women's liberation.

The key, then, to a radical feminist position is the centrality placed on patriarchy as the single most important focus of political struggles. Other social movements (such as those related to class struggle) have been treated with suspicion by radical feminists who believe that they distract attention from the core political goal of dismantling the patriarchal nature of all human societies.

Socialist feminists

In contrast to radical feminists, socialist feminists connect the exploitation of women to the wider field of social relations, shaped by factors beyond unequal gender relations. Socialist feminists consequently do not present patriarchy as an unchanging constant in human society, but rather a system of exploitation that has changed in relation to other aspects of society, in particular the organisation of the economy (see WGSG, 1984: 29). Consequently, and as the name suggests, socialist feminists have been keen to draw connections between gender exploitation and class exploitation. As Gillian Rose (1990: 3) explains:

> [S]ocialist feminism . . . has argued that women are associated with reproductive labour, and that this ideological association is a fundamental aspect of the division of labour in

workplaces, between work and home, and in the home. Feminist geographers insist that reproduction is as important a part of social and economic life as the sphere of production that geographers have traditionally explored, and that the interconnections between the two spheres are central to a fully human geography which acknowledges women as social subjects.

Socialist feminism therefore builds on the forms of radical and Marxist approaches that were adopted by geographers in the late 1960s and early 1970s (see Chapter 4). But this approach extends Marxist critique by both critiquing forms of capitalist exploitation, while simultaneously drawing attention to different experiences of men and women within the capitalist system. As Rose (1993: 113) explains:

> While Marxists examine the uneven development of capitalist production, feminists focus on the relationship between production and reproduction as part of capitalist patriarchy.

As Gillian Rose points out, this draws our attention to many of the sites and practices of capitalism that had formally been ignored as outside the formal economy: nurturing, domestic labour, reproductive activities and informal or part-time work. While drawing attention to these often ignored practices, socialist feminists do not seek to endorse such gendered divisions of labour. Rather, they are seeking to highlight the exclusion of women's experiences from the double exploitation of capitalism and gender inequality, and seek social transformation in both arenas.

Establishing feminist geography

As we have seen, second wave feminism constituted both a political movement aimed at countering women's discrimination in a patriarchal society and an intellectual movement directed towards the gendered ways in which knowledge is produced. This combination of political and intellectual objectives is reflected in the emergence of a feminist geography as a distinct field of the discipline of geography. With a field as broad as feminist geography, there is considerable debate as to when geographers began to adopt feminist concerns, since there have been feminists who have addressed issues of space (while they would not self-identify as geographers). But we would see the upsurge of feminist activity in US and UK geography departments in the late 1970s and early 1980s as the key moment in the adoption of feminist perspectives in geography. Within this broad trend we can identify the activities of the WGSG of the Institute of British Geographers (now the Royal Geographical Society with the Institute of British Geographers) as a key moment in the institutionalisation of feminist geography. This group has produced a series of groundbreaking texts that sketch out the field of feminist human geography, such as *Geography and Gender: An Introduction to Feminist Geography* (1984); *Feminist Geographies: Explorations in Diversity and Difference* (1997) and *Geography and Gender Reconsidered* (WGSG, 2004, self-published and available as a CD-ROM). Through these publications one can trace the approaches, perspectives and concerns of feminist geography as it has established itself as a major field of intellectual inquiry within human geography. Any scholar interested in exploring this field of inquiry is advised to consult these original texts and explore their contribution first-hand. The first step in establishing feminist geography was to increase the visibility of women within the discipline, both as the subjects and investigators of research.

Making women visible

> Feminists writing geography have an ambivalent relationship to the discipline: they share some things with geography but also feel excluded from it. (Rose, 1993: 113)

One of the initial tasks for feminist geographers seems, at first sight, relatively straightforward: to make women, and issues relating to women, visible within the discipline. As prominent feminist geographer Linda McDowell (1992: 400) explains: 'this was a critical moment for many geographers involved in the earliest work in feminist geography, a moment of staking a claim within a discipline that had either ignored women or constructed them as the 'other'. McDowell (1992: 404) goes on to illustrate the areas of inquiry that opened up to feminist geographers:

> [R]edefining geography to include 'women's issues' was one of the major achievements of the first stage of feminist geography. A whole range of new areas become admissible for investigation. These included childcare, domestic power relations, housework, women's life cycle stages (single parenthood or widowhood etc.) and their relationship to spatial behaviour, access to resources, male violence, women's health, friendship networks, the gendering of skills, women's informal labour in a range of societies at different stages of development, women's social mobility, the power relations built into urban symbolism and customs.

McDowell's comments highlight the topics that were brought into the gaze of geographers following the adoption of a feminist perspective on social, economic and political life. But the transformation of the discipline needed to be more profound. As feminist scholars began to explore the nature of women's experience, the very practice of 'doing geography' began to be questioned. Rather than simply denoting new areas of inquiry, feminism challenged the history and practices of the discipline of geography. In particular, Gillian Rose argued that the discipline was masculinist: 'to think geography – to think within the parameters of the discipline in order to create geographical knowledge acceptable to the discipline – is to occupy a masculinist subject position' (1993: 4). Rose goes on to illustrate how women have been marginalised as both the subjects and producers of knowledge, and argues that this is a consequence of a number of unstated assumptions within geography about what men and women do, and reading the discipline to concentrate on the spaces, places and landscapes that it sees as men's (Rose, 1993: 2).

Absence from departments and publications

Rose points to the relative absence of women from geography departments in universities and the corresponding imbalance between men and women in terms of published articles in disciplinary journals. This topic had been a feature of geographical debate for some time: Monk and Hanson (1982) had urged reflection on what 'human' meant in human geography. Their call was for the discipline not to exclude 'half of the human' (Monk and Hanson, 1982). Certainly, women have been under-represented within the geography departments of UK universities, as indicated by Linda McDowell's landmark 1979 survey. One of the interesting aspects of McDowell's study is the decrease in women's participation in geographical study from undergraduate (where women comprised 42 per cent) to full time post-graduate study (where women only comprised 20 per cent). The disparity between men and women was exacerbated at the teaching level, where McDowell found that, in 1978, only 7 per cent of full-time university teachers were women (see McDowell, 1979: 151–152). More recently the WGSG (1997) examined the publication of research articles in the journal *Area* in 1994 and 1995. Their research showed that in 1994 only 8 per cent of journal articles were single-authored by a woman, as compared with 68 per cent single-authored by a man. The situation improved in 1995, although still only 23 per cent of published articles were single-authored by a woman. There are other well-documented mechanisms through which women have been marginalised in the production of geographical knowledge, for example the exclusion of women members of professional societies such as the Royal Geographical Society and Association of American Geographers (see Bell and McEwan, 1996 and Chapter 2 of this text).

The enduring masculinist rationality of geography

As alluded to in Rose's comments on what she terms geography's *masculinism*, the exclusion of women is more than a technical point of exclusion from scholarly positions or representation in professional publications. In arguing that geography is masculinist, Rose is making a much more profound statement that is connected to the form of knowledge that is produced by geographers. Drawing on the feminist scholarship we discussed earlier in the chapter on Donna Haraway, Rose argues that geography has relied on a form of masculinist rationality 'which assumes a knower who believes he can separate himself from his body, emotions, values, past and so on, so that he and his thought are autonomous, context-free and objective' (Rose, 1993: 7). This form of knowledge serves to reproduce the masculinist approach of the discipline: 'the white bourgeois heterosexual masculinities which are attracted to geography, shape it and are in turn constituted through it, imagine their Other in part as feminine. Their Other is associated with all that they deny as part of themselves: the bodily, the emotional, the passionate, the natural and the irrational' (Rose, 1993: 11). Rose identifies a masculinist fantasy figure of Woman, an alternative identification that acts as a repository of non-masculine characteristics.

Divergent strategies of resistance

Gillian Rose's work provides an important starting point for considering feminist geography. Perhaps one of the most intriguing aspects of this work is her consideration for practices of resistance to masculinist power. Rose sketches the political possibilities inherent in identifying the ingrained masculinity of geography and its attendant production of the category 'Woman'. In particular she focuses on two divergent strategies of resistance. On the one hand, Rose argues, feminist scholars could seek to strengthen this identity formation and allow it to speak back to masculinist arguments and provide a political platform for critique. This approach is often labelled *Identity Politics*, since it indicates the political opportunities inherent in gathering under the banner of a single identity label. On the other hand, Rose suggests that it may be more profitable to dismantle this identity formation and challenge the history of oppression and discrimination on which it is founded (Rose, 1993: 11–12). As we will see later in the chapter, this division between the assumption of an identity label for strategic or political purposes and the deconstruction of identity categories is a crucial area of debate in both contemporary feminism and post-colonial studies. We draw attention to this distinction since varieties of this schism are evident through much of feminist geography: whether to seek to address masculinist assumption on their own terms, or to step outside the constrictions of masculinist geographical work to envision a different mode of knowledge creation. In some ways we can trace a connection between this distinction and the different strategies of first wave and second wave feminists: between the desire to seek constitutional change to improve the political rights of women (Women's Social and Political Union, first wave) and the moves to imagine a radically different form of egalitarian social organisation (Wild communes, second wave).

Practising feminist geography

> One of the major, and earliest, achievements of feminist scholars in the social sciences has been to challenge the definition of what is geography, and hence appropriate topics for research, by adding in previously neglected areas. (McDowell, 1992: 403)

As McDowell suggests, one of the first tasks of feminist geographers was to expand the themes and areas of study that were considered relevant for geographical scrutiny. This included rethinking women's participation in formal structures such as employment (WGSG, 1984), development (Boserup, 1970) and the city (McDowell, 1983) as well as exploring the experience of women within a range of settings, such as the workplace (Oakley, 1981) and the home (Miller, 1983). This latter work brought into the geographer's gaze the lived experience of women's oppression, as the WGSG (1984: 86) explains:

> Women's lives themselves hold the key to unlocking the complicated interrelationships between the nature of their exploitation through waged work and their oppression through patriarchal social relations. Rather than our dismissing women's own descriptions and analyses of their lives at home and at work, we should include them in our discussion.

One of the first activities of feminist geographers was to place quantitative studies in context in order to dismantle the image of the 'non-working' woman (Madge *et al.*, 1997). This required placing quantitative datasets in context in order to bring into view the role and activities of women. Madge *et al.* (1997: 91) explain this process with reference to archival research:

> Historical research, for example, shows the ways in which archival sources and archival indexing systems 'silence' women's histories. 'Official' sources of data are shown to contain a male bias. To overcome this bias, archival sources are contextualised and read in conjunction with other sources.

But considering the critique set out above of geography as adopting masculinist rationality, is it possible to endorse qualitative methodologies and simply attempt to 'write women in'? Feminist scholars have considered this point and offered a series of criticisms of quantitative studies – we will focus on two (drawing on Madge *et al.*, 1997: 91). First, the practice of quantitative research suggests a hierarchy between the researcher and the researched, where the researcher sought to establish an objective account of a particular issue. Consequently, the practice of quantitative research could be identified as an element of the masculinist rationality which Rose suggested characterised geographical research. Second, quantitative studies often use pre-existing categories to organise the collection of data. Considering the patriarchal nature of society, these categories are representative of a male-oriented understanding of social life. The example given by Madge *et al.* (1997) is that of the category of 'work'. This category nearly always refers to paid work, thereby erasing other practices of work, such as subsistence farming, domestic labour or childcare, all of which are often undertaken by women.

Considering these criticisms of quantitative work, feminist geographers have attempted to develop a set of qualitative methodologies more attuned to examining the experience of women. The interest of feminist geographers in the lives of women required qualitative methodologies that granted insights into people's lived experiences. In this respect feminist scholars drew on the work of humanistic geographers (see Chapter 3), and shared their interest in the role played by everyday feelings and experiences (WGSG, 1984: 36). By drawing on qualitative methodologies of ethnography, participant observation, extended interviews and focus groups, feminist scholars have attempted to create a different form of rationality to the masculinist rationality that they suggested characterised existing geographical research. To this extent feminist geographers have used qualitative methodologies in order to sketch a *feminist vision of objectivity* (after Haraway, 1991), which McDowell (1992: 413) explains as an attempt to produce 'limited and situated knowledges, knowledges that are explicit about their positioning, sensitive to the structures of power that construct these multiple positions and committed to making visible the claims of the less powerful'.

Following Donna Haraway (1991), feminist geographers have turned to styles of methodology that allow for an appreciation of the 'situated' nature of knowledge production. Put simply: feminist scholars have argued that ideas do not exist in isolation from social contexts in which they are produced. This is a crucial point and reflects the feminist commitment to understand practices, identities and spaces in the real-world environments within which they are shaped. But perhaps just as crucially, the investigator cannot be separated from that research context and placed at an objective distance. As Gillian Rose (1997: 314) outlines, '[r]esearchers are entangled in the research process in all sorts of ways, and the demand to situate knowledge is a demand to recognize the messiness'. The subjectivity of the researcher is, therefore, a fundamental part of the research process. In order to understand research as a subjective and contextual process, feminist scholars have developed a geographer's understanding of qualitative methods, research position and the importance of collaboration. We will consider each of these in turn.

Qualitative methods

As we have seen, some feminist scholars have questioned quantitative methods as masculinist and making false claims of objectivity. Instead, feminist scholars have looked to qualitative methods, where questions of experience, meaning and subjectivity can be addressed, while allowing recipients to speak with their own voice. This does not mean qualitative methods are viewed in unquestionably positive terms. Anne Oakley (1981) criticised interviews as reproducing hierarchical relations between an interviewer and interviewee, and reproduced the notion of an objective observer gathering 'truths' from a particular informant. As you can see, Oakley is criticising certain qualitative methods in the same terms that feminists have criticised quantitative methods. In Oakley's view, feminist research should proceed with a clearer sense of mutuality and collaboration, rather than a strict power hierarchy between subject and investigator of the research.

In order to understand forms of oppression and social difference, feminist scholars have argued that we need to study the intimate and 'peopled' landscape through which social geographies are made. As Judith Stacey remarks (1997: 115), 'most feminist scholars advocate an integrative, trans-disciplinary approach to knowledge which grounds theory contextually in the concrete realm of women's everyday lives'.

This desire to understand the cultures through which gender is produced has drawn feminist scholars to the anthropological methodology of ethnography. Ethnography, literally 'people-writing', involves long-term research in a particular site, or series of sites, where the researcher observes everyday life and often joins in particular events, routines or traditions. This methodological approach has its roots in European colonialism, and has been widely used by anthropologists in order to understand and write about different human communities and cultures. As Cloke *et al.* (2004) identify, ethnography 'treats people as knowledgeable, situated agents from whom researchers can learn a great deal about how the world is seen, lived and works in "real" places, communities and people' (2004: 169). Ethnographic approach therefore offers the potential to release the research process from the strictures of participant and observer, so that, instead, research becomes a shared subjective experience between researcher and researched. The process of 'gathering data' becomes a lived experience where the subjectivity of the researcher plays an active and acknowledged part in the forms of knowledge that are produced about particular communities, places or issues.

As we discuss in Chapter 3, ethnography was initially adopted within geography by scholars adopting a humanistic perspective, such as David Ley's study of black gang graffiti in Philadelphia. But some feminist scholars have argued that women are particularly well-disposed to ethnographic research, as Stacey (1997: 116) explains:

> in ethnographic studies the researcher herself is the primary medium, the 'instrument' of research, this method draws on those resources of empathy, connection, and concern that many feminists consider to be women's special strengths and which they argue should be germinal in feminist research.

This is an interesting point and one that would be challenged by many feminist scholars (including us) who would question the concept of particular enduring or biologically determined traits that are 'feminine'. Such a position undermines the wider feminist scholarship that presents such categorisations of behaviour as socially produced, and therefore reproductive of unequal relations. This is a point we return to later in the chapter in the discussion of gender, since recent scholarship both within and beyond the discipline of geography has sought to question the binary of masculine/feminine as a socially constructed fantasy without any biological basis (thereby challenging the established concept of sexual differentiation between men and women that we discussed earlier in the chapter). Rather than reproducing this binary by suggesting the existence of 'feminine traits' many feminist scholars are seeking to undermine its significance and critique the forms of exclusion and hierarchies which it fosters.

Of course, embracing qualitative methodologies such as ethnography poses a set of new ethical challenges for the researcher. The imagined objectivity of previous qualitative or quantitative studies allowed the researcher to imagine him/herself 'at a distance'.

By embracing qualitative methodologies such as ethnography, this imagined distance is purposefully erased and research is thought of as a shared experience. This means that the researcher has a clear ethical commitment to the research participants, in particular in terms of understanding the effect of the research on the communities within which it is undertaken. One of the key points here is that there will always be a certain power relation between this investigator and the researched, since the investigator has the authority to design the research, draw conclusions and select the means of disseminating the work. In a personal assessment of the ethical dilemmas of this approach, Stacey (1997: 117) argues that 'elements of inequality, exploitation, and even betrayal are endemic to ethnography'. Part of the reason for Stacey's assessment is her experience of being told in confidence about a lesbian love affair between two research participants, one of which was a married woman. When the married woman told Stacey about the affair she asked her to keep it a secret from her relatives, co-workers and friends. Stacey felt this placed her 'in situations of inauthenticity, dissimilitude, and potential, perhaps inevitable betrayal, situations I now believe are inherent in the fieldwork's method' (1997: 117). This critical assessment is not shared by all feminist geographers, although it points to the complex ethical questions posed by qualitative methods in general, and ethnographic methods in particular.

In order to understand and negotiate some of the ethical challenges posed by qualitative methods, scholars have drawn attention to the role of research position and the potential benefit of collaborative research.

Research position

The emphasis of feminist methodologies on the subjective nature of the research process has consequently directed attention to the role of the researcher. As we have seen in the case of ethnographic methodologies, in moving away from a vision of masculinist objectivity the researcher becomes an active agent within the research process. Feminist scholars have labelled this recognition of the role played by the subjectivity of the researcher as *positionality*. Understanding the positionality of the researcher helps identify the range of ethical and methodological challenges posed by the research process, while also drawing into view the power relations between the researcher and the researched. There are two important points to recognise regarding positionality. The first is that one's positionality is neither static nor singular: there are multiple ways in which positionality may be understood and interpreted. Clare Madge (1993: 296) explains how we need to consider the role of the '(multiple) "self"', showing how a researcher's positionality (in terms of race, nationality, age, gender, social and economic status, sexuality) may influence the "data" collected and thus the information that becomes coded as "knowledge"'. Madge therefore points to the rich array of different subjectivities that researchers embody. But this leads on to the second point regarding positionality that has been recognised by feminist scholars: as researchers we are never in possession of perfect self-knowledge (Rose, 1997: 306). We cannot fully understand how particular aspects of our perceived identity shape the research process, since these are subjective responses experienced by research participants or collaborators and perhaps never voiced. The perpetual uncertainty over the influence of researcher positionality on the process of research is illustrated in the story

Box 6.2
The challenge of positionality

Prominent feminist geographer Gillian Rose (1997: 306) opens her paper with a discussion of the challenge of thinking through research as a subjective process, shaped by the particular positionality of the researcher and the subsequent need to write these issues into the research. Rose sees this as a challenging prospect, and she relates a tale of her research experience through which to identify the challenge of positionality:

> The event that brought my difficulty home to me was a joke made by one of my interviewees. We were sitting in the cafe of an arts centre talking about his work, with my tape recorder sitting on the table between us. He's Scottish and working class. As a friend of his, another worker at the centre, walked past us, he laughed and said, 'look, I'm being interviewed for Radio 4'. She laughed and so did I, and the interview – a long and very helpful one for me – continued. But that joke has bothered me ever since; or, rather, my uncertainty about what it meant has bothered me. Was it just a reference to the tape recorder? Was it to do with his self-consciousness at being interviewed? But Radio 4 is a national station of the British Broadcasting Corporation, which means in effect it's English, so was his joke a reference to the middle-class Englishness of my accent? If so, was the joke a sign of our different 'positions'? But does he like Radio 4's Englishness? And how do any of these possibilities relate to how the interview went? I don't know the answer to these questions, and this, I felt, was my failure. Indeed, now I think about it, I can't even be sure he said 'Radio 4' and not 'Radio Forth', which is a regional commercial station, which would raise some but not all the same questions, and some more besides. Or not. I don't know what the joke indicates about our position, let alone how to write it into my research.

Source: Rose, 1997: 305–320

told by Gillian Rose at the start of a influential academic paper on the topic of situated knowledges (see Box 6.2).

One of the innovations that has emerged from feminist geographers' interest in positionality is a recognition of research as an *embodied* and *performed* social practice. Addressing first the question of embodiment, scholars have therefore drawn attention to the materials through which subjectivity is enacted: that is, the body. The opening discussion in this chapter concerning breastfeeding in a public space drew attention to the significance of the female bodily form and its policing through regulation and social norms. But the desire to explore research as an embodied practice is more profound than simply recognising the significance of bodies to social life. For feminist scholars, understanding research as embodied has been a crucial part of challenging masculinist visions of objectivity which have been based on a repression of the body (Rose, 1993: 33). By focusing on the body, scholars have begun to consider how questions of sexual desire, emotion, insecurity and health shape our understandings of space. But bodily practices do not simply exist; they are temporary dispositions. In this sense we need to think of them as performed. The concept of performance is useful on a number of levels. It first draws attention to the dynamic nature of any social act, including research. Performances are fleeting and ever changing. Second, the notion of performance highlights the presence of a performer and an audience. This relationship is itself extremely complex, but it reminds us that particular aspects of one's identity are understood by others on their own terms. Recalling the insight concerning positionality made above, we cannot fully know how an audience will receive and interpret particular performances. The relationship between the body and performance is illustrated by Gregson *et al.* (1997: 196):

> The process through which bodily performances are constituted through particular spaces are beginning to be explored in more detail by some feminist geographers concerned with the process of subjectivity. The ways in which bodies themselves are imagined as spaces, and the spaces which they are imagined as inhabiting, are being examined by some feminist geographers in relation to a range of subjective, emotional and psychic processes.

As we will see both later in this chapter, and in Chapter 12, the emphasis placed on embodiment and performance has extended beyond studies of feminist geography into, in particular, the study of geographies of sexuality and non-representational geographies.

Collaborative practice

In seeking to establish a *feminist vision of objectivity*, feminist geographers have promoted collaborative and participatory approaches to knowledge production. These approaches have shaped the way in which scholarly research is both conducted and disseminated. In terms of collaboration, the emergence of feminist geography as a field of intellectual endeavour was itself a collaborative exercise. The WGSG presented its ideas as a fully collaborative venture; in its 1984 text the chapters were not assigned to individual authors, but instead the entire authorship of the book was attributed to the WGSG. This style reflects a more cooperative and less individualistic approach to the academic endeavour, where the collaborative nature of the writing process is brought to the fore. The innovations of the WGSG extend beyond collaboration; in their 1997 text a significant part of the second chapter is devoted to testimonies of feminist geographers concerning their engagement with feminist geography. The results are personal and often emotional discussions of the background to individual interests in feminist geography and the political motivations for their intellectual interests. This personal narrative approach demonstrates the significance that this style of inquiry places on positionality and context.

In addition to promoting collaborative practices between academics, feminist geographers have been at the forefront of moves to foster more participatory methods of research. This approach – or more correctly set of approaches – attempts to break down the unequal power relation between researcher and researched, by bringing 'research participants' into the process of designing research, collecting data and drawing conclusions. This approach takes seriously the notion that research is 'inter-subjective': a reflection of a shared experience, rather than an 'objective' process of gathering data (McDowell, 1992: 406). This participatory move has both ethical and political motives. From an ethical standpoint, a participatory approach can limit the exploitative nature of qualitative research, where research participants have little voice in how their lives or ideas are portrayed in the research process. In political terms, a participatory approach may ensure that research objectives are closely linked to the interests and desires of research participants. These aspects of the participatory research are illustrated in the formation of the Participatory Geographies Research Group (see Box 6.3).

Box 6.3
The Participatory Geographies Research Group

The shift towards participatory geographies is reflected in the emergence of the Participatory Geographies Research Group (PyGyRG) as part of the Royal Geographical Society with the Institute of British Geographers. Here we reproduce the key information of the PyGyRG, published on their website (http://www.pygyrg.org/).

Participatory Geographies Research Group
The Participatory Geographies Research Group (PyGyRG) is a research group of the Royal Geographical Society (with the Institute of British Geographers). Recently there has been a surge of interest in the study and application of participatory research methods. Whilst a number of geographers have used participatory approaches and methods for many years, there are a number of reasons for the more recent interest across all fields of human geography. These include:

- A growing feeling that geographic research should have benefits for those affected by the social, economic and environmental issues which are at its heart.
- The belief that groups outside the academy have meaningful contributions to make to setting agendas, project design, analysis, interpretation and writing outputs of geographical research.
- Disillusionment with the ability of many mainstream quantitative or qualitative approaches and their sets of ethical principles to effect this, or to contribute to significant change, even where findings are disseminated to policy-makers or (non-participatory) action research frameworks are applied.
- Discontent with the increasingly elitist and exclusionary nature of the structures of higher education and UK geography, including the RAE [Research Assessment Exercise], which privilege forms of research which are highly theoretical in nature, have a narrow audience and few political impacts (in contrast to theoretically aware 'action'-oriented work, with a focus on examining the difficulties in trying to undertake such work).
- Development of critical debate over participatory approaches, including those promoted in public policy programmes.

Rethinking gender

Like many binaries in recent years under the philosophical turn of post-modernism, the categories of 'masculine' and 'feminine' have come under increased scholarly scrutiny in recent years. Feminist geographers have contributed to this process of reflection, concerned that the categories on which feminism has relied are themselves exclusionary and oppressive. Some scholars have subsequently identified a broad shift in feminist scholarship, from the study of women and women's oppression, to the study of the production of gender categories and the political implications of their continued existence. This more reflective turn has been met in some quarters with a sense of scepticism, as theoretical and philosophical reflection replaces the certainties of first and second wave feminism. As the geographer Linda McDowell (1992: 412) explains, 'in some hands . . . uncertainty about the use of gender as an analytical category has led to a profound pessimism about the future of feminism as either a theoretical category or as a political movement'.

But renewed scrutiny of the nature of identity has been extremely productive and, in many ways, broadened the political scope of feminist geographies. This approach is indebted to the work of leading feminist thinker Judith Butler. It is difficult to overstate the significance of Butler's work in thinking about the nature of gender and it is important to study her original works in some detail to grasp the complexity of her argument. Her work *deconstructs* gender, meaning that she does not see gender as a stable or essential identity but rather seeks to uncover the practices and assumptions through which prevailing norms of gender identity are reproduced. One of her key arguments in the text *Gender Trouble* (1990) is that gender labels are *performative*, in that they do not simply describe an identity but bring identity into being. This insight draws our attention to the role of wider social structures and practices in normalising particular gender identities. But crucially, and perhaps most controversially, Butler rejects the idea of a biological underpinning to gender labels of 'femininity' and 'masculinity'. Butler takes an *anti-foundationalist* position, arguing (as the name suggests) that there is no biological or essential foundation of identity that prefigures action. Instead, seemingly knowable sex categories of 'male' and 'female' are themselves fundamentally unstable discursive productions that serve to make masculinity and femininity intelligible (see Nayak and Kehily, 2006: 460). Butler's work has therefore focused attention on how gender is performed in everyday life: the styles of clothing, bodily comportment and speech acts that produce, confirm or contest prevailing gender norms.

One of the key advances in feminist thinking prompted by Butler's work has been a turn to explore the construction and variations in gender identities. We would draw particular attention to three new areas of scholarship. First, scholars have drawn on post-colonial literature (see Chapter 11) that has critiqued feminist scholarship for privileging the attitudes of elite Western women over the experience of women elsewhere in the world. The question was therefore posed whether it is possible or desirable to speak of an over-arching category of 'women' when the experiences of women vary between times and places. Geographers have been particularly well-placed to trace and incorporate the experience of women in the Global South into wider feminist thought. The geographer Nina Laurie (1999), a member of the WGSG discussed earlier in the chapter, has conducted extensive research into the experience of women in Peru through their involvement in work/welfare programmes. Laurie explores a specific governmental intervention in Peru called PAIT (Programme of Support and Temporary Income), an initiative that was aimed at addressing the country's prevailing economic crisis of high inflation and high unemployment. This programme was designed to provide unskilled labouring work to unemployed men at the minimum wage, work that involved 'constructing access roads, irrigation channels, walls and foundations, reforestation, installing basic sanitation and painting frontages of schools and health centres' (Laurie, 1999: 69). Laurie goes on to explore gender identities through a number of lenses: first that though the PAIT programme of work was designed for men, it was predominantly taken up by women (around 80 per cent). This 'feminisation' of PAIT was a reflection of the opportunity for men to enter the informal labour market and receive higher wages, an option that was not open to women. Using qualitative methods, Laurie explores how participation in PAIT changed some gendered and sexist attitudes towards women held by husbands and male co-workers, namely that women would not be able to work hard and complete the challenging manual labour. In addition, Laurie traces the fractured nature of gendered identities as women and men working within the PAIT programme played different roles and formed different allegiances. Laurie does not talk in universalised ways of gender identities, and from these read off particular political strategies, but rather explores the range of practices and strategies employed by different gender groups. But Laurie also traces the ways in which entrenched ideas of different gender roles endured through PAIT: particularly evidenced by the Peruvian government's move to re-label the policy as a 'welfare' as opposed to 'work' programme, despite the fact that the work remained the same (hard labouring). This positions women as merely 'welfare recipients within a paternalistic state system' (Laurie, 1999: 81). Therefore, reflecting the emerging work that is aiming to re-think gender, Laurie's research talks more about the construction and solidification of gender identities than simply the oppression of women.

Second, feminists saw the binaries between 'masculine' and 'feminine' as exclusionary towards transsexuals who do not fit neatly within these gender norms. The feminist scholar Kath Browne (2004) has suggested that the assumption of two neatly defined genders should be understood as 'genderism', a form of prejudice that shapes political and intellectual life to the exclusion of those who are not deemed to 'fit in'. This argument seems to develop Judith Butler's ideas of **performativity:** gender categories are not descriptive of a external reality, but they are instead *bringing into being* a particular understanding of the social world. Therefore they are not innocent labels, but are rather *performative* of exclusionary categorisations. Browne illustrates this argument by documenting the feelings of insecurity felt by transsexuals and transvestites when using public toilets:

> Toilets, as sites that are separated by the presumed biological distinction between men and women and their different excretionary functions, can be sites where individuals' bodies are continually policed and (re)placed within sexed categories. (Browne, 2004: 332–333)

Consequently, Browne laments the lack of work exploring 'gender disidentification', referring to those who do not see their gender identity reflected in the prevailing binary between 'masculine' and 'feminine'. This work illustrates the distance that feminist geographies have travelled from early studies directed towards addressing the masculinist nature of society, to recent work that is looking at the production of categories of *both* masculinity and femininity (see McDowell, 1992: 400). Expectations of gender roles, practices and spaces is a central aspect of geographies of sexuality, discussed in the following chapter.

Conclusions

> The history of feminist theory, like all histories of thought one of contested positions and contradictions, is a story of great, though perhaps immature, certainty now being replaced by a period of mature doubt. In little less than a decade feminist scholarship seems to be in danger of swinging from wild optimism about the prospects of the construction of a body of explicitly feminist thought associated with empowerment through knowledge, to extremes of self doubt about the validity of the category of 'woman' itself. (McDowell, 1992: 409)

The profound influence of feminist ideas on the geographical discipline means that the approaches and ideas they have developed extend beyond a neatly defined field of 'feminist geography'. Feminist scholars have challenged and transformed the way we conduct geographical research, confronting the fallacy of political neutrality and, more recently, the very categories used to order the social world. The transformative nature of feminist work in geography can be traced to its roots in first and second wave feminist movements in the US and Western Europe. The excerpt from Linda McDowell above illustrates the ways in which this early politicised work was imbued with a sense of optimism: that there was a clear target for feminist scholars, namely the masculinist nature of society at large, and the geographical discipline in particular. But, as we have seen, recent theoretical turns to the work of scholars such as Judith Butler have served to question the category of 'women' as a stable and fixed identity. Instead this recent work has suggested that gender identity is a fiction constructed through social norms and practices.

The legacy of feminist geographical work is reflected throughout the later chapters of the book. Theoretically this legacy is expressed through work that seeks to challenge the categories used to order the social world; for example, geographers have recently begun to question labels such as 'heterosexual' and 'homosexual' in work examining the relationship between sexuality and space. Empirically, feminist work forces us to think through our research positionality and the influence this has on the approaches we use and the topics we address. Think through your own research interests: to what extent is this a consequence of your background, identity or stage of life? But the significance of positionality extends beyond this process of self-reflection: it also urges us to explore the embodied nature of ideas that we encounter through field research. This is a point developed in Chapter 10 on critical geo-politics.

Summary

- Feminist geography emerged during the 1970s as political stances developed in first and second wave feminism were brought into the geographical discipline.
- Feminist scholars have argued that geography is a masculinist discipline that has marginalised women.
- Feminist geography challenged the supposed neutrality of social research, encouraging scholars to reflect on the 'embodied' nature of knowledge production, both in terms of the positionality of the researcher and the social context of research participants.
- Recent work in feminist scholarship has questioned gender categories, suggesting that these are social constructions that should be challenged rather than reproduced. This is intellectually liberating, but critics are concerned that it undermines the original political project of feminism: to seek equality between the sexes.

Further reading

Blunt, A. and Wills, J. (2000) *Dissident Geographies: An Introduction to Radical Ideas and Practice*, Harlow: Prentice Hall.

Browne, K. (2004) Genderism and the Bathroom Problem: (Re)materialising sexed sites, (re)creating sexed bodies. *Gender, Place and Culture*, 11: 331–346.

Butler, J. (1997) Gender Trouble, Feminist Theory and Pyschoanalytic Discourse, in L. McDowell and J. Sharp (eds), *Space, Gender, Knowledge: Feminist Readings*, London: Arnold, pp. 217–261.

McDowell, L. (1983) Towards an Understanding of the Gender Division of Urban Space. *Environment and Planning D: Society and Space*, 1: 59–72.

McDowell, L. (1998) Elites in the City of London: Some methodological considerations. *Environment and Planning A*, 30: 2133–2146.

Rose, G. (1993) *Feminism and Geography: The Limits of Geographical Knowledge*, Cambridge: Polity Press.

Rose, G. (1997) Situating Knowledges: Positionality, reflexivities and other tactics. *Progress in Human Geography*, 21: 305–320.

Women and Geography Study Group (1984) *Geography and Gender: An Introduction to Feminist Geography*, London: Hutchinson.

Women and Geography Study Group (1997) *Feminist Geographies: Explorations in Diversity and Difference*, Harlow: Longman.

Geographies of sexuality

Learning objectives

In this chapter we will:

- Explore how scholars in cultural geography have drawn on feminist perspectives to examine the object of geographical research.
- Introduce work that has argued that society and the geographical discipline are heteronormative; they actively endorse certain lifestyles and sexual preferences, while presenting others as deviant and abnormal.
- Explore the geographies and politics of sexuality: examining, in particular, the emerging geographical scholarship that identifies emerging forms of sexual citizenship.

Introduction

On 5 December 2009 the Episcopal Diocese of Los Angeles voted to elect Reverend Mary Glasspool (see photo) as an assistant bishop. This vote prompted global comment and in some quarters criticism, since Reverend Glasspool was in a sexual relationship with another woman. The sexuality of Reverend Glasspool was the focus of intense scrutiny, in particular stressing that she was 'openly gay' and that she had been in a relationship with her partner 'since 1988'. For example, Reverend Kendall Harmon of the more traditionalist Diocese of South Carolina, US, viewed the appointment of a lesbian assistant bishop as corrupting Christian values, suggesting 'this decision represents an

The Reverend Mary Glasspool
Source: Press Association Images / Mark J Terrill

intransigent embrace of a pattern of life Christians throughout history and the world have rejected as against biblical teaching' (BBC News, 2009).

The appointment of Reverend Glasspool was not the first time that sexuality had been placed at the forefront of the selection process of the US Episcopal Church. Six years earlier the first bishop in a relationship with another man, Gene Robinson, was elected in New Hampshire. The appointment of Bishop Robinson caused a rift in the global Anglican Church, as traditionalists refused to recognise the possibility of a gay man occupying such a position of moral leadership. In particular – though as we have seen by no means exclusively – Church leaders from Africa condemned Bishop Robinson's appointment and declared a 'broken communion' with the Episcopal Church in the US, meaning that they refused to attend church meetings with American representatives or accept missionaries from the US.

On the basis of the rift caused by Bishop Robinson's appointment, the election of Reverend Glasspool was met by a rapid response by the leader of the Anglican Church and the Archbishop of Canterbury, Dr Rowan Williams. In a statement on his website Dr Williams suggested that the election 'raises very serious questions not just for the Episcopal Church and its place in the Anglican Communion, but for the Communion as a whole'. Although the statement makes no explicit criticism of the election of Reverend Glasspool, Dr Williams makes it clear that the appointment still requires confirmation by the diocesan bishops and committee, and he urges a period of 'gracious restraint'. The question of sexuality is therefore posing another threat to the unity of the Anglican Church. There is a fear, reported by BBC News (2009), that the appointments will 'mean that the US Anglican Church is consigned to the second of a two-tier global communion'.

This example is alive with issues relevant to the geographies of sexuality. Most mundanely, the reaction to the elections of Reverend Glasspool and Bishop Robinson points to an assumption that there are certain spaces that are reserved for heterosexual people, such as churches, pulpits, church congresses and so on. More profoundly, the reactions to the elections are based on the assumption that individuals may be defined by their sexual preference, that it becomes part of their identity. This, in turn, suggests that certain identities (in this case Christian) are exclusive of others (homosexual). On an international scale, the differential reaction to the appointment of gay clergy across the world illustrates differential histories and geographies of tolerance to sexual practices that deviate from an imagined heterosexual norm. All of these points are subsumed within a prevailing context that there is a 'normal' approach to sexual desire within the Anglican faith (heterosexuality) and other approaches are deviant and should be cast out as they pose a threat to the very moral fabric of the Church itself. Illuminating, challenging and reworking this complex mix of desire, identity and politics has been at the heart of work within geographies of sexuality.

This chapter is divided into five sections. In the first we examine how the emergence of research on sexuality in geography has demanded engagement with the 'object'. This section examines a key concept within geographies of sexuality: **heteronormativity**. This term refers to the portrayal of heterosexuality as the appropriate form of sexual disposition in contemporary society. It is worth spending some time examining this point, since it not only explains the marginalisation of alternative sexual preferences but it also points to the broader exclusions of questions of sex and desire from scholarly work. The majority of this section is taken up exploring three key figures – Sigmund Freud, Michel Foucault and Judith Butler – who in different ways have attempted to question heternormativity as a natural condition of society, and explore its production through the family, social institutions and cultural practices. The second section examines the emergence of scholarship on the geographies of sexuality. It traces the development of this work, from attempts to 'map' residential and leisure spaces populated by homosexual, lesbian and bisexual inhabitants, through to work that has explored the plural and fluid nature of the constitution of sexual identities and their expression in both public and private life. This work has challenged some of the key binaries that circulate within geographical inquiry: notably the concepts of public and private space, **essentialism** and constructionism, and the gap between researcher and researched. In the next section we develop these points by looking at the politics of this work. Specifically we are interested in developing two points. First we examine the ways in which scholars have sought to challenge and transform the discipline of geography – in particular undermining its perceived heterosexist nature. Second, we examine how work within geographies of sexuality has sought to highlight links between scholarship and activism, and in doing so illustrates new modes of citizenship that challenge heteronormativity. This engaged scholarship poses more questions about the relationship between theory and practice and the ability for researchers to enact political change. The final section of the chapter provides a conclusion, pulling out some key points and developing questions for future inquiry.

Engaging with the object of research

> [G]ay and lesbian geographies have remained largely outside squeamish academia, and have not been subject to the same depth of analysis as other marginalised groups, despite claims that homosexuals are the last minority and the most oppressed people in society.
>
> (Bell, 1991: 327)

Since David Bell wrote the above words, there has been a remarkable growth in geographical scholarship engaging with the nature of the relationship between sexuality and space. This area of study has drawn on history, sociology and psychology to develop a set of theoretically and empirically rich accounts of how sexual identity is constituted spatially and, conversely, how space constitutes sexual identity. This rapid growth in interest is reflected in university geography curricula that are more likely than not, at least in the UK and US, to explore geographies of sexuality, often within wider debates on social geographies. In research terms, this increased interest is reflected in the establishment of research groups within the two largest geographical societies, the Association of American Geographers and the Royal Geographical Society with the Institute of British Geographers, and the publication of two influential edited collections bringing together leading scholars in the field (see Bell and Valentine, 1995; Browne *et al.*, 2007).

In this chapter we want to present the spirit of this research and give an introduction to some of the key theoretical and political concerns of geographers working in this field. We share with others a desire to move away from presenting this work as an adjunct to work within feminist geography. We are aware that geography textbooks often introduce the geographies of sexuality towards the end of a discussion of geography and gender. This approach can lead to an assumption that work on the geographies of sexuality emerged as a coherent subset of feminist geography at some point in the 1980s. This is, of course, unsatisfactory since it underplays the significant independent contribution of scholars of sexuality and space, while also masking the ways in which scholarship on feminist and sexual geographies can disagree. Instead, we want to draw attention to aspects of the shared history of feminist approaches with scholarship on the geographies of sexuality, while also charting out this area of work as a distinct field of geographical inquiry and political practice.

In a sense, this discussion will draw on perspectives and forms of analysis that have been raised in earlier chapters. In particular, much of the work within cultural geography after the cultural turn (see Chapter 5) has been concerned with challenging assumptions about the identity of the subject *and* object of research. In terms of questioning the subject, as we have seen, geographical inquiry is no longer simply concerned with mapping, charting, labelling and exploring space, but also in addressing research as a human process that both engages with, and is undertaken by, real people. This has allowed a number of different aspects of human identity to come to the fore – questions of class in Marxist geography (see Chapter 4), race (see Chapter 8) and sex and gender (see Chapter 6). At the heart of this work has been the realisation of the social creation of these identity labels and categories. They do not present essential biological truths: there is nothing 'natural' about race, class or, as we have seen, even sex. We have therefore learned that the act of naming is not innocent; it creates the categories that it seeks to represent. Hence the work of geographers has been to understand and unsettle the spatial effects of these processes.

This work also unsettled the *object* of research: a process that can be imperfectly divided into two parts. First, scholars have sought to turn their attention to the role of the researcher and the wider disciplinary context in which she or he is embedded. As we saw in the previous chapter, feminist geography has been at the forefront of work in this area, arguing that geography is a masculinist discipline that privileges a particular style of 'objective' and 'rational' knowledge production, which was simultaneously marginalising alternative styles of inquiry (see Rose, 1993). Geographers of sexuality have also turned their attention to the nature of the discipline, arguing that there are barriers to working on questions of sexuality within geography. This was a point made by David Bell (1995: 127) when he discussed having the title of a conference paper altered to remove a swearword:

> Maybe we can dismiss this as a trivial point . . . But when it is added to a growing list of censoring and discriminatory practices within our discipline – having our articles pulled from library collections, gaining negative press coverage whenever we get 'public money' to do our work, having secretaries refuse to type up papers, not to mention all the whispering and all the silences from colleagues – we begin to uncover many more of the limits of working in geography on issues of sex and sexuality.

Scholars working on the geographies of sexuality have therefore sought to attend to this discrimination, and explore the different styles of research and dissemination that are prompted by taking sexuality seriously within geographical study.

The second – and related – part of unsettling the object of research is to collapse the distance between researcher and researched. Again, feminist scholars have been at the forefront of work that has highlighted the importance of understanding positionality, a term that refers to the role of the researcher's identity, history and geography in shaping the outcome of scholarship. By taking positionality seriously, feminist scholars, and others working in different parts of the geographical discipline, have refuted the notion that research constitutes 'a view from nowhere', highlighting the researcher as an active participant within the research process. The key implication of this argument is that it illustrates that research is not a technical exercise that could be undertaken by a machine – it is a human process through which different individuals articulate their ideas, practice and desires. Therefore research is necessarily a political practice, since it involves negotiations between researchers and researched (who occupy different positions of power) and it involves advancing particular arguments about the nature of society and the possibilities of transformation. This point has been at the heart of much work within geographies of sexuality, scholarship that has sought to challenge homophobic attitudes and the privileging of heterosexuality within the discipline and wider society. As we will see, the political objectives of enacting change can often come into conflict with the more theoretical desire to embrace difference (see Binnie, 1997).

Heteronormativity

There is a tendency within popular culture and media to present heterosexuality as normal. This is not a controversial statement: look at adverts for a new housing development, where a young heterosexual couple clink glasses of champagne in a newly fitted kitchen; look at holiday brochures for tropical resorts that show young families walking along a sandy beach; or take a big-budget Hollywood film almost at random and the leading (usually) male will look to 'get his girl' (see Nast, 1998: 192). Taking this last example, the confirmation of the expectation of heterosexual relations in film was explored by Sacha Baron Cohen in his film *Brüno*, which used the homosexual camp of the eponymous hero to unsettle viewers' (and unwitting film participants') expectations of romantic behaviour – including the public exhibition of homosexual practices and sadomasochistic uniforms. Cohen's comedy emerges from playing with the boundaries of acceptability within modern society – boundaries that serve to help delineate between acceptable and unacceptable sexual behaviour.

These assumptions of heterosexuality extend beyond popular culture: the formal political structures of the state are oriented to favour heterosexuality and often discriminate against alternative sexual identities. This is most starkly illustrated in the case of legislation banning sex between two men in the UK until it was decriminalised at the 1967 Sexual Offences Act. But the liberalisation of laws of sex has not been matched by access to all careers within the state, for example within the military. The US Army prohibits anyone who 'demonstrates an intent' to commit a homosexual act from serving in its forces, a piece of legislation that is often referred to as the 'Don't Ask, Don't Tell' Act on account of its emphasis on individuals concealing their sexual identity. In the UK context, the ban on homosexuality in the military was lifted in 2000, following a judgment by the European Court of Human Rights in 1999 that declared the ban broke laws of privacy. But, despite this legislative advance, forms of everyday harassment towards homosexual and bisexual soldiers still act as barriers to participation.

In less direct terms, state structures also tend to encourage heterosexuality through favourable tax conditions for married couples living in nuclear families, for example through married tax credits or child benefits. This norm has been eroded recently, as certain states (beginning with Denmark in 1989) have legally recognised civil partnerships between same-sex couples. However, the legal and spatial ambiguity of this process is encapsulated by the US, where there is no federal law providing for civil partnerships, leaving states to make their own legislation. This has resulted in a highly uneven legislative landscape, where certain states permit same-sex marriage (such as Indiana, Vermont and New Hampshire), some states grant limited legal rights to same-sex partnerships (for example California and Washington State), while in other states their constitution and/or statute ban same-sex marriage (for example Texas and Florida).

Despite these legal advances, heterosexuality remains an assumed norm within social, cultural and political life. Consequently, scholars talk of the prevalence of heteronormativity: a term which indicates that heterosexuality is presented through cultural and political life as the appropriate mode of sexual orientation. The dominance of heteronormativity has occupied minds of scholars across the natural and social sciences for the last three centuries. Initially these studies were confined to the discipline of biology: in the post-Enlightenment period (after the 18th century) the quest for scientists was to discover a biological basis of sexual behaviour (Blunt and Wills, 2000: 144). This work focused, in particular, on the role of procreation within sexual practice, and thus the biological instinct to reproduce presented heterosexual identities as 'natural' (Valentine, 1993: 238). In the 19th century new scholarly approaches to sexuality began to question this biological narrative. Rather than mapping desire neatly on to sex, work within psychology, and later in history, sociology and geography, attempted to understand *how* specific sexual desires were formed. Extending the work of geographer Phil Hubbard (2002), we can isolate three distinctive approaches to understanding human sexuality: psychoanalytic, discursive and performative approaches. These approaches do not sit in neat opposition, as contemporary scholars often draw on all three in order to understand heteronormativity. However, they do offer distinct frameworks for analysing the emergence of heteronormativity, and thus their central traits need to be illuminated.

Psychoanalysis: Sigmund Freud

Sigmund Freud (1856–1939) was a psychologist working in Vienna in the later 19th and early 20th centuries. He is famous for inventing **psychoanalysis**, a way of exploring human psychological behaviour and its role in establishing self-understanding. One of the key components of Freud's pioneering work was its focus on the role of the nuclear family (two heterosexual adults and children) in reproducing sexual identity. Psychoanalysis explored the ways in which subconscious decisions about sexuality are established through infant development within the family. By using psychoanalysis, Freud moved away from the biologically determinist views of earlier scientists, arguing that individuals are not born with a sexual identity; it is learned through social interactions with immediate family members, in particular parents. As the geographer Gill Valentine (1993: 238) has recognised, this does not equate to a total break from biology; Freud continued to stress the importance of biological sex in framing individual interpretations of individual sexual roles.

Freud used his psychoanalytic approach to develop a theory of infant sexual development known as the Oedipus complex. This theory is named after the mythical Greek king who unwittingly killed his father and married his mother, a tale recounted in the play *Oedipus Rex* by Sophocles. In this theory, Freud argues that the key to understanding the development of an infant's sexual orientation lies in the repression of subconscious desires. Freud argued that all humans are born bisexual but at as they grow up their 'sexual identities and gender roles become intimately bound up together' (Blunt and Wills, 2000: 142). Specifically, Freud argued that sexual development centred on the repression of desire for the parent of the opposite sex. Freud articulates a material and psychological motivation for the repression of alternative sexual desires, focusing in particular on castration anxiety (among boys) and penis envy (in girls). Ultimately children learn to defer and displace sexual desire onto parent substitutes. As Heidi Nast (1998: 194) explains in relation to the Oedipus complex in boys, '[r]epression of desire for the Mother is, according to psychoanalysis, what a Son must do if he is to be properly socialised, or, more specifically, oedipalised'.

This analytical framework may appear somewhat obscure, or at least of marginal interest to the study of geographies of sexuality. But it is important to stress the significance of Freud's work for developing a relational understanding of sexual development: that is, that a sense of a sexual self does not emerge organically, but is shaped through relationships with others. As the geographer Phil Hubbard has explained (2002: 366), Freud's work helps us understand how sexual identities emerge 'through interaction between self and other, with feelings of attraction and repulsion entering the unconscious only to be projected back on to Others who become objects of desire or disgust'. Methodologically, Freud's focus on the subconscious allows scholars to explore the imaginative geographies of sexuality, identifying how individuals imagine their socialisation with others and how this shapes senses of desire and the Self. But also Freud's approach has been 'scaled up' to explore 'socio-spatial repression' (Hubbard, 2002: 371), where non-heterosexual identities and desires are banished from material landscapes through architecture, town planning and building design. In this case the language of psychoanalysis – in particular the centrality of repression – serves as a metaphor for understanding the dominance of heterosexual relationships in the configuration of urban space.

Freud's work has not been without criticisms. First, and perhaps most significantly, there have been suggestions that there is an underlying universalism within Freud's psychoanalytic framework, whereby the Oedipus complex serves as a narrative for understanding *all* infant sexual development. As we will see in later discussions of the work of Michel Foucault, Freud's approach is therefore seen as underplaying the significance of social and historical contexts by looking for shared human experiences of sexual development. Second, Freud's framework can be seen as supporting heteronormativity through the suggestion that heterosexuality is the *normal* outcome of child sexualisation. Homosexuality is therefore presented as an incomplete or imperfect process of sexual development and a deviation from an idealised norm. Third, Freud's focus on the subconscious can make it difficult to explore empirically these ideas in practice. In banal terms: how do we identify the development of subconscious ideas of sexuality when the individual involved has little or no conscious recognition of the operation of these processes? Freud himself used a process of 'free association' where the subject of analysis would voice a stream of consciousness, in doing so voicing apparently unconnected aspects of their past, the relevance of which Freud would discern through analysis (see Blunt and Wills, 2000: 141). This approach places a great deal of emphasis on the analytical observer in discerning the relevance of social contexts for the formation of sexual identities, at the potential expense of the subjects themselves articulating how particular sexual preferences have been established.

Despite these criticisms, Freud's work continues to shape scholarship working on the relationship between sexuality, space and place. As we will see throughout this chapter, scholars often draw on Freud's ideas or language in combination with other scholars – notably Jacques Lacan (1901–1981). Readers who are interested in Freud's ideas are urged to explore his major works, in particular his *Three Essays on the Theory of Sexuality* (2000 [1905]) and *Beyond the Pleasure Principle* (1990 [1920]). We direct you to explore Freud's work yourself on the basis that – like most great thinkers – his ideas were not static, but changed throughout his life, as he developed his theory of psychosexual development.

Discursive: Michel Foucault

French philosopher and historian Michel Foucault has had a profound influence on the development of recent geographical thought (see, in particular, Chapter 5). One of the key attractions of Foucault's work for geographers is his rejection of biological essences or truths underpinning identity claims. Instead of searching for biological essences, Foucault looked to the historical and spatial contexts (for example prisons, hospitals and schools) through which particular ideas of the individuals as subjects and populations emerged. In this way, Foucault's historical work is not so much an exploration of the past (although this is necessarily involved); it is more concerned with constructing a 'history of the present', an exploration of how certain discourses (constellations of power and knowledge) of identity or status come into being (Weeks, 1991: 162–163). Through his work, Foucault presented a theory of power that deviates from dominant thought within political science that located power centrally as a repressive dynamic connected to state or class interest. As the geographer Linda McDowell (1995: 78) explains, Foucault provides a rather different picture, arguing that:

> . . . power, rather than being a totalising system, is diffused throughout the whole social order, and exists at all levels, from the micro-scale of the body, the home and the workplace to the structural institutions of society. The control, discipline and surveillance of bodies is particularly important in the production of what Foucault terms, 'docile bodies', which conform to the historically- and spatially-specific ideas of what is normal and appropriate forms of the presentation of self and daily behaviour in particular spaces . . .

Foucault brought this understanding of power to the study of sexuality. His contribution to this field has been profound, encapsulated by his three-volume *The History of Sexuality* written between 1976 and his death in 1984. This work challenged Sigmund Freud's ideas concerning the development of sexual identity, arguing that sexuality was not simply a result of sexual desires or acts, but was the product of a specific discourse of power (Howell, 2007). This approach took the focus of analysis away from the nuclear family as the key site of the development of sexual identity, and looked instead to wider social, political and scientific practices in order to explain how certain notions of sexual desire and identity are perceived as 'normal' or 'natural'. As the geographers Alison Blunt and Jane Wills (2000: 144) have argued, Foucault was particularly critical of supposedly 'scientific' studies of sexuality suggesting that rather than uncovering 'true' and essential meanings about sexuality, 'such analyses act as techniques of power, and that such techniques of power come to be internalised by people who regulate and discipline their own behaviour accordingly'. Thus, for Foucault, dominant scientific attempts to 'uncover' the essences of sexuality do not expose pre-existing truths, but through their claims to superior knowledge (through academic credentials, for example), they *bring these sexualities into being.*

Let's look at Foucault's idea of the 'production' of sexuality in a little more detail. In the following excerpt from *The History of Sexuality Volume 1* Foucault argues that the establishment of dominant heterosexism since the 18th century has required the enrolment of law, medicine and education in order to marginalise and criminalise alternative sexual identities and practices:

> Through various discourses, legal sanctions against minor perversions were multiplied; sexual irregularity was annexed to mental illness; from childhood to old age, a norm of sexual development was defined and all the possible deviations were carefully described; pedagogical controls and medical treatments were organised; around the least fantasies, moralists, especially doctors, brandished the whole emphatic vocabulary of abomination. Were these anything more than the means employed to absorb, for the benefit of a genitally centred sexuality, all the fruitless pleasures? All the garrulous attention which has us in a stew over sexuality, is it not motivated by one basic concern: to ensure population, to reproduce labour capacity, to perpetuate the form of social relations: in short, to constitute a sexuality that is economically useful and politically conservative?
>
> (Foucault, 1976: 36–37, cited in Del Casino Jr, 2007: 44)

We can draw out two points from this excerpt from Foucault's work. First, in drawing attention to the relationship between heterosexuality and labour capacity he seems to be advocating what could be considered a Marxist approach to sexuality: foregrounding the concept of heterosexuality as crucial to the reproduction of capitalism. But we must temper our assumption of a convergence between Foucault and Marx, since Foucault expressed a different opinion of the nature of power. And this leads on to the second point: Foucault was insistent that power was defused through micro-situations, individual and practices, and therefore could not be reduced to being 'held' by the capital-owning class. Through his detailed study of the operation of institutions and individuals, Foucault sketches the ways in which particular discourses of sexuality penetrate society, producing norms and marginalising 'deviant' sexual practices. For Foucault, this is the mechanism through which heteronormativity was established in Western European states.

Foucault's work provides a theoretical backdrop to much of the later writing in geography on the relationship between sexuality, space and place. While we must be careful not to overstate his contribution (see Weeks, 1991: 7 for a critique), and remember the continued attachment to psychoanalytic theories in some strands of the study of sexuality (see Nast, 1998; Hubbard, 2002), there is an implicit attachment to Foucault's theoretical framework in much of the writing on the geographies of sexuality in the 1980s and 1990s. We can draw out, in particular, three themes of Foucault's work that have provided particular stimulus for scholars in geography:

1 The anti-essentialist nature of Foucault analysis allows geographers to study the spatial practices through which sexual identities are created and reproduced. This has involved studying the home, the workplace and public space as locations where certain sexual identities are normalised and others presented as deviant or illegal. In this way we can see a development in Foucault's work beyond that of Freud: Foucault was interested in looking beyond the private sphere as the locus of sexual development, to draw attention to public sites as formative spaces of sexuality.
2 Foucault's approach is attentive to the expression of power, allowing scholars to explore the ways in which certain practices are dominant and the forms of resistance that are undertaken in order to unsettle these prevailing norms. As we have seen, power is not a uniform exercise in domination within Foucault's work; it is productive of myriad forms of political expression that have the capacity to challenge social norms.
3 Building on these first two points: Foucault's work is attentive to the politics of studying sexuality. Scholarship in the field of geographies of sexuality is rarely descriptive or detached, but rather emerges with a strong normative element focusing on changing dominant practices that have marginalised non-heterosexual forms of desire, lifestyle and sexual practice. Just as the emergence of feminist geography is closely tied to the wider feminist political movement, so geographers examining the production of sexual identities and spaces have been inspired by wider gay rights campaigns over the late-20th century.

Performativity: Judith Butler

Throughout the discussion of the work of Freud and Foucault, we see a challenge being posed to essentialist claims that sexuality is simply a natural or biological trait. Instead, in various ways and extents these scholars – among others such as Fausto-Sterling (1985) and MacKinnon (1987) – argued that sexuality was socially constructed. This work draws gender and sexuality together as the geographer Gill Valentine (1993: 238) explains:

> Constructionists argue that there are no essential biological characteristics but that, although biological differences do not have any inherent meanings, a whole set of complementary hierarchical characteristics and behaviours have been attributed to being male or female. Sexuality and gender identities cannot therefore be separated.

Crucially for our discussion, these scholars argue that 'socially constructed and interlinked gender and sexual identities are not fixed but are constructed and reconstructed over time and space' (Valentine, 1993: 238). This allows us to focus on the practices, structures and sites through which particular sexual identities are expressed and how heteronormativity may be challenged.

The philosopher Judith Butler has been at the forefront of understanding the social construction of sexuality. We have encountered Butler's work during the discussion of gender, and her discussions of sexuality extend these arguments. Butler (1997) argued that sexuality and gender are not fixed statuses but should more readily be understood as sets of performances that reify particular prevailing understandings and hierarchies of gendered and sexualised identity. Through this focus on performance, Butler argues that sexuality is a social artifice, reflecting nothing that exists externally in nature (Nussbaum, 1999). As the geographer Phil Hubbard (2002: 366) explains:

> Focusing on the performativity of sex/gender, Butler's writing disturbs any neat [. . .] correspondence between sexuality, biology and corporeality [the body]. Arguing that sex and gender categories are brought into being through bodily acts which are labelled as male/female or straight/gay, Butler denaturalizes what she terms the 'heterosexual matrix' the grid of discursive intelligibility which normalizes bodies, genders and desires.

By introducing the idea of the 'heterosexual matrix', Butler ties gender to sexuality: the very notion of a binary of two sexes/gender is mirrored by the notion of opposite-sex attraction. Sustaining the idea of one (two coherent and stable sexual identities) provides the basis for the other (that heterosexuality is considered 'natural'). As we will see, Butler's work has been hugely influential to geographers who have felt liberated by her approach of denying all essences, focusing instead on the production of imagined stable identities through social, cultural and political life. Consequently Butler's work is often cited as one of the origins of queer theory, a theoretical position inspired by post-structuralist ideas which argues that identity is never fixed 'but always in a process of becoming' (Kitchin and Lysaght, 2003: 490).

There is something intellectually satisfying about the notion of no fixed categories, instead looking at the production of categories themselves and the forms of knowledge and identity they privilege. However, this approach has also been the focus of criticism, particularly in relation to the political implications of Butler's argument. Butler suggests that resistance mainstream understandings of gender or sex may be mobilised through parodies of dominant practices, alterations to expected gender or sexual roles that may challenge established categories and ideas. The philosopher Martha Nussbaum (1999) explains:

> Thus the one place for agency in a world constrained by hierarchy is in the small opportunities we have to oppose gender roles every time they take shape. When I find myself doing femaleness, I can turn it around, poke fun at it, do it a little bit differently. Such reactive and parodic performances, in Butler's view, never destabilize the larger system. She doesn't envisage mass movements of resistance or campaigns for political reform; only personal acts carried out by a small number of knowing actors. Just as actors with a bad script can subvert it by delivering the bad lines oddly, so too with gender: the script remains bad, but the actors have a tiny bit of freedom.

This notion of 'parodic performance' is encapsulated for Butler (1997) by drag acts, who challenge expected notions of dress, comportment and language in established gender roles (see Browne, 2007). This leaves the question of whether such performances offer the hope of transforming established – and unequal – relations of sexuality. On this point Butler seems unsure, which leads critics – not least Martha Nussbaum (1999) – to suggest that this work offers an 'empty politics' that ignores pressing normative questions as to the nature of society that Butler envisages through the practice of parody. This is a very complex debate that is fuelled by different understandings of what politics and political action mean, and what forms of political agency are possible. Rather than seeing this as a settled argument, we would urge students to examine the primary texts to engage with the complexity of these different positions on the performance of sexualised and gendered identities.

Geographies of sexuality

Since the early 1980s, geographers have attempted to challenge the dominance of heteronormativity and expose the forms of discrimination and marginalisation that it masks. This work was prompted by the increased political pressure in both Europe and North America to recognise the rights of gay, lesbian and bisexual men and women. A formative moment in this process was the 1969 riot that followed a police raid on the Stonewall Inn, a gay bar in New York City. While raids had become a commonplace form of discrimination against homosexual and bisexual activity, the physical response by members of the lesbian, gay and bisexual community marked a turning point in terms of staging a visible and physical resistance to threat and intimidation. The riot and its commemoration acted as focal points for gay, lesbian and bisexual communities, and its annual commemoration in cities across the world has transformed into annual gay pride events (see photo). The very naming of this movement serves to emphasise resistance to persecution and discrimination of non-heterosexual communities and to 'reclaim' public space through public display of gay, lesbian and bisexual lifestyles.

This political context serves as the backdrop to increased scholarly focus on the geographies of sexuality. In some ways the disciplinary developments mirror those of feminist geography, as initial scholarly work sought to 'write sexuality in', rendering the notion of sex, desire and sexuality as appropriate topics for academic debate. This work evolved over the 1990s to shift away from 'mapping sexualities' to explore the forms of subjectivity and spatial practice that could be illuminated through an understanding of sexuality. This second phase of work sought to introduce an analysis of the multiple spaces through which sexual identity is expressed and challenged: the home, the street and the workplace. The

Gay Pride march

Source: Press Association Images / Cuccuru / Milestone / Empics Entertainment

third area of study has been to deploy the theoretical and empirical tools developed to understand homosexuality and bisexuality in order to analyse heterosexuality. Rather than assuming heterosexuality to be a unitary and unchanging form of dominant sexual practice, this work has looked to expose the multiple forms of heterosexuality and the implications of these on spatial practices and conceptions of citizenship. Over this section of the chapter we will examine each of these bodies of work in turn, and illuminate some of the emerging work in these fields.

Mapping sexuality

As we have seen, one of the effects of heteronormativity is that it renders invisible other sexual identities and practices. Extending this point, scholars have argued that heterosexuality is normalised in society to the extent that sexuality itself is rendered invisible. Scholars such as Heidi Nast (1998) argue that these social conditions influence the nature of scholarly inquiry, where heterosexuality is considered so normal and unremarkable as to be viewed as benign or even equivalent to asexuality. Early sexualities scholarship in geography sought to challenge this dominance, at first by seeking to map the distinct geographies of sexuality in the city, often focusing on the location, landscapes and form of 'gay ghettos' (see Lyod and Rowntree, 1978; Weightman, 1981). Perhaps the most high-profile work in this field is that of urban sociologist Manuel Castells (1983), who studied lesbian and gay spaces in San Francisco. The geographers Jon Binnie and Gill Valentine (1999: 176) provide an overview of his approach:

> This work was most easily and unproblematically 'spatial'. The boundaries of specific gay male neighbourhoods and commercial districts were marked and defined by dots on maps – the dots being lesbian and gay facilities such as bars and other businesses gleaned through business directories. Castells argued that the geography of gay men and the geography of lesbians reflected their respective gender roles and gendered behaviour. Gay men acted primarily as men and were therefore more territorial, had more disposable income and desired the visible spatially defined commercial scene. Lesbians acted primarily as women, were not territorial, were reliant on informal networks rather than commercial facilities, were more politicised than gay men and created lesbian space within feminist networks.

Despite criticisms – which we will explore below – there is no question that Castells's approach was popular in the emerging arena of studying the geographies of sexuality. In simple terms, it acted as a catalyst of interest in the relationship between sexual identity and urban space. In particular it drew attention to the role of gay men in processes of gentrification, work that explores the relationship between sexual identities and economic class (see Knopp, 1987, 1995). In more theoretical terms, Castells's work provoked interest in the relationship between sex and gender identities. As we have seen, his work suggested that there were core differences in the spatial and economic practices between gay men and lesbian women, a point that has been explored – and challenged – in later scholarship (see in particular Valentine, 1993, and the discussion below). Finally, Castells's work develops a Marxist argument in attempting to draw connections between sexual identities and space, suggesting that the nuclear family is crucial to capitalist reproduction and therefore connections can be drawn between sexual and economic marginalisation. In this way Castells's work prompted an interest in the relationship between housing design and the reproduction of heteronormativity, where the design and financing of housing serves to reinforce the norm of the nuclear family (see Watson, 1986).

But, despite pioneering an interest in the study of gay and lesbian lifestyles and spaces, Castells's work has also attracted criticism. Perhaps most fundamentally, Castells's writing suggests that gay men and lesbian women live lives 'distinct from each other and from straight society' (Binnie and Valentine, 1999: 176). First, this concept of ghettoised existence has been challenged in more recent work that has illustrated that gay and lesbian identities are formed through experience of living and being discriminated against within heterosexual environments. Second, and in relation, Castells's research and other studies that it has inspired have been criticised for focusing specifically on 'public' spaces – bars, streets and shops – and not accounting for the role of other sites and spaces such as the home and the workplace in shaping gay and lesbian lives. As the geographer Gill Valentine (1993: 246) explains:

> By focusing in this way on how the public expression of gay sexual identities affects the character of space at a local level, geographers have ignored the fact that the majority of lesbians and gay men do not live openly gay lifestyles in gay defined environments but instead spend most of their time living, working and socialising in heterosexual environments.

Consequently Castells's work could be accused of reproducing a heteronormative approach to academic scholarship that does not engage with sex and desire, but focuses instead on public practices that are perceived as separate from such bodily and sensual practices. In a sense, can Castells's work be seen as reproducing geography's 'squeamishness' (in the words of Bell, 1991)? Finally, Castells's work can be criticised for its methodology, which centred on survey data gleaned through secondary sources, rather than in-depth primary data collection. The complex and dynamic nature of sexual subjectivity means that it can be difficult to grasp these processes through quantitative survey; instead qualitative methodologies such as interviews and participant observation are required, as more recent work in this field has demonstrated (see in particular Valentine, 1993, below).

Sexuality and space

Over the course of the 1980s and 1990s, geographers developed new analytical and methodological approaches to the study of the relationship between sexuality, space and place. In particular, scholars drew on the theoretical perspectives developed in feminist and cultural geographies to challenge the notion of sexual identity as a fixed and stable condition. Refuting the possibility of plotting the geography of gay, lesbian or bisexual communities, this work has developed more textured accounts of sexual life and citizenship in order to understand how forms of discrimination and marginalisation are reproduced. The geographer Gill Valentine has been at the forefront of these endeavours through a range of papers, chapters and books over the 1990s and 2000s, and we will begin to illustrates this body of work through an exploration of her research into the nature of lesbian existence in a southern English town (Valentine, 1993).

In this paper Valentine (1993) poses a challenge to Castells's suggestion that sexual identities and practices may be mapped and fixed in space. Instead, Valentine (1993: 238) draws our attention to the ways in which 'lesbians negotiate multiple sexual identities over space and time and the strategies used to manage conflicts between these often contradictory identities'. In order to illustrate this conception of plural sexual identity, Valentine conducted 40 interviews with young women aged between 18–29 and older women aged 30–60. One of Valentine's key observations is that previous scholarship has focused on openly expressed sexual identity, ignoring the fact that 'many lesbian and gay men conceal their sexualities and so "pass" as heterosexual at different times and places' (1993: 237). This dynamic conception of sexual identity clearly challenges ideas of 'coming out' as a simple temporal event that distinguishes between closeting sexual identity and carefree public expression. Using the qualitative material gathered for the study Valentine (1993: 240) portrays a more complex picture:

> This process of 'coming out' is usually conceived as a duality: a gay person is either 'out' and living a completely open lifestyle amongst the gay community, or 'in the closet' and completely secretive about their sexuality and isolated from other gay people. Alternatively, it is thought to be an unfolding narrative as the person 'comes out' in more and more spheres of their life. But in fact the lesbians interviewed perceive that different people and organisations will react differently and therefore they negotiate different and contradictory sexual identities in different time/space frameworks.

Valentine goes on to identify the different sites and events that caused the participants in the research to alter their sexual identity in order to 'fit in' or avoid discrimination. One of the key spatial practices that Valentine (1993: 243) identifies is the idea of establishing a temporal and geographical 'break' between different parts of the participants' lives:

> Because women are raised by their parents as if they were heterosexual or actually have heterosexual relationships before they identify as gay, one of the most common sites of potential conflict is the local neighbourhood where others know them as or assume them to be heterosexual. Therefore when women first identify themselves as lesbian one of the most frequent ways to reduce the risk of family and friends finding out is to *establish geographical boundaries between past and present identities* by moving away from places where they have an established heterosexual identity and creating a lesbian identity (at home, work and in the community).

We see in this account the ways in which identity may be understood as performed (after Butler, 1990, 1997): identity does not pre-exist performances, rather it is through these practices of concealment or open expression that dynamic and plural lesbian identities come into being.

But Valentine's analysis of the performance of sexual identity is not limited to reflections on the practices of others. In line with the methodological and theoretical advances afforded by work on sexuality and gender (see above), Valentine has also drawn on these reflections concerning the social construction of identity to explore her own positionality. What is unique – and very distressing – in Valentine's reflections on her own identity is that they were prompted by a prolonged episode of harassment, involving silent phone calls and, later, threatening letters. This event is detailed in a 1998 paper in the radical journal of geography *Antipode* and we would suggest engaging with this work in order to explore how a pioneer in this field articulates the challenge of writing about discrimination and the performance of their identity. Echoing the discussion of lesbian lives in the earlier study Valentine (1998: 307) discusses how she has performed her own sexuality in different time/space contexts:

> . . . while I have been held up as a 'lesbian-geographer' who is assumed to be 'out' both publicly and privately, I have been actually performing a very different identity to my family, creating a very precarious 'public'/'private,' 'work'/'home' splintered existence. To add to the irony, it was, of course, writing about how other women manage just such multiple subjectivities that has helped to create and therefore exacerbate the cleavages in my own life.

Valentine's work on the 'splintered' nature of sexual identity has been picked up by others who have explored the ways in which identity is performed in a range of settings. One of the key sites for such discussions is the home. The geographers Stewart Kirby and Iain Hay (1997) examined the role of the home in constituting sexual identities in Adelaide, South Australia. This work challenged accounts of the home as a refuge, a site that is wholly 'private' and beyond the (heteronormative) gaze of wider society. Instead they use interview data to examine the differential experiences of gay men towards the 'home', detailing how some are driven to 'de-gaying or straightening up' (1997: 297) their homes when heterosexual family or friends are visiting. Others who lived with family felt senses of loneliness and alienation – a feeling of being 'out-of-place' – since they were unable to exhibit their sexuality in such environments. In both these cases we see a reflection of Foucault's work of the power of heteronormativity to shape behaviour and establish social 'norms'. As Kirby and Kay (1997: 297) note, 'even the home, that "free" heartland of individual identity, is penetrated deeply by the primacy of heterosexism'.

One of the conclusions of Kirby and Hay's (1997) work is that the practices of gay men in changing their environment and practices in order to conform to heterosexual norms can, in fact, assist in the reproduction of such norms. In particular Kirby and Hay (1997: 303) argue that the practice of hiding gay identity serves to reinforce the spatial supremacy of heterosexuality in at least two ways:

- First, Kirkby and Hay argue that the enforced and reproduced dominance of heterosexuality perpetuates the invisibility of gay men in everyday spaces, sometimes pushing the expression of gay identities into gay-specific areas.
- Second, the authors argue that it is difficult for gay men to identify and meet others except in gay-specific spaces. By taking on sexual facades in different spaces and places, gay men may not be able to develop authentic relations with others, for fear of destroying the credibility of their act.

These points draw us back to the work of Michel Foucault and his discussions of the way in which particular discourses (in this case the discourse of heteronormativity) shape the behaviour of individuals and leads to self-disciplining behaviour. A similar point is made by David Bell (1991: 326) in his discussion of the ways in which the heteronormativity of workplaces leads gay men to conceal their sexuality. We must be clear that this does not mean that gay, lesbian or bisexual individuals can be held responsible for the reproduction of heteronormativity, but rather it points to how deeply ingrained heteronormative assumptions are in home, public and work life that this disciplining behaviour is deemed necessary to avoid discrimination. It also should be noted that this raises a question concerning the nature of agency – to what extent and how individuals challenge and reshape heterosexual norms and offer alternative sexual discourses. We will confront this question of politics in 'The politics of sexuality' in this chapter.

Heterosexuality

So while dominant notions of sexual citizenship serve to spatially exclude lesbians and gays from visible space, this interpretation has latterly been extended to consider the way that a *particular* notion of heterosexuality is implicated in the process. Specifically, it has been stressed that the distinction between good and bad sexuality is not simply drawn between heterosexuality/homosexuality, but also between monogamous/polygamous, procreative/commercial and polite/perverted hetero-sex. (Hubbard, 2002: 368)

The geographer Phil Hubbard's statement draws our attention to the need to question all conceptions of fixed, or stable sexual identity, including heterosexism. While much of the early work within geographies of sexuality sought to 'write gay/lesbian/bisexual' geographies in, this project has now included examining heterosexuality as a differentiated and unstable form of identity. This work often draws on Butler's ideas and, more broadly, on **queer theory** – a theoretical position that denies any fixed sexual identity (see Butler, 1990). One of the key criticisms raised by this emerging body of work is that absence of analysis of heterosexual geographies reproduces the notion that heterosex is equivalent to no sex, with 'normal' heterosexuality seemingly unworthy of explanation (Hubbard, 2000: 206). Recent studies have begun to explore heterosexual spaces such as bars (Lebrun, 1998), workplaces (Hall, 1989) and housing (Watson, 1986).

Linda McDowell's study of the reproduction of heterosexual norms in the workplace provides an example of this body of work. McDowell conducted qualitative research – structured around extended interviews – in merchant banks in the City of London, in order to explore the extent to which these institutions conformed to a 'heterosexist stereotype of aggressive masculinity' (McDowell, 1995: 75). One of the sites that McDowell is interested in is the body: examining the role of the '(hetero)sexed body and its significance in shaping power relations in the workplace' (McDowell, 1995: 76). Drawing on the ideas of Foucault, McDowell (1995: 77) examines the ways in which bodily comportment is disciplined by the corporate environment:

> One of the ways in which interactive service workers are selected and controlled is through emphasis on and careful surveillance and disciplining of their bodies; explicit rules about weight, permitted hirsuteness, and styles of dress and implicit rules about sexual identity, or at least its transfer into workplace performances, are enforced to produce a corporate image of the worker.

Using the interview responses, McDowell explains how individuals assume a workplace persona 'constructed as false in opposition to a real self that was to be hidden in the workplace' (1995: 84). While these practices were exhibited by both men and women, McDowell is keen to stress that these performances differ between genders and age groups, for example that younger women felt compelled to adopt particular styles of femininity in order to conform to heterosexual norms, whereas older women felt confident exaggerating or parodying feminine characteristics, a strategy that is reminiscent of the forms of resistance suggested by Judith Butler (1990, see above). Similarly, the workplace demanded 'a particular heterosexual performance of men, exacerbated by excessive "macho" behaviour in dealing and trading and a particular paternalism in corporate finance' (McDowell, 1995: 85). Through this argument we see the interface between the study of gender and sexuality: that heteronormativity simultaneously demands particular gender roles and assumptions concerning 'appropriate' styles of sexual attraction and desire.

The politics of sexuality

As we have seen, the stimulus for much of the work on geographies of sexuality is a political commitment to challenge heteronormativity and the forms of discrimination it fosters. From early work emerging from the reaction to the Stonewall riots through the public reaction to the AIDS epidemic in the 1980s all the way to challenging the banning of gay, lesbian or bisexual recruits to the army in the UK and US, work within geography and affiliated disciplines has used scholarly inquiry to illuminate the ways in which certain sexual practices and identities are portrayed as deviant or undesirable. But the politics of this work is not straightforward or singular, since a number of divergent causes has been pursued through the study of sexuality, space and place. In this section we want to challenge two of these: the politics of the discipline and the advancement of sexual citizenship.

The politics of the discipline

One of the central themes of work in geography on the spaces of sexuality has been a reflection on the practices, assumptions and institutions of the discipline itself. Just as Gillian Rose's (1993) work illuminated the masculinist nature of geographical knowledge production, over the 1990s scholars argued that geography is heterosexist (Binnie 1997). While scholars have adopted sophistical theoretical tools, such as queer theory, in order to examine the production of sexualised identities, rather less attention has been paid to the assumptions that underpin the wider discipline of geography. As Jon Binnie and Gill Valentine (1999: 183) have suggested, 'what we need is not so much a queer reading of space, but a queer reading of the discipline of geography itself'.

David Bell's interventions (1995, 2007) have been at the forefront of challenging the heterosexist nature of geography. He has called for a 'queer turn' in geography that could unsettle the assumptions and discourses that frame the production of geographical knowledge. He argues (2007: 84) that the methods and theories of geographical inquiry can silence certain voices that are not deemed to be framed in appropriate language or exhibit the 'correct' credentials:

> This 'queer turn' was all about intervening in the discipline, all about testing, pushing, seeing what could be said where: where the fuck could you say fuck? Trying to expose the boundaries of 'proper' geographical discourse, and also expose the policing of boundaries, was a big part of the agenda: who is being silenced, what cannot be thought, what is at stake?

It would be wrong to read such an intervention as simply a desire to test where profanities may be published – and the question could be asked whether the ability to swear is really an emancipatory or transformative political act in the first place (Bell asks this himself later in the essay). More fundamentally, Bell's point urges us to explore both the nature of discipline of geography and scholarship on sexuality. Within the discipline, Bell points to its enduring 'squeamishness', a desire to sidestep discussions of sexuality and, in doing so, reproducing heterosexual norms. The geographer Jon Binnie's (1997: 224) work supports this point by illustrating how the methodologies employed by geographers can be used to distance questions of sexuality:

> Geographers as social scientists have been trained to uphold the clear distinction and distancing between the reality out there (which we map), and the in here (our bodies or selves). The wish to uphold an objective stance may partly reflect a certain discomfort with one's own body and (hetero)sexuality.

The work that we have explored in this chapter has exhibited how scholars of sexuality have sought to challenge this 'distancing', by drawing on in-depth qualitative methods, by taking seriously the commitments of positionality and by exploring the body as a site through which sexual identity is performed and contested.

But Bell goes on to suggest that criticism of discomfort with the body can be extended to work within geographies of sexuality. He makes the point that it has been 'easier to write about sexual identities and politics than sexual practices' (Bell, 2007: 84). This point is reflected in the thematic choices made on scholarship of sexuality: exploring the urban geographies or the lived experiences of gay, lesbian and bisexual respondents. Recent work has begun to challenge this point, by exploring the politics and representations of sexual morality surrounding bondage, discipline and sadomasochistic sexual practices (Hubbard, 2001). But this has not been straightforward, since embodied accounts of sexual practice would require forms of autobiographical writing that could be harmful to the author's academic career. This is a point made by Jon Binnie (1997: 224) when he explains that he has practised self-censorship in the past and avoided writing an embodied account of a bar pickup for fear of harming his chances of getting an academic job. He explains:

> Embodying geography is easier in theory than in practice. I am also mindful that the autobiographical will seem cringe-making in years to come. That I felt constrained by the bondage of conventional academic discourse to disembody my own work [. . .] reflects the extent of homophobia within the discipline.

Therefore a challenge remains: within studies of sex and sexuality embodiment is predominantly something that scholars have done to others, rather than explore the embodiment of the self.

Sexual citizenship

Much of the work within geographies of sexuality is inherently political – since it is attempting to illuminate how structures of power shape the sexual subjectivity of individuals, and exploring the strategies that are undertaken to resist and contest these structures in everyday life. One of the most explicit ways in which these strategies of resistance are made visible is through the articulation of styles of sexual citizenship – exploring the relationship between the individual and the state. This work has often drawn attention to the role of gay, lesbian and/or bisexual communities in political activism: work that has explored the insurgent forms of citizenship that seek to make claims to justice and human rights that are often denied by the state. In doing so new spaces of sexual citizenship emerge, both materially through the site of protest and metaphorically through the establishment of networks of trust and communication between gay, lesbian and bisexual activists.

The geographer Gavin Brown has explored these conceptions of sexual citizenship through a study of the *Queereruption* gatherings. These events emerged over the course of the 1990s through queer anarchist networks: consequently these were occasions that sought to extend political mobilisation beyond sexual identity to challenge other economic, political and social structures of society. From 1998 onwards *Queereruption* gatherings have taken place annually or thereabouts, and have used appropriated spaces as venues such as squats and disused tenement buildings in New York, San Francisco, Berlin, London and Vancouver. The *Queereruption* gatherings are designed to construct autonomous spaces of expression:

> Freed from the sexual and gender constraints of the quotidian world, participants in these queer autonomous spaces often find themselves questioning the social relations that normally restrict the free expression of their desires. Immersion in these autonomous spaces can be an intensely emotional and cathartic experience, with the consequence that when they end many participants experience a real 'come down'. (Brown, 2007b: 203)

Reflecting its anarchist roots, *Queereruption* sought to challenge *both* dominant forms of heterosexual and homosexual behaviour. In particular the activist participants saw expectations of mainstream homosexual behaviour as closely allied to capitalist consumption:

> these radical queer spaces are important because they provide a constructive and practical attempt to offer a non-hierarchical, participatory alternative to a gay scene that has become saturated by the commodity. (Brown, 2007b: 205)

The form of political mobilisation of *Queereruption* illuminates the diversity of political strategies adopted by queer activists. These events sought to imagine a more inclusive social order through the creation of spaces set apart from mainstream society, within disused and derelict spaces. This form of spatially-segregated activism reflects a shift away from the identity politics that marks early queer politics: where the existence of a coherent and unified image of gay identity offered a powerful banner under which groups may organise.

While this example does not mean that unified action has been surpassed – notice the increasing prominence of gay pride marches across the world – it illustrates a central schism of queer politics, often presented between academic and activists. As we have seen, academics have been drawn to the performative accounts of Judith Butler to present gendered and sexualised identity as a social fiction recreated through repeated performances. These identities may be undermined by parodying these social constructs and thereby summoning into existence new forms of social order. In some ways, this perspective is illustrated in the approach of *Queereruption* that seeks to challenge labels such as 'gay' and 'straight', seen as oppressive structures that limit human expression, and create new forms of socialisation within bounded spaces. On the other hand, other activist groups have sought to retain identity politics, in this case gay, lesbian and bisexual political movements, and use these labels to organise under a collective banner to challenge social marginalisation. So rather than focusing on a scholarly concern for the fiction of sexual identity, these groups have used these identities as coherent labels under which to mobilise. As Jon Binnie (1997) points out, for academics identities such as 'gay' or 'straight' may be fictions, but for activists they are often seen as very real and empowering.

Conclusions

This chapter has argued that geography of sexuality is not just a subject that geographers study: it challenges the very practices, assumptions and make-up of academic geography. This area of inquiry and political action emerged from the engagement of geographers with key figures from psychology and social history: Sigmund Freud, Judith Butler and Michel Foucault. As we have seen in other chapters of the book, the influence of these figures extends beyond geographies of sexuality, and their influence within geographies of sexuality extends beyond what we can cover here. We would encourage students to engage with the original works by these authors and to that end we have included references to their work in the further reading below.

Over the course of the chapter we have outlined some of the main areas of research in the study of geographies of sexuality. In some ways there is a narrative arc to this work: starting with studies that sought to 'write in' gay, lesbian and bisexual geographies into our understanding of (mainly urban) space. This work developed into a wider range of studies that sought to challenge the notion of fixed sexual identity and, in doing so, opened up a range of sites within which to study sexuality: the home, the workplace, leisure spaces and so on. This work has introduced a more complex understanding of sexuality as performed in different ways and in different times. In doing so it has prompted greater reflection on construction of heteronormativity. One of the main contributions of this work is that it brings the study of sexuality away from examining minority or marginalised groups and explores how sexual desire and assumptions of heterosexuality shape practices in elite institutions – such as merchant banks. There is obviously a political danger in such a transformation of studies of sexuality: as work enters the mainstream is there a risk that its radical or transformative potential may be diluted? In order to address this question we explored the politics of this work and how it has sought not simply to understand marginalisation, but challenge and reverse these forms of discrimination.

Summary

- Scholarship on sexuality in geography is indebted to a long lineage of work on psychology and society. We have drawn out three notable influences: the work of Sigmund Freud, Michel Foucault and Judith Butler.
- We can trace a development of the work of geographers, from mapping sexual identities through to exploring how sexuality shapes conceptions and practices of space.
- Geographies of sexuality should not be seen as simply another topic that geographers 'do', but rather a field of inquiry that challenges how geographical study is organised and practiced.

Further reading

Bell, D. and Valentine, G. (eds) (1995) *Mapping Desire: Geographies of Sexualities*, London: Routledge.

Binnie, J. and Valentine, G. (1999) Geographies of Sexualities: A review of progress. *Progress in Human Geography*, 23: 175–187.

Brown, G. (2007) Mutinous Eruptions: Autonomous spaces of radical queer activism. *Environment and Planning A*, 39(11): 2685–2698.

Brown, G. (2009) Thinking Beyond Homonormativity: Performative explorations of diverse gay economies. *Environment and Planning A*, 41(6): 1496–1510.

Browne, K., Lim, J. and Brown, G. (eds) (2007) *Geographies of Sexualities: Theory, Practices and Politics*, Aldershot: Ashgate.

Butler, J. (1997) Gender Trouble, Feminist Theory and Pyschoanalytic Discourse, in L. McDowell and J. Sharp. (eds), *Space, Gender, Knowledge: Feminist Readings*, London: Arnold, pp. 247–261.

Foucault, M. (1981) *The History of Sexuality, Vol. 1*, Harmondsworth: Penguin Books.

Freud, S. (1962) *Three Essays on the Theory of Sexuality*, New York: Basic Books.

Valentine, G. (1998) Sticks and Stones May Break My Bones: A personal geography of harassment. *Antipode*, 30: 303–332.

Geography, ethnicity and racialisation

Learning objectives

In this chapter we will:

- Interrogate the concept of race and show how a belief in this idea has led to biological and cultural expressions of racism.
- Consider the imperial relationship between geography and race.
- Investigate how social and cultural geographies of race came to emphasise the social construction of race categories and the power of racism.
- Discuss urban and rural forms of racism.
- Consider the role of whiteness and anti-racist geographies in the discipline.

Introduction

In his popular book *Beyond the Boundary*, exploring class, colour and cricket in the West Indies (see photo), C.L.R. James examines how different sports clubs continued to organise through a fictitious notion of race imposed during the colonial era. Cricket clubs posed as 'black', 'brown', 'mixed', 'white' or 'coloured' in what appeared a seemingly endless spectrum of distinction. And yet the author reminds us that such divisions are arbitrary and that the idea of race is a modern invention.

> . . . historically it is pretty well proved now that the ancient Greeks and Romans knew nothing about race. They had another standard – civilized and barbarian – and you could have white skin and be a barbarian and you could be black and civilized. (James, 1973: 124)

The 1933 West Indies cricket team

Source: Alamy Images / Lordprice Collection

The suggestion here is that being seen as 'civilised' or 'barbarian', itself a problematic social construct, did not rest with the colour of one's skin but was dependent upon other physical or cultural markers. If the pre-modern societies referred to above did not employ race as a way of understanding human populations what are we to make of this idea and why is it of powerful geographical significance? In response to this question this chapter seeks to interrogate the concept of race and dispel any lingering belief in its reality. We argue that the idea of race gives us little and takes too much. The chapter begins by outlining biological and cultural expressions of race-thinking to expose how these myths are formative of 'commonsense beliefs'. We then explore different geographical approaches to the study of race, beginning with the discipline's problematic relationship to empire (see Chapter 1). This is followed by a discussion of an early post-World War II tradition and its obsession to map, monitor and spatially locate patterns of settlement and segregation in the pursuit of a robust, quantitative, social geography. The contribution and limits of these techniques are debated.

The chapter then turns to newer approaches concerned with dismantling – or deconstructing – the idea of race itself. This tendency is frequently found in social and cultural geography accounts of race and place, where there is a representational focus upon race as a sign, symbol or discourse that is wilfully made and unmade. These geographies form part of a paradigm shift, from looking at race as a knowable object to focusing upon racism as a set of social relations inscribed with power. More recently, however, some of this work has been critiqued for its obsession with the metropolis and the black and minority ethnic communities that reside there. To counter this approach we consider some of the recent work on rural racism that examines how the countryside and other spaces are geographically imagined. One of the drawbacks of examining the racialised experiences of ethnic minorities is a failure to address the racial privileges afforded to those who identify as 'white'. In response some geographers are producing critical accounts of whiteness to disclose how it shapes and structures how race is lived. In exploring geographical accounts of whiteness we seek to illustrate its connections to modernity and explain why it should be a necessary component of critical race thinking. Finally, the chapter concludes by reflecting upon the way anti-racist geographies can be utilised to transform the theories, practice and overall constitution of the discipline – from the inside out.

The idea of race

Race is arguably one of the most controversial concepts in the English-speaking world. For these reasons it is often parenthesised in inverted commas, omitted from discussion, or conflated with other seemingly less contentious terms such as ethnicity or culture. But what do we mean by 'race' and why does it have such lasting appeal?

Broadly speaking race is premised upon a belief in innate human difference and the abiding conviction that biological distinctions exist between social groups, delineating, for example, between 'Europeans', 'Africans' or 'Asians'. To speak of the world's 'different races' gives credence to the false notion that such a thing as race exists. This also implies an illusionary level of homogeneity within these societies and the suggestion that race is a timeless and enduring set of characteristics passed down through our genes to successive generations. There is, however, little evidence to support the idea of any essential racial characteristics, let alone anything approaching the existence of definitive or unique species within humankind. We are all *homosapiens* and where any minor genetic differences can be traced within parts of a national population scientists claim this is a consequence of environment, regional diet, health and close intermarriage over time. This means that genetic variation between, for example, a Nigerian and an Englishman, melt into insignificance when compared with the differences within each of these national populations. Race, then, is anything but a biological fact.

Race can be seen as a modern myth – a human invention used to divide up and make sense of a complicated social world. In this sense, when one speaks generically about 'Africa's attitude to children' or 'whether blacks make good workers', such statements elide the enormous cultural differences that may exist between Tanzania, Morocco, Nigeria or Kenya. Such remarks also conceal the multitude of differences that exist within these populations, as it is highly unlikely that the wealthy share the same understandings as the poor, or that women enjoy the same privileges as men. And while generalising is part and parcel of how we communicate, this becomes problematic when particular ideas or values become *racialised*, i.e. where specific forms of behaviour become attributable to particular social groups. This can lead to popular racist stereotypes concerning how the Germans, French, Jews, Muslims, blacks or Irish 'are'. While many of us may recognise the inadequacy of racial **typologies**, it is often difficult to fully escape the pull of racist stereotypes and the powerful way in which they manifest in the imagination. In this vein cultural geographers such as Peter Jackson have intimated that it is preferable to 'speak of *the idea of race*, rather than of race *per se*, to emphasise its socially constructed as opposed to biologically given character' (1995: 133).

Box 8.1
Race and space: dividing practices in South African apartheid

One of the most graphic displays of how race and space intersect can be seen during the apartheid era in South Africa which finally came to an end in 1991. The word apartheid literally means 'apartness' and it came to be the defining project of the prevailing South Afrikaner regime. Designed to keep people in 'their' allotted place, individuals were demarcated through irreducible race markers, coming to be defined as 'white', 'black', 'brown', 'Cape coloured', 'mixed' and so on. Ethnicity and tribalism was a further point of distinction as separate spaces were identified for Indians, Malays, Sotho, Tswana, Xhosa, Zulu and Others. Rather than reflecting pre-existing race categories, apartheid operates as a pernicious dividing practice that actually produces and upholds these distinctions. The fiction of race is made to appear solid through this spatial politics of separation. Interestingly, despite longstanding antagonism between British and Dutch South African descendants, white ethnicities were not segregated from one another, leaving 'whiteness' as a mode of racial classification intact – an issue we shall return to later in the chapter.

Race is then concretised in space. Apartheid seeks to produce 'pure spaces' comprising a single group located within clear boundaries, allotted zones and demarcated territories. Inevitably these communities are more diverse than terms such as 'coloured' would suggest. The racial encoding of space and its frenetic regulation shows how the arbitrary, fictitious idea of race can come to settle upon particular places to locate them within a colonial imaginary that carries racialised meaning – for example the 'bush', 'Chinatown', township, ghetto or 'hood'.

Above all apartheid exemplifies how white elites can effect power through the racial ordering of space. Legally defined through racial absolutes, apartheid enabled the white minority prime access to the nation's resources. It also allowed whites to move unimpeded through spaces of privilege while the majority black populations were stuck in place. This geography of race extends further as whole cities, regions and the nation become divided along lines of ethnicity, class, heritage and colour. Through this racialised cartography markedly different 'scapes' of existence come to be inhabited by white urban elites, rural landowners and the majority of the population living in the predominantly black townships.

The effects of a geography of apartheid can be seen at a macro-scale where large differences between whites and blacks may exist when it comes to nutrition, wealth, education, employment levels, health indicators, disease and life expectancy. Commonplace claims that black South Africans are 'ignorant', 'criminal' or 'carriers of disease' then have little to do with the genetics of race and all to do with being victim to the exercise of colonial authority. In turn these ideas serve to bolster the spatial practice of segregation as it is applied to residential areas, schooling, public transport, beaches, washrooms, public playgrounds and so on. Despite the release of Nelson Mandela from prison and the subsequent election of the African National Congress (ANC) in 1994, the conflict and contestation over space continues in the 'new' South Africa.

What this case study reveals is: (i) race and space are mutually constitutive relations; (ii) apartheid imparts a geography of exclusion that allows the idea of race to be realised in built architecture, segregated living and everyday practices, and (iii) racism is spatially expressed as exemplified in apartheid, but also operates across local, national and global scales when it comes to contemporary debates on borders, territories, migration, asylum, citizenship and belonging.

The recognition of race as little more than an 'idea' offers an important intervention in critical race thinking. It suggests that what we might consider to be race is little more than a fallacy. But if race is no more than a modern myth why is it such a compelling fiction to behold? Part of the explanation rests with the divisive way in which race has been used in the past, as seen in the arbitrary territorial delineations made in the name of race, as our example of South Africa shows (see Box 8.1).

The invasion of distant lands, the acquisition of local resources and the butchery of indigenous peoples have all taken place in the name of race. Racism, as an ideology inflected with power, has given way to the elucidation of an imaginary hierarchy of human 'types' upon which ideas of civilised/primitive, superior/inferior, developed/undeveloped, clean/dirty were rapidly ascribed (see Chapter 1). Biological forms of racism were a recurring motif in imperial expansion and remain a feature of modern-day genocides. For example, when Belgian colonists arrived in central Africa in 1916, they considered the Tutsis to be a biologically superior tribe, an attribute Tutsis were keen to embrace as they achieved better jobs and privileges over the next two decades. What had once been a mixed society, in which intermarriage was commonplace, gave way to growing resentment and sporadic conflict, culminating in the slaughter of nearly 20,000 Tutsis by Hutus in 1959; as many fled to the neighbouring countries of Burundi, Tanzania and Uganda. Three years later, Rwandan independence was granted as Belgium withdrew from Africa, but more recent genocide demonstrates the power of race to return and haunt the post-colonial present, an issue which we address in Chapter 11. The phantom history of race is a feature of modern times. During World War II the Nazi Holocaust witnessed the mass extermination of Jewish people, premised on the belief that they were inferior beings whose blood could only weaken and 'pollute' the 'purity' of the Aryan race. Recent world examples of biological racism can be seen with the break-up of the former Yugoslavia, where conflict between Serbs and Croats resulted in 'ethnic cleansing' and the attempt to purge 'impure' people from the body of the nation state. Such atrocities, carried out in the name of race, have led many people to seek refuge and asylum in other countries.

What is striking in each of these violent examples is the will to make race difference appear *real.* Here phenotype or somatic markers, how we might look or sound, come to stand in for who we 'really are'. Differences in skin tone, hair texture or other embodied markers which do exist are falsely burdened with the weight of race. As Anderson (2007) documents, race continues to remain a point of 'crisis' in most modern civilisations that struggle to reconcile their borders with a global humanity. We would concur by arguing that race as a cultural force is an *organising principle* in modern society; it is the grammar through which nations are constructed, territories are designated and claims to human rights, citizenship and belonging are iterated. This presents something of a theoretical conundrum – race may not be real but, once enacted as a meaningful category, its effects certainly are.

Although race may be an 'empty' category, it has achieved an almost cinematic appeal. We are often drawn to participate in this fiction and may all too easily put into motion a stock of familiar racialised images. One of the common ways in which people 'see' race – and render it as-if-real – is through skin colour, what the post-colonial critic Homi Bhabha (1994: 42) identifies as 'the most visible of fetishes'. Differences in pigmentation have led people to think of race as natural, yet the application of race signs through colour is both an arbitrary act and, as the opening quote from C.L.R. James revealed, a modern phenomenon. Seen in this light the seemingly permanent signs of racial authenticity, inscribed most potently by way of skin colour – black, yellow, red or brown – need to be understood as socially constituted and relationally connected categories. Furthermore, these signs change over time and place, taking on different meanings and new significations. As such, there is no 'inner truth' to the values accorded to skin pigmentation, only a set of socially ascribed

Box 8.2
President Barack Obama: post-race identifications

The election of Barack Obama as the US Democratic President in 2008 has been deemed by many a watershed moment in American history. The media has focused upon his appointment as the first 'black' candidate to take office in the White House. The son of a Kenyan father and mother from Kansas, Obama's racial identity has been subject to unstinting scrutiny. Raised by his white mother and grandparents in Hawaii and Jakarta before studying law at an Ivy League University, Obama 'fits no mould of black American life', according to self-identified black British journalist Gary Young (*The Guardian*, 1 March 2007). Young qualifies his remarks by identifying the uncertainty that surrounds Obama, his mixed-heritage and his steadfast Midwestern delivery in a country where 'black' and 'white' have come to be seen as poles apart. He explains,

> Some refer to Obama as a racial category all to himself: in the eyes of those who praise him he is a new, different and (by implication) better kind of black politician that those who have previously strode across the American political landscape. To others his background suggests a lack of authenticity. He's black alright, but simply not 'black enough'.
>
> (http://www.guardian.co.uk/world/2007/mar/01/usa.uselections2008).

Young reminds us that such debates tell us much more about America's obsession with race than they can ever do about Obama. But they also inform us of the 'burden of representation' (Hall, 1993) endured by ethnic minorities; that is that they are seen to be 'representative' of 'their' supposed 'race'. This burden is seldom endured by whites who pass simply as 'individuals'. In contrast political commentary on Obama's campaign continually speculated whether he could appeal to 'Joe the plumber', a symbol of white working-class masculinity. In his campaign against Hilary Clinton to lead the Democratic Party, parts of the US electorate questioned whether someone with the name of Barack Hussain Obama was a threat to national security, in light of anti-Islamic sentiments and fears concerning conflict in Iraq and Afghanistan after September 11, 2001. This shows how skin colour, nomenclature and other markers of difference can quickly become racialised and linked as black bodies and global terroristic events are fused together.

The racialisation of Obama is a double-edged sword and as Gary Young intimates, for some minorities he just 'ain't black enough'. Indeed, another journalist and well-known race commentator Yasmin Alibhai-Brown (*The Independent*, 9 June 2008) reflects how the US tendency to designate Obama as 'black' is as much a denial of his white upbringing and mixed heritage. Identifying Obama as 'black' does not allow space for his multiple identifications, placing him securely within a familiar racial typology. At present, post-race subject positions are still some way off, although the rise of popular individuals such as Barack Obama, the golfer Tiger Woods or the Formula One driver Lewis Hamilton begin to question what we might mean by such fixed terminology as 'black', 'white' and so forth. Ultimately, the example of Obama and his designation as 'black' informs us of the impossibility of race categories to adequately describe who we are. At the point at which race becomes intelligible, it stands to be erased. Obama is black, yet he is not.

attributes or deficiencies. In this respect we might consider race to be a hollow sign, what Roland Barthes (1972) would term a 'myth'. The mythic properties of race as an imaginary concept are opened out in our discussion of US President Barack Obama, where the impossibility of racial classification is made evident (see Box 8.2).

Finally, although race may not be real, racism presides as a lived reality. For this reason many nation states have legislation outlawing racial discrimination and pass laws advocating human rights equality. This may be targeted at perpetrators of racist violence; explicit discrimination in employment, housing, education, health and welfare services; political organisations looking to incite racist hatred; institutional forms of racism related to the police service, armed forces, schools and other state organisations; religious forms of discrimination, as well as much more. Already we can see through these examples a complex array of racisms emerging that continue to exist despite legislative attempts to counter their proliferation. To understand this widespread appeal of racism throughout modern society we shall now consider perhaps its most prevalent form of prosecution: cultural racisms.

Cultural racisms

If biological notions of racial superiority are greeted today with a healthy scepticism from most scientists and specialists, recent years have witnessed a proliferation of *cultural* forms of racism. In language, images, tropes and a dense scattering of representational signs, the lie of race can be made to appear a 'reality'. It is through an assemblage of words, filmic images, music and other registers that race is able to masquerade as an intelligible 'regime of truth'. These 'texts' perform the work of race. The myth of race is then sustained through its constant iteration. Discourses communicate meaning and in producing the objects of knowledge can in time come to form commonsense beliefs. Discursive ideas about race are frequently carried in language where casual references to 'black muggers', 'Muslim fundamentalists' or 'bogus asylum-seekers' are embedded in contexts that make race appear an inescapable truth, or at least, a normalised category. This is seen in the newspaper headline shown below, which simultaneously equates asylum-seekers with terroristic activities and 'sponging' of the state, conveniently absenting the fact that it is illegal for those seeking asylum to work.

DAILY EXPRESS

The World's Greatest Newspaper

How Patsy Kensit has fought back to be the new Queen of Emmerdale

Now schools minister wants to bring back classes of 60

BOMBERS ARE ALL SPONGEING ASYLUM SEEKERS

THE SCHOOLBOY: Said Ibrahim, circled, in a 1994 class photograph

Britain gave them refuge and now all they want to do is repay us with death

By David Pilditch and Cyril Dixon

Scare-mongering headline from the *Scottish Daily Express*, 27 July 2005

Source: John Frost Historical Newspapers

Cultural racisms then centre upon the idea of immutable ethnic or cultural differences. Here, **ethnicity** refers to the everyday practices associated with a nationality or culture, such as our choice of food, music, dress, kinship relations and symbolic rituals. There is a danger, though, in ascribing ethnicity and culture in immobile ways to national populations. If we take the example of food one might consider how the British diet has changed markedly in the last 30 years to accommodate new 'traditions', such as the popularity of 'beer and a curry', that now forms something of a national staple.

Ethnicity is not then a 'bounded' category but is always in the making. Cultural racisms, however, not only treat ethnicity as a timeless, closed set of practices tied to particular national populations, but also may impute that some cultures are superior to others – being 'modern', 'progressive' and 'civilised' as opposed to 'traditional', 'backward' or 'primitive'. In such instances the idea of race is displaced from the biological to the cultural sphere.

An illustration of the prevalence of cultural racisms can be seen where contemporary Muslim youth are frequently demonised as 'fundamentalists', 'suicidal bombers', drug-dealers, 'victims of the veil' or rucksack-carrying terrorists (Alexander, 2000; Amin, 2002; Hopkins, 2007). Such representations are circulated through the high-tech spectacle of global media and appear to suggest an inevitable, geo-political, 'clash of civilizations' (Huntington, 1996). At these moments racism appears a product of one's upbringing and elides with an implacable notion of cultural and religious difference. For example, in Denmark discrimination against those of Muslim faith and new refugees reveals cultural racism to be prominent (Wren, 2001). Social constructionist thinking has been especially incisive in challenging cultural racisms by imploding the concept of race to disclose how it is brought into being through discourse which is infused with *power*. Indeed, the idea of race and how it is socially practised has come to shape educational opportunities, labour market structures, residential housing patterns and school choice around the world. At a cultural level the effect of race thinking can be seen in the popular attitudes that circulate around minority ethnic people and their abilities or deficiencies with regard to sport, crime, education, parenting or mental health.

So, let us briefly recap on why we should consider race a fictitious construct. Race does not precede discourse, but it is the consequence of it. The bringing-into-being of race through discourse is one of the most pernicious ways in which race thinking is sustained, kept alive and made new again. While the concept of race may be theoretically bankrupt, it retains a high material currency in the everyday world. Drawing upon the ideas of the anthropologist Claude Lévi Strauss, we might denote race is a 'floating signifier' that moves between – and occasionally settles upon – particular bodies, spaces, objects and things as encapsulated by the newspaper headline in the *Scottish Daily Express* (see the photo on the previous page). Ultimately race is a modern invention, an arbitrary sign that has become imbued with particular meaning in late modernity. Having debunked the myth of race, the remainder of this chapter will discuss the different approaches used by geographers to research this slippery concept. This includes the constitution of imperial geography, post-war mappings of immigration, social and cultural geographies documenting urban and rural racisms, studies of whiteness and examples of anti-racist praxis.

Geography and the shadow of empire

At an inaugural meeting of the Scottish Geographical Society in 1884, the renowned explorer H.M. Stanley proudly remarked how the discipline, 'has been and is intimately connected with the growth of the British Empire' (1885: 6). As we witnessed in Chapter 1, geographical investigation, in large part, was validated as a consequence of being applicable to empire building and expansionism. It was nothing less than a 'sternly practical' pursuit (Livingstone, 1992: 216). During this period geographical expertise was deployed in a plethora of ways: to map the non-Western world; to gauge the conviviality of climates and the peoples who resided within these zones; to identify suitable areas for colonial settlement; and to locate reservoirs of resources for extraction and production.

Unsurprisingly the interlacing of geographical knowledge with the imperial drive meant the category of race loomed large throughout this period, continuing to cast a shadow on the constitution of the discipline in the present day (see Chapters 1 and 11). As the social geographer Alastair Bonnett explains, 'Geographers interested in issues of "race" saw their task as the elucidation of the hierarchy of the world's "races" and the provision of informed speculation on the implications of White settlement and colonial governments' (1997: 193). In this respect geography, not unlike much mainstream anthropology (see Rich, 1987), was a prime contributor to models of racial classification. The picture from *Maury's New Complete Geography* (1906) reveals how scientists looked for cranium or

The making of race types (*Maury's New Complete Geography*, Maury, M.F., 1906)

Source: GNU Free Documentation License: Wikipedia

anatomical differences across populations in a feverish attempt to assert the false premise of race. Yet beneath the signifying envelope of skin, our flesh and bones are 'racially' indecipherable.

The disciplinary production of race can be seen where the RGS, established in1830, was content to sponsor a series of presentations in which the concept of racial difference was legitimated. As we saw in Chapter 1, race was fundamental to the conceits of empire, and empire was propped up by colonial discourse. Indeed, a number of 19th-century geographers were convinced that climate had a deterministic influence upon the racial disposition of native groups. Essays on 'The Connection between Ethnology and Physical Geography' by John Crawford in 1863, or by James Hunt on the topic of 'ethno-climatology' in the same year, are testimony to pseudo-scientific tendencies to determine human characteristics through phenotype, the environment or the physical landscape. Griffith Taylor's various essays on racial geography (see Livingstone, 1992: 227–230) made spurious links between the environment and migration movements that resorted in him depicting blacks as primitive people akin to Neanderthals.

What is noticeable in these representations is a flawed attempt by geographers to use environmental hypotheses to explain what *they perceived to be* biological differences within humankind. Charles Huntington's famous cartographic illustrations in the *Character of Races* in 1924 provide further evidence, if any was needed, of how the geographical imagination came to be imbued with an imposing sense of white superiority. In Huntington's damning depiction, the defining characteristics of black Africans, 'are those which unspecialised man first showed when he separated from the apes and came down from the trees. It is not to be expected', he postulated, 'that such people would ever rise very high in the scale of civilisation' (cited in Livingstone, 1992: 231). Although some geographers did depart from Huntington's thesis to varying degrees, what transpires is an assumption of race as a 'reality'. In this sense race became the *leitmotif* of 19th century geographical discourse where it manifests in debates on human development and colonial acquisition of territory. Significantly, along with the themes of economic resources and urban planning, race was a central pillar of geography from the 1830s. As we will go on to see, these foundations are not easily dislodged.

Mapping and monitoring race

With the decline of British imperialism after the 1930s race was relegated from the centrepage of geographical analysis to a mere footnote. Within the British context it may well have stayed there too had it not been for the large-scale migration of people from former Commonwealth countries, Ireland and Pakistan, succeeding World War II. The timing is not coincidental for, as Bonnett succinctly remarks, 'The issue of "race" returned to British geography when the Empire "returned" to Britain' (1997: 194). In other words, the presence of visible minorities in the urban landscape prompted a new swathe of geographical interest. This itself is interesting, for it reveals the manner in which race becomes conflated with 'colour' in modern society.

Under the grip of the quantitative revolution (see Chapter 2), geographers sought to legitimate their status as 'spatial scientists' to the extent that measuring and mapping immigrant communities was common currency from the 1950s to the mid-1970s. Two significant tendencies worthy of our attention in this paradigm concern the methodology and epistemology evident within early accounts. At a *methodological* level geographical work concerned itself with empirical data, quantitative methods and spatial analysis. For example, in the collection of essays that comprise Peach's (1975a) early edited volume *Urban Social Segregation*, 'Its basic hypothesis is that the greater the degree of difference between the spatial distributions of groups within an urban area, the greater their social distance from each other' (p. 1). Influenced by the concentric urban models developed by quantitative researchers at the Chicago School, Peach and his collaborators sought to use various mathematical indices to map and interpret the levels of segregation and integration endured by different 'immigrant groups'. This work has been enormously influential in providing a mass of data from which to understand social relations and developing evidence-based policy.

An *epistemological* aspect deriving from this approach, however, is the tendency to treat race as a 'proper object' that can be quantified, mapped and located. In such positivist studies race appears a discrete and stable unit for analysis – an issue challenged by later urban cultural geography. This approach led to **essentialist** representations of race in which it appears an unchanging biological, genetic truth. This primal tendency is seen where Stea (1965: 13) argues territoriality to be 'as pervasive among men as among their animal forbears'. What is noticeable here is that structural explanations related to poverty, availability of affordable housing, employment opportunities and so forth are eschewed in favour of pseudo-biological ones. The stigma of essentialism remains where it is often said that some minority ethnic communities 'stick together' when it comes to population settlement and residence, yet in most cases this remains a consequence of a lack of choice and mobility (Phillips, 2006). These loose descriptions were particularly applied to the popular work undertaken on the poor as well as studies on what became known as 'coloured neighbourhoods' and the 'Negro ghetto'.

Although statistical analysis continued to remain the dominant mode through which geographers studied race, by the mid-1970s the theoretical and methodological limitations of mapping race were becoming apparent. The focus of early race studies on labour markets, housing, crime and education served to perpetuate the idea that race was a 'problem' requiring forensic analytic investigation (see CCCS, 1982; Gilroy, 1987). The quantification of race also raised some uncomfortable questions: if race does not exist as a 'real' category, is collecting data upon it a fruitless endeavour, or worse still an insidious mechanism that enables race to materialise in the first place? It is interesting to note that some countries such as France and Belgium do not collect this data from their citizens, in compliance that race is an erroneous category. However, it could be argued that such statistics are vital for monitoring inequalities and under-representation in such spheres as the media, universities, legal professions and so on. At this point you may wish to consider the photo on p. 183, 'The Race Map of Britain' published on 6 October 2006 on the front page of the *Independent* newspaper. It offers a simplified version of the types of statistical maps many geographers have produced on race, migration and settlement within cities. What is the value of mapping race in this way? What might be the limits of approaching race in this way? How reliable do you think the data are, and what can we learn from it? What other techniques or approaches might enrich this portrait?

Despite some glaring deficiencies not all positivist approaches are consumed by the essence of race. In *Social Interaction and Ethnic Segregation* (1981), editors Peter Jackson and Susan Smith would similarly draw on the insights of Robert E. Park and other researchers at the Chicago School. Unlike Peach (1975a) who was compelled by Park's conceptual relationship between social distance and spatial association, Smith and Jackson were drawn to the manner in which Park offered scope for an anti-essentialist, cultural account of race. 'His contribution was significant', they attest, 'for turning attention away from biological theories of race towards cultural theories: "instinct" was replaced by "attitude", a product of contact and communication between groups; the concepts of "race" and "racial differences" were themselves revealed to be subjective and problematical' (Smith and Jackson, 1981: 2). By moving away from a belief in race difference towards the structural power of racism, social geographers have gone on to expose its material dimensions. Smith's (1987) examination of race and residential segregation led her to question whether the geography of English racism in the housing market represents 'the highest stage of white supremacy' (p. 39), an issue that is especially pertinent to the US (see Stanton, 2000). These processes of racialisation are not only attributable to the housing stock, but come to mark the structure of the labour market (Miles and Phizacklea, 1979), schooling choice (Dwyer, 1993), neighbourhood territories (Ley and Cybriwsky, 1974) and the everyday 'landscapes of power' (Zukin, 1993) we move between. As Jackson remarks, 'Many forms of racism have an explicitly territorial dimension that requires us to examine the complex interweaving of social relations and spatial structures' (1987: 14).

Interestingly, recent years have witnessed a revitalised quantification that has emerged in the field of race and ethnic studies. This is partly due to advances in technology such as the development of Geographical Information Systems (GISs) and other more sophisticated computer software. It also reflects a trend towards evidence-based policy along with a need for transparency and statistical clarity in what remains a politically contentious and often turbulent field. A good example of how such work can be used to implode race myths can be found in recent social geography research. When Trevor Phillips, the equality chief for what was the Commission for Racial Equality, announced in 2005 that 'Britain is sleepwalking to segregation', he depicted a country divided by colour and risking becoming ghettoised in a fashion after the US. However, social geographers Daniel Dorling and Bethan Thomas's report *People and Places: A 2001 Census Atlas of the UK,* published in 2004, uses index formulae to demonstrate that it is social class rather than 'race' generating social division. Such portraits offer clear, concise and policy-relevant empirical data that can be used to dismantle popular race myths about mass immigration or deep-seated segregation.

The evidence presented through quantification has also been supported in mixed-method race research undertaken with British South Asian Muslim populations in Northern England (Phillips, 2006; Phillips *et al.*, 2007). This work has the added advantage of not only showcasing what is happening in particular neighbourhoods, but offering some explanation as to why these patterns are prevalent. This important work by Deborah Phillips and her colleagues has helped debunk the notion that Muslim communities are 'self-segregating' and wilfully choose to live 'parallel lives' that run alongside but never intersect with other residents. The popular perception that Muslim communities are homogeneous, untrustworthy and insular is imploded by including their perspectives into the research. These lines of distrust figuratively 'thicken' when particular events are condensed, such as the destruction of the Twin Towers in 2001, terrorist strikes in Bali, the 2005 London bombings and the on-going war in Iraq and Afghanistan. However, such out-of-scale images are far removed from the prosaic struggles many Muslim young people face with regard to religious intolerance, racism and segmented neighbourhood experiences (Dwyer, 1999; Fortier, 2007; Hopkins, 2007).

An innovative feminist approach to quantitative mapping can be seen in a study by Mei-Po Kwan (2008), which combines oral testimonies with GIS to examine how the travel routes of Muslim women in the US changed in the aftermath of the attacks on the World Trade Center in New York City on September 11, 2001. The narrative and visual method affords opportunities for mapping movement and understanding how racism generates fear that impedes mobility. Such examples showcase the value of mixed methods and the significance of incorporating the voices of respondents in the production of critical understandings of race and racism.

A current force giving renewed impetus to the quantification of race is the globalisation of migration. With cheaper and faster transportation, more people of diverse ethnicities and nationalities are travelling, working and living in other countries. The enlargement of the European Union (EU) in 2004 granted people in new member states freedom of movement to work and reside across the EU. Global conflict has also led to new migratory movements as people are forced to seek asylum and refuge elsewhere. The paucity of information that may exist at the time when such upheaval occurs offers opportunities for population geographers and empiricists to provide valuable data from which to build future research and policy on. It is also interesting to note that the 1991 national Census was the first of its kind to document 'ethnicity' in Britain. A decade on and the ethnic categories deployed in the 2001 national Census were more diverse, including references to 'mixed' identities (such as 'White and Black Caribbean') and a splintering of the once homogeneous category 'White' to delineate those of white-Irish or -Scottish heritage for example. In the last Census religion is also documented as a point of identification and more open-ended classifications such as write-in options for 'White Other' or 'Other British' have emerged. The suggestion is that race and migration are more complex issues than may have previously been imagined and existing ethnic categories need to be more flexible to capture at least some of these changing dynamics.

Empirical approaches must then continue to be open and reflexive to recent theoretical developments on race and changing global flows of migration. They must also include the needs and perspectives of the minority ethnic people so often sidelined (and silenced) in these studies. This entails quantitative geographers acknowledging that 'they are merely making a limited, problematic, yet strategically useful, contribution to the wider debate on "race" and place' (Bonnett, 1996: 868). Such work remains significant for its potentially wide audience and effective policy impact (Holloway, 2000). Carefully managed, the production of statistics has been used to challenge fears concerning immigration, myths about asylum-seekers and popular tendencies to equate migration with the question of 'colour'. This data can also provide valuable information on new population movements and patterns of settlement. To do so it must recognise that it can only present a 'flash-bulb' image of what is a mobile and mutable set of relations.

70p

Friday 6 October 2006
www.independent.co.uk
NUMBER 6,232

THE INDEPENDENT

NEWSPAPER OF THE YEAR

Arts& Books Review
32-page magazine inside

THE RACE MAP OF BRITAIN

• UNPRECEDENTED VIEW OF A DIVERSE NATION, PAGE 2

SCOTLAND AT A GLANCE

• The total minority ethnic population is just over 100,000. This is an increase of 62.3 per cent in the past decade.

• The largest non-white group in Scotland is people of Pakistani origin, with 31,793 people – two thirds of one per cent (0.63 per cent).

• The second largest non-white group in Scotland is people of Chinese origin – 16,310 people – one third of one per cent (0.3 per cent).

• 28 per cent of people in Scotland describe themselves as not religious.

• 65 per cent of the population describe themselves as Christian. 45 per cent of people in the Church of Scotland denomination are aged 50 or over, compared to 32 per cent of Roman Catholics.

• The second largest religious group is Muslim, despite accounting for less than one per cent of the Scottish population.

• Muslims have the youngest age profile of any religion in Scotland, with 31 per cent aged under 16 years.

• Nearly half of the 6,400 Jews resident in Scotland live in East Renfrewshire.

NORTHERN IRELAND

• There are 14,279 non-white people living in Northern Ireland from a population of 1.68m – less than one per cent

• There are 737,412 members of the Catholic community as defined by upbringing (43.8 per cent)

• Protestant faith communities, defined by upbringing, number 895,377 (53 per cent)

• Chinese are the largest non-white community, numbering 4,145 people.

• Asian communities – including Indian, Pakistani and Bangladeshi – number 2,679, while black communities number 1,136.

• There are 234,800 people who either speak, read, write or understand Irish.

LEAST ETHNICALLY DIVERSE COMMUNITY
If you bump into someone in the street in Easington, Co Durham, there is just a two per cent chance they will be from a different ethnic group

HIGHEST CONCENTRATION OF PAKISTANIS
Frizinghall, Bradford has the biggest concentration of Pakistanis in England and Wales, making up 73 per cent of the population

HIGHEST CONCENTRATION OF JEWISH PEOPLE
Broughton in Salford has the highest concentration of Jews outside London – they make up 49 per cent of the population

HIGHEST CARIBBEAN CONCENTRATION
One in four of the population of Rusholme, Manchester is Black Caribbean, the highest concentration outside London

HIGHEST CONCENTRATION OF MIXED ETHNICITY
More people in Princes Park, Liverpool, describe themselves as mixed race than anywhere else – 11 per cent

HIGHEST PROPORTION OF MUSLIMS
Three quarters of the population of Moseley and Camp Hill, Birmingham follow Islam. One third of all Muslims live in the West Midlands

LOWEST CONCENTRATION OF CHRISTIANS
Only 13 per cent of people describe themselves as Christian in Moseley

MOST ETHNICALLY DIVERSE AREA
In Brent, London, there was an 85 per cent chance that any two people chosen at random in the street would belong to different ethnic groups

BIGGEST CONTRAST IN NEIGHBOURS
In one ward in Camberwell, London, Black Africans make up 41 per cent of the residents – yet in the neighbouring ward covering Herne Hill they account for just two per cent

Index of ethnic diversity

- 0.50 to 0.85
- 0.26 to 0.49
- 0.12 to 0.25
- 0.02 to 0.11

Experts devised a "diversity index" – based on the probability that any two people chosen at random from a particular area would be from different ethnic groups, even if neither of them are white.

A score above 0.5 means there is more than a 50 per cent chance of this happening – and the area is classed as highly ethnically diverse

MOST ETHNICALLY DIVERSE PLACE OUTSIDE LONDON
In Slough there is a 62 per cent likelihood that any two people will come from a different background

LEAST RELIGIOUS TOWN
Four out of ten people in Brighton described themselves as having no religion

MOST RELIGIOUSLY DIVERSE COMMUNITY
In Harrow there is a 62 per cent chance that any two people will practise a different religion

Graphic: Kristina Ferris

The race map of Britain (*The Independent*, 6 October 2006)

Source: John Frost Historical Newspapers

Urban cultural geographies of 'race'

Some of the most incisive and theoretically adept investigations of migration and settlement can be found in richly nuanced geographical accounts of race and place. What is apparent in these studies is the move away from plotting 'immigrant groups' to a focus upon the discursive production of race within the landscape. This work considers how the idea of race is brought to life in time and space, and how it may come to be concretised in place. The various examples of urban cultural geography discussed here are then indicative of a shift from race as a knowable category to the competing practices of *racialisation.* These accounts are marked by a turn away from the mapping, monitoring and segregation familiar to early social geography reports, in favour of an opening out of the conflicting representational strategies and signs that surround particular racialised places and events. Concurrent with this is a shift from looking at race and 'racial groups' to a focus upon how this concept is tenuously constructed, contingently utilised and multiply appropriated by different actors and agents.

Where statistical reports on race had once been the dominant disciplinary paradigm, discursive accounts are now thoroughly represented. This 'turn to culture' (see Chapter 5) is exemplified in a recent collection *New Geographies of Race and Racism* (Dwyer and Bressey, 2008). Such studies challenge the power-geometries of everyday mobility and the territorial racialisation of space frequently endured by ethnic minorities. International examples of race cultural geographies include Jackson's (1993) study of race and crime in Canada, Jacobs's (1993) work on landscape use and symbolism in Australia, Davis's (1991) research on race, gangs and the 'political economy of drugs' in the US, Wren's (2001) report on cultural racism in Denmark, Robinson's (1996, 1998) analysis of the changing dynamics of race and space in South Africa, Saldanha's (2007) study of whiteness on the Asian Goan Trance scene and Pred's (2000) research of the spectre of racism 'even in Sweden'. These studies finely illustrate how cultural geographies of race and place are locally and globally materialised. In this section we will consider some examples of this type of urban cultural geography which frequently draw upon ethnographic, historical and textual forms of representation. In doing so, we take a detour away from race towards understandings of *racism* as an expression of power.

David Ley's (1974) pioneering exploration of racial dynamics in a Philadelphia neighbourhood provides an early example of an ethnographic and cultural geography approach to the field. Over the course of six months, Ley became attuned to the rhythm of everyday life in the Monroe district of Philly, walking, bowling, shopping, worshipping and attending meetings in the urban quarter. Drawing upon quantitative surveys of residents, participant observation of 'gang territories', mobility maps and newspaper articles pertaining to the neighbourhood, Ley archives a rich array of information culminating in the publication of his book *The Black Inner City as Frontier Outpost* (1974). Ley's main finding is that the perceived meaning of an environment is ultimately more important than its physical characteristics when it comes to accounting for the routes people travel and their decision making. Ideas about community are seen to be socially constructed, while a neighbourhood and its borders tend to be defined by those inhabiting these spaces. Ley discovered that the repetition of familiar patterns of movement may over time lead to racially demarcated zones. These ideas may come to be further entrenched through popular associations of particular areas as gang 'turf' that are symbolically realised through graffiti and 'tagging'.

In this way the boundaries of community are always being contested and redrawn through everyday practice that may lead to racial and class segregation.

The spatialised representations of race and their material, symbolic and historical effects are captured in a series of urban studies. Jacqueline Burgess's (1985) early account of media representations of British urban uprisings in the 1980s shows how particular race discourses around the space of the inner-city evolve. She examines how print media construct particular places through the lens of race. To this effect, 'a myth is being perpetuated of *The Inner City* as an alien place, separate and isolated, located outside white, middle-class values and environments' (1985: 193). What is noticeable in this approach is the keen focus on representation. Deploying semiotic methods – that is the study of signs – Burgess considers how the repeated use of particular words and phrases, such as the popular terms 'riots' or 'ghettoes', come to inform and speak back to a broader repertoire of race thinking and race making devices. These techniques are richly enhanced in Michael Keith's (1993) interpretations of the spatialised construction of race and riots in 1980s London. Keith's argument, that 'places are moments of arbitrary closure' (1991: 187), is used to disclose how particular groups are involved in different productions of space. For example, what may be deemed a 'no-go' area to the police and outsiders may be a 'safe space' and 'home' to various resident black communities. In other words, places are assigned multiple and contested meanings by different people (the police, the media, local residents, those living in neighbouring areas, etc.) and their status is always in a process of negotiation.

To exemplify this tendency we can turn to the work of Kay Anderson (1993a, 1993b) who has focused upon the development of 'racial' meanings in the 'Aboriginal suburb' of Redfern in Sydney, Australia. The multiple ways in which people are positioned in relation to place is detected when Anderson found an appeal to 'Aboriginal land rights' by indigenous communities could be used to counter 1970s racist depictions of Redfern as a 'white-only' space. What we can see happening in the ensuing years is a discursive struggle over 'racial' belonging and the meaning of place. Using archive and interview material Anderson shows how politicians and officials 'drew on an established (pejorative) set of images of Aboriginality, not out of any simple 'prejudice', but in order to win the support of local white residents' (1993a: 86–87). Thus Redfern became a central and disputed category in a socio-spatial conflict between white preservationists and black activists/sympathisers, forming part of the contingent openings and closures that Keith identifies in the British metropolis.

Doreen Massey's (1996) celebrated and colourful description of life on Kilburn High Road in North London further suggests that local places can be re-invented through an open-ended and transnational, 'global sense of place'. Indeed, Chinatowns are perhaps a case in point where many of these once vilified spaces are now central to our 'sense of place' of cities such as San Francisco, London and New York, as our case study of the cultural geographies of Chinatown reveals (see Box 8.3). Massey suggests that such new attachments and practices are continually forming to enable a 'progressive' sense of place to emerge. Massey's point is that places do not have to be seen in a parochial way, tied to an earlier imagined 'way of life' but can be re-invented as potentially inclusive and cosmopolitan in outlook. Although the multicultural idea that people from diverse parts of the world may rub along together is insightful, aspects of her approach are critiqued in Jon May's (1996) empirical study of another London neighbourhood, Stoke Newington.

May found that many middle-class residents enjoyed living in an area with tree-lined avenues that exuded a certain English 'village charm' and were equally attracted to the idea

Box 8.3
Cultural geographies of Chinatown

In a series of studies undertaken in Vancouver's Chinatown district in Canada, Kay Anderson (1987, 1988, 1991) produces an important account of the racialisation of place. As a cultural geographer Anderson explores how these spaces have long been seen through an 'Orientalist' gaze (see Chapter 11) and were subject to salacious newspaper reports which focused upon opium dens, vice and crime. The representation of Chinatown and its residents is produced through a familiar repertoire that distinguishes it from other 'Western' spaces of the city. Anderson reminds us that these spaces 'grew out of, and came to structure, a politically divisive system of racial discourse that justified domination over people of Chinese origin' (1988: 146).

Today, many Chinatowns across the globe are more likely to function as exotic sites for a cosmopolitan 'tourist gaze' (Urry, 1990) where food, artefacts, massage, acupuncture and ornaments may all be on offer for the Western consumer. Anderson is also interested in the way signs and symbols are used to designate and racialise space. She explores the way in which Chinese lettering, lanterns, coloured pillar boxes and gated entrances are used to signify and demarcate the Chinatown district. What Anderson's account ultimately reveals is that such signs are rarely markers of authenticity and much of what is thought of as Chinese is largely part and parcel of the Western imagination. Hence, many goods are marketed to tourists rather than those of Chinese descent. Similarly, a number of restaurants may have a separate menu for Chinese customers indicating how these places are often involved in the presentation and staging of ethnicity. Together discourse and representation may function to transform a space into a place by assigning it to a particular 'community' as seen in the case of Chinatown, Little Italy and 'ethnic neighbourhoods'. The point is that these representations are culturally given meaning to – and may be resisted, embraced or even alter over time – so are rarely stable coordinates for understanding what we might think of as race.

Anderson's writings suggest that power lies at the heart of the various constructions of Chinatown and that these ideas always collide at the crossroads of contestation. Needless to say, there are no Chinatowns in China. However, the recent development of 'Thames Town' at the edge of Shanghai, complete with mock-Tudor pubs, Victorian buildings and village greens, targeted at a rising Chinese middle-class, further suggests that such communities are far from passive recipients of processes of racial representation. Through such mimicry Occidentalism – as a selective and playful understanding of 'the West' – is then a fitting parody that enables the Orient to strike back.

that the neighbourhood lay within easy reach of an eclectic patchwork of multicultural shops, restaurants, cafes and clubs. But May's interviews with residents found this was not the 'progressive sense of place' Massey is reaching towards, but an *aestheticisation of difference* where both the 'traditional' and the 'exotic' are sought out. In other words, difference is commoditised where Stoke Newington residents may enjoy the benefits of living in a quaint 'old English village' while taking weekend expeditions to the nearby exotic and hip multicultural areas. Although residents may speak about the value of multiculturalism, May's study suggests that in practice this is largely a consumer experience. This enables mobile white, middle-class residents to 'have it all' as difference is fetishised and transformed into a bourgeois form of cultural capital. In contrast ethnic minorities and white working-class residents alike are rooted to their allotted place and unable to enjoy the privilege afforded to middle-class whites in Stoke Newington. The racialised modes of consumption May identifies are more about segmentation than multiculturalism; an act of splitting, that serve to hold visible minorities and white working-class subjects apart from everyday living. Sensitively developing this critique of Massey's 'global sense of place' McGuinness

(2000: 225) remarks how it is all too easy to discover 'colourful empirical demonstrations of . . . cultural **cosmopolitanism**' in metropolitan areas yet such analyses rarely extend to the seemingly mundane estates, suburban areas and New Towns that make up much of British life. In response the following section moves beyond the urban to examine race and racism in seemingly prosaic 'out of the way' places. For if there is a criticism of much urban cultural geography on race, it is that it has been largely metropolitan-centred and has tended to focus upon visible minorities at the expense of a critical engagement with whiteness and its position in 'race relations'.

Geographies of rural racism

The bulk of geographical research on race has concentrated on urban areas, and in particular the spaces and neighbourhoods where ethnic minorities reside. As Paul Watt remarks, such accounts are in danger of reproducing 'the hegemonic status of the inner-city discourse in relation to race and space' which he claims serves to 'reinforce the notion that racism only spatially occurs where black people live' (1998: 688). Urban cultural geographies may even give rise to a presumption that rural and mainly white areas are immune to racism. Reports on racism in the countryside and rural areas emphatically challenge these preconceptions (Jay, 1992; Troyna and Hatcher, 1992). They suggest that places are popularly racialised in different ways through a particular set of geographical imaginaries. For example, cities can be seen as energetic places, with a 'buzz' and cosmopolitan vibrancy. The other side of this racialised portrayal, though, is to conceive of inner-cities as crowded, polluted and dangerous spaces, long given over to immigrants. Contrastingly, the countryside is envisioned as the epitome of whiteness, Englishness and nationhood. Consider the touristic staging of national heritage through literary and filmic traditions, such as the Yorkshire landscape's rendering as 'Brönte country', the Lake District's familiarity with Wordsworth and the pastoral poets, or the Peak District's connections with Jane Austen's *Pride and Prejudice*, where scenes were filmed around Chatsworth House. The timeless appeal of rural Englishness is captured in numerous television series such as *Last of the Summer Wine*, *Heartbeat* or *Peak Practice* and signified through 'rolling green fields, winding lanes, cream teas, chocolate box villages' (Neal, 2002: 443) and other pastoral, antiquated iconography.

Ideas of nationhood are then inscribed in landscapes. However, it is possible to deconstruct these homogeneous visions of the nation, as geographers have shown in critical studies of rural and Christian 'divine Englishness' (Matless, 1998; Palmer, 2002). In subjecting dominant national identities to analytic scrutiny, we can begin to address issues of rural racism, social inclusion and the need to make the countryside more accessible to black and minority ethnic communities, who, at least in Britain, are emphatically equated with urban spaces (Kinsman, 1995; Ageyman and Spooner, 1997; Neal, 2002; Garland and Chakraborti, 2006). The symbolic reworking of rural landscapes is found in the photography of Ingrid Pollard (see photo), a black British artist whose images offer a counter-point to the prevailing geographical imagination (see Pollard, 1989, 1994). Through sepia-tinted postcard images and provocative captions such as, 'I wandered lonely as a black face in a sea of white', Pollard's photographs of the Lake District bring out into the open the manner in which ethnic minorities are emphatically constituted as 'out of place' in the English countryside and filled with 'dread, unease'. This experience is a reminder of how these landscapes are repeatedly constituted through an upper-class prism of whiteness, Englishness and Christianity.

Ingrid Pollard reworking the relationship between race, place and rurality
Source: Ingrid Pollard

To gain insight into the way in which the English countryside performs as white, we can turn to Wendy Joy Darby's (2000) ethnography of country rambling associations. Darby's anthropological observations reflect upon the 'geographies of nation and class in England' wherein 'Black and Asians were seen to be urban' (p. 244) and the rural landscape is largely bleached of ethnic minorities. Occasionally these cultural imaginings of race and nation give rise to crude racist beliefs such as those held by one hill-walking leader. On learning that the author is an anthropologist, he suggests she should be 'eating meat and chewing bones with the natives'. When Darby challenges these racist perceptions and the idea of race difference – once part of anthropology's past roots in eugenics and colonialism – she is informed in no uncertain terms, 'I'm not going to believe that I've got the same DNA as the jungle bunnies' (Darby, 2000: 239). Despite the enactment of whiteness, Englishness and racism in the countryside, it remains possible to construct alternative cultural geographies from which to rethink the rural. For example Divya Tolia-Kelly (2006) has shown in cultural geography work with South Asian women that the English countryside can reacquaint diasporic communities with memories of their own agricultural heritage and village past. Many respondents sketched images related to their rural upbringing in the Indian sub-continent and in doing so could figuratively connect to the English countryside and render places meaningful to their own mobile biographies. We suggest that the valuable insights generated by Pollard, Tolia-Kelly and other minority ethnic artists and academics are indicative of the way these landscapes can be reclaimed from an imperial imagination and reworked in a post-colonial context as described in Chapter 11.

Although studies of constructions of race and racism in the English countryside have developed considerably in recent years, geographers have largely overlooked the everyday spaces that lie in between the racially demarcated urban/rural binary and the seemingly 'ordinary' spaces where whiteness abounds (McGuinness, 2000). A counter to this is Paul Watt's (1998) study in the 'escalator region' of South East England, exploring perceptions of space and territory among 70 Asian, black and white young people. Interestingly, Watt found South Asian youth held the strongest attachments to their locality, while white youth living in predominantly 'white places' were likely to regard these areas as non-place spaces. However, this does not render white places unproblematic or devoid of racialised meaning. For example, using visual ethnography Anoop Nayak examines the geography of racist violence in the English suburbs (Nayak, 2010). The study demonstrates how a skinhead 'gang' were able to transform their neighbourhood into a 'white space' and a 'no-go' zone for other youth living outside the area. At a macro level this is partially achieved through economic transformations in the region, 'white flight' from the inner-city and historical patterns of settlement throughout the conurbation (Rex and Moore, 1967). However, the constitution of the estate as 'white' is further secured at a local micro-political level through intimidation, harassment and the perpetuation of far-Right racist graffiti. When black and Asian families were moved into housing on the estate they would frequently be subjected to racist threats and burglary that would result in them choosing to relocate back to urban areas. In other words, it is through everyday practices, repetitive performances and their affects that the idea of race becomes 'real', as it slides from imminence to emergence.

The relationship between whiteness and suburbia is further documented in US geographical research. Audrey Kobayashi and Linda Peake (2000) reflect on the press and media representation that followed the shooting of students at Columbine High School in Littleton, Colorado, by members of a Goth friendship group known as the 'Trench Coat Mafia'. In these readings Littleton is repeatedly represented as a 'safe' (white) space when juxtaposed to the vividly racialised inner-city 'hood' and ghetto. Although the story would have had dramatic potential in any context, Kobayashi and Peake note how the narrative is given added potency as it presents against a backcloth of white normalcy; the school is wealthy with a largely white cohort and the perpetrators were white and supposedly from 'ordinary' suburban families. Through these unseen markings of whiteness the shooting is rendered inexplicable. In the absence of familiar racialised signifiers of bodies and spaces the media came to depict the perpetrators as 'abnormal, *individuals* who have deviated from the established norm as individuals, not as products of a particular social context' (Kobayashi and Peake, 2000: 395). The material and symbolic place of suburbia in the US geographical imagination is seen when Frances Winddance Twine (1996) examines the lives of 16 female university students of African-American descent. She notes how these women were 'raised white' on the outskirts of America, as a consequence of a 'parachuted plurality in the suburbs' (Peach, 2000: 622) that has seen affluent minorities relocate to the outer-ring. For these 'brown skinned white girls', as Twine (1996) describes her respondents, the suburban privileges of whiteness are seen to act as a type of 'comfort zone' (p. 215) which insulates them from at least some of the forces of race hatred by enabling their inclusion into 'white' spaces and institutions. Extending race inquiry to rural, suburban and coastal areas is then a means of bringing 'white places' into critical focus to understand how the power of whiteness is produced and negotiated across space, an issue we shall now turn to.

Turning to whiteness

Although geographers have occasionally considered patterns of white European migration and settlement, an interrogation of whiteness as a social norm and hegemonic form of power has been largely omitted in these accounts. Where cultural geographers have sought to expose the social construction of race, they have concentrated mainly upon visible minorities and the manner in which their bodies – and the spaces they inhabit – become marked. Absenting whiteness in this way means that it is rarely held up to scrutiny, but comes to operate as the blank canvas of experience upon which Other ethnicities are so vividly painted. This section will argue that whiteness is an unspoken form of privilege inhabited by ethnic majorities in the Western world. We demonstrate how geographers are engaging with whiteness and suggest they can add much to the study of whiteness by developing a critical sense of race as a global project and providing international accounts of white identity.

There is now a number of geographical accounts concerned with the relationship between race, place and whiteness. Ruth Frankenberg's (1994) interviews with 30 feminist white women living in Southern California expose the racial geography of their belonging and the unacknowledged privileges accrued when identifying as 'white'. Frankenberg reveals how whiteness is a normative category in the lives of her respondents who may regularly identify through gender but rarely through ethnicity. To this extent whiteness is frequently unseen in individual accounts yet its power to structure identities through bestowing various forms of advantage cannot be dismissed. In a neighbourhood account of consumption practices in the Wood Green and Brent Cross areas of London, Peter Jackson (1998) found many elderly residents expressed feelings of nostalgia that occasionally elided with romanticised constructions of Englishness. 'It is here', Jackson declares, 'that one begins to approach a localized construction of "whiteness", expressing concerns about the effects of national decline and social change at a neighbourhood level' (1998: 103). This theme is pursued in Slocum's (2007, 2008) study of alternative food practices and farmers' markets where whiteness coheres in the complex acts of consumption, differentiation and the sense of 'doing good' through middle-class involvement in ethical trade, organic purchases and locally sourced products.

In a multi-site youth ethnography exploring race, class and post-industrial change in contemporary Britain, Anoop Nayak (2003a, 2003b) considers the making and unmaking of whiteness. In this account whiteness operates through a complex circuit of local-global 'flows', where it is interwoven within the discursive and material histories of place relations, and comes to be embodied by different subjects in different ways. For example, young working-class *Real Geordies* may draw upon a masculine background of skilled labour that had once facilitated 'white flight' from the city, home ownership and other symbols of white respectability. It is argued that their industrial past enables them to present as authentic whites, through such emblematic metaphors as being 'backbone of the nation' or 'salt of the earth'. This embodiment allows the *Real Geordies* to perform as arbiters of ethnicity and exert a superior form of whiteness over *Charver Kids* who herald from families where long-term unemployment exists. It is suggested that *Chavs* or *Chavers* occupy a precarious relationship to whiteness – Nayak depicts them as 'not-quite-white' – which is derived from geographies of dense urban living, a proximity with ethnic minority and asylum-seeking families, historical associations with crime and the 'black economy', symbolic connections to Gypsies, Romany and Traveller communities and a distinct 'rough' subcultural style exhibited through dress, music, accent and behaviour. At the other end of the spectrum are *B-Boyz* and *Wannabes* who, through an outward commitment to hip-hop, basketball and a repertoire of global signifiers, were displacing the meanings of whiteness in complex and often contradictory engagements with 'black cool'. What each of these subcultures reveal is the place-based 'doing' of whiteness – how it is precariously recuperated, negotiated, done and undone. This suggests that whiteness is far from homogeneous, is situated in relations of power and is performed in a myriad of ways.

If there is one prominent area where geographers can make a sustained contribution to the study of whiteness it concerns the development of critical global perspectives. Research on whiteness is largely derived from the US and opportunities to explore understandings of whiteness in international contexts remain. The complex intertwining of whiteness to national cultures is evident in 19th-century Barbados, where Lambert (2001) focuses upon the plight of poor whites and former slaves from the colonial order who became freedmen. Lambert argues that both these social groups function as 'liminal figures' (p. 335) as they each occupy a position of 'in-betweenness', an ambivalence that means their status is constantly in flux. The issue of post-colonial ambivalence is discussed in Chapter 11. Lambert goes on to remark upon a broader project of 'racial reinscription' (p. 337) which sought to recuperate poor whites within the pantheon of whiteness and relegate black freedmen to 'slave-like marginality' (p. 337), a process that failed as it attempted to overwrite the complex and competing ways in which race signs are constantly produced, circulated, resisted, mis/recognised, consumed, adapted and so forever incomplete. In the modern context it has been demonstrated how black British-born return migrants to Barbados can be interpreted as 'both black and symbolically white' (Potter and Phillips, 2006) where an English education, accent, dress and deportment afford particular opportunities that would only be open to whites. According to Potter and Phillips's (2006) study of 51 'Bajan-Brits', respondents were ascribed qualities of being 'hard working' and 'trustworthy' in 'an extension of white hegemonic power . . . shared by Barbadians and non-Barbadians alike' (p. 919) leading to race privileges in employment, banking services and so on.

In Sydney, Australia, Wendy Shaw (2006) discloses how an area known as 'the Block' was either ignored by estate agents and promoters of the 2000 Sydney Olympics or is discursively represented as a 'black ghetto' or 'no-go zone' by the Australian media and white population. The mass gentrification that accompanied the Olympic games witnessed a 'new version of Sydney urbanity' that sought to write-out indigenous communities and 'to disregard the local, its various pasts and presents, by promoting a vision of a stylish, urbane, and very white Manhattan' (Shaw, 2006: 863–864). In this way one may argue that whiteness, gentrification and corporate capital operate to have the effect of producing areas like the Block as 'dead spaces' in which indigenous communities can appear 'abject bodies', haunting the contemporary post-colonial landscape. The high currency of whiteness on the international stage is also found in Arun Saldanha's (2007) vivid ethnography of the Goan trance scene, taking place in the Indian village of Anjuna. Amid the tropical sands, sounds, sunlight, drugs and massala tea we may expect to observe something approaching a post-race communal hippy paradise. What Saldanha reveals instead is a sharply stratified, racially demarcated terrain upon which Western rave tourists are able to deploy the privilege of whiteness and a knowing cosmopolitan consumerism. In these new renderings, race is a viscose sign that carries affective intensities as it attaches itself to bodies, spaces and actions. Such embodied and emotional geographies of difference now offer future possibilities to extend the 'radical' impulse of social constructionism for new geographies of race, place and landscape – a theme addressed in Chapter 12. Geographies of whiteness as described above, exploring Europe, Latin America, the US, Caribbean, Australia and South Asia, suggest pioneering routes for examining whiteness locally and globally (see Box 8.4).

Box 8.4
Whiteness and geography: internationalising the debate

Inspired by debates taking place in American labour history as well as literary and cultural studies, Alastair Bonnett is the first geographer to develop an analysis of whiteness within the discipline. In an early paper he remarks how white people appear devoid of history or geography; indeed, 'White appears as the Other of ethnicity' (1993: 175–176). Elsewhere Bonnett (1997) elaborates on the impact this myopia has where whiteness can elude geographical fixity and the ensuing processes of racialisation. He suggests this evasion is actually a form of race privilege that enables whiteness to present as a homogeneous category where those classified under this rubric appear 'forever white'. But as Bonnett (2000a) and numerous other scholars have demonstrated, whiteness is continually being resituated over time and place. For example, the ancient Chinese had once identified as 'white', while other social groups such as the Irish, Gypsies or Jews were displaced from this marker of privilege.

The impact of Bonnett's work can be seen in Peter Jackson's earnest attempt to redress the absence of whiteness in his own studies of racial inquiry. The beginnings of an international approach to whiteness can be traced in Bonnett's book *White Identities* (2000a) which considers understandings of whiteness in Japan and parts of Latin America. In particular he reveals how business, global media and advertising are transforming the meaning of whiteness. This can be an unpredictable process where local 'identities also make use of whiteness as something to aspire to, to define oneself against or, as in the case of Japan, in a more complex and tense relationship of assimilation and refusal' (p. 75). This is seen in Bonnett's case study of Xuxa, an infantilised, blonde young woman who hosts a 'zany' children's show for the Brazilian Globo television network. Drawing upon Amelia Simpson's (1993) comprehensive analysis of a programme sponsored by Coca-Cola, Bonnett shows how whiteness has become synonymous with 'sexiness' and full-throttle 'American-style' consumption. In this respect, whiteness is portrayed as the high-water mark of modernity and progress.

Anti-racist geographies: subverting a white discipline

Despite the rise of new cultural approaches to race, geography has been, and continues to remain, an ostensibly white discipline. Whiteness is embedded in its historical foundations (see Chapter 1), its institutional apparatus and the networks of university students, teachers and researchers that constitute the 'geographical community' (see Chapter 11). Although a number of geographers from the 1970s onwards started to think through race as a part of society, this was often treated as something *external* to them. Even now few geographers delve deep into their complicity with whiteness and white society – its knowledge, hierarchies and forms of cultural capital. In this section we consider how the discipline has come to comprise mainly white practitioners, what the implications of this composition are for the formation of geographical knowledge and anti-racist strategies for developing more inclusive forms of learning.

Where feminist geographers have long contested and subverted the masculine basis of the discipline (see Chapter 6), the whiteness of geography has only recently been subject to discussion. For example, Delaney wryly comments upon the suspicion that geography 'is nearly as white an enterprise as country and Western music, professional golf, or the supreme court of the United States' (2002: 12). The whiteness of geography is then a deeply fraught paradox for a subject that has illusions to being the 'world's discipline' but is entangled in the academic production of 'geographies of exclusion' (Sibley, 1995). Some anti-racist geographers suggest that the only way to penetrate the white heart of geography is through cultivating a critical mass of minority ethnic scholars (Kobayashi, 1994; Pulido, 2002). This theme is addressed in a special issue of the human geography journal *Gender, Place and Culture* (2006), in which a number of essays appeal to the development of anti-racist geographies of inclusion in order to displace its inherent whiteness. As Kobayashi remarks, 'the very fact that geography remains such a white discipline shows that, at least at this time in our history, colour *does matter*' (2003: 533). Given this effect in most areas of social life many minority ethnic people choose to identify as 'black' more as a political marker of resistance than an accurate descriptor of skin colour.

Given the overbearing whiteness of geography it is necessary to ask, how does this shape geographical thinking today? We might suggest that the whiteness of geography enables certain truths and norms to achieve ascendancy. This in turn means that other geographical knowledge is obscured or subordinated: it is left 'hiding in the light'. This issue will be discussed in finer detail in Chapter 11 where it is argued that a post-colonial approach not only serves to open up alternative geographies but in doing so can 'provincialize Europe' (Chakrabarty, 2000), showing how our own paradigms, canons and taxonomies of knowledge are peculiar to their history. It would then amount to an academic conceit to presume that just because the discipline of geography is now interrogating its imperial legacy through what Smith and Godlewska term 'critical histories of geography' (1994: 1), that it can simply emerge unblinking from the shadow of race and empire.

To illustrate this point we might consider how white Australian acquisition of Aboriginal lands has displaced many indigenous communities from these territories and subjected them to the forces of modernity. In doing so a new, dominant white geography unfolds and like an avalanche conceals all before it. Where whites have sought to 'master' the lived terrain through land ownership, the naming of streets, building of transport infrastructures and the accumulation of property, indigenous populations do not set people and place apart. Instead many Aboriginal people's existence is spiritually interwoven into the landscape and intimately connects with bird-song, animals, bush, wind and rivers. Through this alternative geographical imagination, 'the bush' cannot simply be understood as empty space waiting to be filled by the masculine and racialised penetration of corporate capital, but as part of humanity's spiritual and ancestral being. To many of us this may seem an archaic and pre-modern disposition to adopt, not least because 'whiteness works as an *epistemology*, that is, as a particular way of knowing and valuing social life' (Dwyer and Jones III, 2000: 210). The recursive privileging of white racial norms effectively means that geographical knowledge is a product of the West, constituted as it is through the white eye of power. But if we consider how Aboriginal communities such as Native Americans have been displaced from their land – and at times provided with reservations – it becomes difficult to ignore the fact that Western modernity has evoked a geography of violence upon native subjects.

One way in which we might challenge the hegemony of whiteness then is by becoming aware of it as a racial norm that privileges a few at the expense of others. Recognising this position and how it shapes our research and ideas is a starting point for effecting change. As the feminist geographer Gillian Rose admits, 'my whiteness has enabled my critique of geographical discourses' (1993: 15), having allowed her access into particular spaces that may not otherwise have been held open to her. Developing this theme, Kobayashi and Peake remark how 'the lives of dominantly white geographers, are sites for the reproduction of racism, but they also hold the potential of being strategic sites of resistance' (2000: 399).

Nevertheless, what is particularly striking for Bonnett, is the failure of anti-racists to engage with this site of potential contestation (see Bonnett, 1993, 1997, 2000a, 2000b). To counter this we will now consider some of this potential by turning to anti-racist practices and the challenges to racism posed in contemporary human geography.

Anti-racist geographies, in their many diverse forms, seek to make a meaningful political intervention in race debates. Broadly speaking, **anti-racism** 'refers to those forms of thought, and/or practice that seek to confront, eradicate and/or ameliorate racism' (Bonnett, 2000b: 4). Within academia this could include, for example, the establishment of anti-racist policies across universities; the inclusion of work by black writers and anti-racist activists on course syllabi; attempts to recruit black and minority ethnic students and teachers into the discipline; the organisation of campaigns against race discrimination; the development of race awareness across the campus, and so forth. Tackling the culture of institutional racism in the workplace is a difficult but pivotal aspect of anti-racist practice. The lack of minority ethnic staff and students in many geography departments mean such issues are rarely addressed. However, recent reports by the Policy Studies Institute and the Association of University Teachers (AUT) in the UK found ethnic minority lecturers are more likely to be employed on casual fixed-term contracts, have heavier teaching loads, are paid lower salaries and have less chance of gaining promotion or financial increments compared with white colleagues across the university sector (Carter *et al.*, 1999; AUT, 2005). In US geography departments, the disciplinary picture is equally dispiriting where the number of ethnic minority tenured and senior staff remains well below those figures in congruent fields such as history, sociology and anthropology (see Mahtani, 2004).

In order to effect an anti-racist geography we might start with exploring our own institutional practices. In a study of whiteness in the University of Georgia, US, Inwood and Martin reflect how 'Campus landscape memorials simplify or ignore race as a social mediator, thereby obfuscating deeply embedded racialized identities and tensions on campus' (2008: 374). If you were to take a geographical tour of your own campus, exploring any statues, prominent photographs or buildings named after individuals, you may increasingly be drawn to the manner in which whiteness can be embedded in institutions. You may further wish to reflect upon geography in your own institution with regard to the ethnic composition of academics, support staff, students, cleaners and other service workers. Is race an 'organising principle' in your institution? Does your geography department adequately reflect the diversity of the city it is situated in? Does your curriculum equally engage with the work of 'Third World' scholars and other minorities? And how is whiteness produced across your university campus, in built architecture and everyday practice? The point is that white norms can be traced in many geography institutions, including our own, where a prominent public lecture series entitled 'Tales from the Bush' echoes the idea that territories beyond the West are wild, primitive and uncivilised – a fantasy space from which pith-helmeted fieldworkers can return with hair-raising anecdotes of adventure and valour. As Richard Phillips (1997) suggests, these geographical narratives are part of the imaginative mapping of masculinity and empire.

An area where anti-racist practice could be utilised is in the scrutiny of everyday geographical materials and practices. Some critics claim that geography textbooks rarely include the perspectives of racialised others (Sanders, 2006) but all too frequently position them and the nation states they are seen to inhabit as the problematic subjects of geographical inquiry. For example, while famine, disease, violence and poverty feature in many parts of the global south and are important areas for geographical research, there remain relatively few alternative representations of these places or enough attempts to show how these issues connect to those of us in the global north. However, work published on behalf of the Geographical Association in the UK attempts to provide teachers with a practical guide from which to develop meaningful anti-racist geographies in the classroom (see Morgan and Lambert, 2003). Here, discourse and representation are also sites for the production of anti-racist geographies (see Nash, 2003).

The potential of this might be found in the practice of geography international fieldtrips common in schools, colleges and universities. Where geography was once the discipline through which racialised Others were 'discovered', seldom are these 'expeditions' placed within a post-colonial framework that could enable a more rigorous anti-racist geography to transpire generating more reflexive forms of knowledge. Derived from personal experiences of leading undergraduate fieldtrips to the Gambia, Abbott (2006) suggests that a post-colonial understanding of whiteness can edge us towards a 'political analysis of fieldwork' that begins to grapple with the issues of power and positionality. An anti-racist project of this kind might ask what it means for students from Western universities to visit developing areas and former colonial spaces. Is education itself a whitening force, that leads to an extended 'cultural voyeurism' through exotic travel? And, as international fieldtrips are a marketing device for many geography institutions, what relations of power are evoked in the entanglements between money, race, class and Western privilege?

A core underpinning of anti-racism is direct action and the politics of activism. An explicit example of this was seen at the Association of American Geographers (AAG) annual conference in 1971, arguably the largest and most prestigious public forum for geographical debate. Here, a number of geographers called for the US to withdraw its troops from Vietnam, a struggle which was viewed as an act of imperial conquest. As Marxist ideas spread within the discipline, many US campuses became sites for political protest, which included a number of geography staff and students giving their support to the Civil Rights Movements (see Chapter 4). This anti-racist zeal is captured in William Bunge's (1969: 35) caustic remarks to what he termed, 'armchair geographers'. Bunge effused, 'Geographers should form expeditions to the poorest and most blighted areas of the country, contributing rather than taking resources, planning with people rather than for them, incorporating local people . . .' and so forth. What Bunge advocates essentially is a type of 'geography for all' in which local people and indigenous communities are part of the solution as opposed to being the geographical objects of study. Today such activism may include participatory geographies that work with refugee and asylum-seeker groups in the voluntary and community sector where the focus is upon meeting their needs rather than deriving publications. The rise of the British National Party and English Defence League in the UK and other far-Right political organisations across Europe suggest that there continues to be a need for anti-racist action. Echoing the work of early radical geographers, Peake and Kobayashi (2002) view such activism as an essential component of anti-racism. As such, anti-racist teaching, political protest and direct action can come to form part of the 'dissident geographies' that Blunt and Wills (2000) identify as a radical feature of the geographical tradition.

Of course, anti-racism works on many levels and to be effective invariably has to be policy relevant. Here the meticulous statistical data discussed earlier can offer a robust empirical base from which to challenge racial mythologies (see Dorling and Thomas,

2004; Phillips, 2006; Phillips *et al.*, 2007). Anti-racist geographers contend that while race may be a social construct it is imbued with material effects that cohere over time and space to produce social inequalities (Jackson, 1987; Jackson and Penrose, 1993). Early studies on institutional racism and the discourses of anti-racism (Phillips, 1987), as well as work on the difference that place makes to anti-racist practice (Bonnett, 2000b), reveal this geographical potential. Some geographers have attempted to develop a 'pedagogy of place' that draws upon local histories of racism and anti-racism in order to make these issues relevant to children and young people living in mainly white areas where different forms of intervention may be appropriate (Nayak, 2008). This approach can be enhanced by utilising biographical techniques and oral histories to show how even white students have a complicated 'geography of belonging' which challenges the enduring relationship between whiteness and the Western nation state. For example, in the predominantly white nation of Scotland, Penrose and Howard (2008) recently explore the '*One Scotland, Many Cultures*' campaign, demonstrating how perceptions and identifications with 'Scottishness' inflect attitudes to anti-racism and belonging in diverse ways. Engendering an anti-racist awareness can then inspire the production of alternative geographical imaginations.

Occasionally spaces for radical dialogue can be opened up within the academy. Many universities are sites for anti-racist campaigns against military activity, global poverty, or the plight of indigenous communities cowed by the whip-hand of Western global forces. The refusal by certain human geographers to contribute chapters to *The Encyclopedia of Human Geography* – until the publisher Reed Elsevier eventually withdrew their sponsorship of the arms trade – is a small but meaningful gesture that suggests possibilities for setting an anti-racist agenda based on a global understanding of human rights. Of course many geographers involved in this project chose to put their individual careers before collective action and will continue to do so, often entering into a tortured theoretical calisthenics to justify injustice. At such critical moments geographers' complicity with Western institutional practices and their personal ambitions are all too often nakedly revealed. In contrast good anti-racist practice is grounded in the materiality of daily life (Jackson, 1987; Jackson and Penrose, 1993). It is alive to its political context and the manner in which race operates as a 'dividing practice' in late modernity. In this respect anti-racist practices are rooted in their own moral base and speak back to social theory, making us rethink how we do what we do and why. This involves a greater reflexivity and openness to the idea that 'geographical knowledge and activism are synonymous' (Kobayashi, 2003: 554) and that addressing racism should never be a purely intellectual pursuit. Anti-racist geographers contend that while race may be a fictional construct when it is enacted as real, it produces grave inequalities: this needs to be politically, theoretically and methodologically challenged.

Conclusions

This chapter makes the concept of race problematic. It demonstrates how a fictitious belief in the 'world's different races' can lead to forms of biological or cultural racism. This assumption was embedded in early geographies of empire where some scholars sought to categorise human difference through distinct racial forms of classification, that are today seen as erroneous. Although discussion of race in geography was to disappear after the age of Empire, it returned in later debates on immigration and settlement in post-war Britain and the USA. In the 1980s, geographers were moving away from treating race as an objective truth, to recognising it as a social construct tied to enduring forms of power. Work on urban and rural environments illuminated the contingent relations between race and place. However, it is only more recently that geographers have begun to consider their own racially located forms of privilege and the power that can be accrued through whiteness. In doing so more critical, anti-racist geographies have transpired that are capable of challenging the overbearing whiteness of the discipline, its ethnocentric assumptions and way in which race continues to remain an enduring presence in geographical thought and practice.

Summary

- Distinct 'races' within humankind do not exist, suggesting race is an invented social construct.
- Geography's imperial roots enabled the myth of race to flourish by making false divisions between the human population.
- Early social geographers concerned with mapping the spatial aspects of integration and segregation tended to treat race as a 'proper object'.
- Cultural geographies of race and place encouraged a representational shift from race to racism.
- Studies on the geographies of whiteness can critique normative racialised categories and demonstrate how white identities differ across time and place.
- Anti-racism contests the whiteness of geography and offers a political means for subverting its racial authority.

Further reading

Amin, A. (2002) Ethnicity and the Multicultural City: Living with diversity. *Environment and Planning A*, 34: 959–980.

Bonnett, A. (1997) Geography, 'Race' and Whiteness: Invisible traditions and current challenges. *Area*, 29(3): 193–199.

Bonnett, A. and Nayak, A. (2003) Cultural Geographies of Racialization – the Territory of Race, in K. Anderson, M. Domosh, S. Pile and N. Thrift (eds), *Handbook of Cultural Geography*, London: Sage.

Dwyer, C. and Bressey, C. (eds) (2008) *New Geographies of Race and Racism*, Aldershot: Ashgate.

Holloway, S.R. (2000) Identity, Contingency and the Urban Geography of 'Race'. *Social and Cultural Geography*, 1(2): 197–208.

Jackson, P. and Penrose, J. (eds) (1993) *Constructions of Race, Place and Nation*, London: UCL Press.

Kobayashi, A. (1994) Colouring the Field: Gender, 'race' and the politics of fieldwork. *Professional Geographer*, 46(1): 73–80.

Nayak, A. (2008) Young People's Geographies of Racism and Anti-racism: The case of north east England, in C. Dwyer and C. Bressey (eds), *New Geographies of Race and Racism*, Aldershot: Ashgate, pp. 269–282.

Nayak, A. (2010) Race, Affect and Emotion: Young people, racism and graffiti in the postcolonial English suburbs. *Environment and Planning A*, 42(10): 2370–2392.

Phillips, D. (2006) Parallel Lives? Challenging discourses of British Muslim self-segregation. *Environment and Planning D: Society and Space*, 24: 25–40.

Slocum, R. (2008) Thinking Race Through Corporeal Feminist Theory: Divisions and intimacies at the Minneapolis Farmers' Market. *Social and Cultural Geography*, 9(8): 849–869.

Watt, P. (1998) Going Out of Town: Youth, race and place in the south east of England. *Environment and Planning D: Society and Space*, 16: 687–703.

Representation and post-representation

Post-modern geographies

Learning objectives

In this chapter we will:

- Introduce the concepts of modernism and modernity.
- Describe post-modernism as a period in time.
- Explore post-modernism as a radical critical approach to knowledge and practice.
- Consider post-modernism as an artistic and stylistic movement.
- Reflect upon critiques of post-modernist geographies.

Introduction

Hip-Hop music and post-modern geographies may seem unlikely bedfellows. Yet despite their many differences, there are a few surprising connections to be made between these forms. These relations will become clearer throughout the chapter, but for now it is worth noting some specific features of Hip-Hop that are characteristic of broader post-modern tendencies.

Hip-Hop is derived from black oral culture, religious sermons and urban street rhyming games familiar to African-American neighbourhoods. It is then the result of a type of scattered stylistic 'borrowing' through which a new urban form is created. In order to understand how genres such as Hip-Hop come into being it is helpful to turn to a

Hip-Hop artist Eminem – the real 'Slim Shady'?
Source: Rex Features / Giovanni Canitano

French metaphor, that of **bricolage**. As a post-modern term this phrase suggests that a pick-and-mix 'collage' can be composed of different materials through the interplay of images, sounds, signs, symbols and iconography. The sticking together of different components in the form of a collage can be a highly creative post-modern act. It may also offer a challenge or contrast to earlier form and convention. If we follow the example of rapping as a popular mode of expression in Hip-Hop we might glimpse how this artistic deployment challenges the highly orchestrated Western narrative convention of singing, melody and performance. Contrastingly many DJs and rappers improvise and utilise 'old skool' beats, sampling Soul legends such as James Brown or baselines derived from the genres of Rock, Pop, Electro and Motown. The integration of 'retro' sounds and materials into the present is also a post-modern characteristic. The eclectic performance of Hip-Hop and Dance is given a further twist when DJs literally 'cut-n-mix' rhythms on vinyl decks, drop basslines and use sequencers to produce new sounds in the heady moment of a dance event. We might then consider Hip-Hop and Dance music – at least at the level of form – as a post-modern *bricolage* where the production is creatively and endlessly reassembled in the moment through a strategic use of riffs, beats, baselines and melodies often plundered from other artists and periods.

In this way Hip-Hop DJs and performers are involved in a post-modern practice of 'symbolic creativity' (Willis, 1990); they are adept 'textual poachers' and cultural pioneers, capable of disrupting musical convention and genre. Furthermore, Hip-Hop does not only involve the mixing of historical genres and musical form; it is frequently marked by what we might term a *geography of transnational becoming*. Many Hip-Hop notes and rhythms are 'borrowed' from Africa, Jamaica, Europe, America and elsewhere and may later intermingle with national and local referents, as seen in the production of artists as diverse as Dizzee Rascal and Eminem. Dizzee Rascal's black British Cockney patois and Eminem's Southern American 'white trash' drawl demonstrate how place and locality can be used to inflect what is an already mixed genre to give it endless new meaning through additional signification. We would suggest that these global, national and local iterations should be of profound interest to geographers in that they remind us of the intersecting scales at which culture operates and the significance of local particularity.

Hip-Hop is then a fragmented, post-modern and 'culturally hybrid' (Bhabha, 1994) form composed at the cusp of what the post-colonial theorist Paul Gilroy (1992) has termed the 'Black Atlantic'. Through this phrase Gilroy seeks to denote the often brutal geography of trade that connected Africa, Europe and America as jewels, spices, commodities and slaves flowed through these routes. For Gilroy these transnational routes remain important arteries for understanding contemporary processes of creolisation. It is through these flows that genres such as Hip-Hop are spawned along with modern Yoga, chicken baltis, street carnival, body-piercing fashions and countless other practices that are a consequence of a global intermixture. Through these examples we can locate an array of seamless creativity that dismantles any notion of something being original or authentic. With this in mind, rather than trace where the primary *roots* of Hip-Hop might lie – a point of much contestation – we might be better to utilise the possibilities of 'hybridity' and '*bricolage*' to consider instead the contemporary *routes* of Hip-Hop culture in an increasingly interconnected world. This search from anchored *roots* to mobile 'routes' is suggestive of a more 'rhizomatic' (Deleuze and Guattari, 1987), post-modern form of understanding that seeks to move beyond dichotomous explanations.

At the risk of stretching our example of Hip-Hop to its limit, one last connection with post-modernism might be indulged. This concerns the issue of identity. On the surface, Hip-Hop culture places an emphasis on authenticity, 'keeping it real', being an OG ('Original Gangsta') or 'home boy' and affirming connections with the 'hood', 'ghetto' or 'block'. Yet if we look closer at Hip-Hop music videos, lyrics and performances, there is something exaggerated or 'hyper-real' about this world. The respective lyrics, embodied movement and musical video performance of artists such as Dizzee Rascal and Eminem are ingrained with a high level of **pastiche** and cartoon irony. Street-names, 'Dizzee' or 'Slim Shady', also signal a post-modern play with identity that is alive in Eminem's track, 'Will the Real Slim Shady Please Stand Up'. Here the 'real' Slim Shady is as much an illusion formed by the mediated impressions of others and remains a persona open to fantasy and duplication as the accompanying video display of Eminem clones demonstrates. In live performances Eminem also dons a Jason mask derived from the *Halloween* horror movies. In this way post-modern notions of pastiche, *bricolage*, masquerade or hyper-reality offer valuable tropes through which to consider the stylistic affects of post-modernism.

However, it is important not to take the connections between post-modernism and Hip-Hop too literally. We might just as easily have used Hip-Hop to begin a rather different discussion in this book on post-colonialism, gender, sexuality or Marxist alienation. What we have tried to do here is simply open with an example that shows how aspects of what is characterised as post-modernism – hybridity, creativity, fragmentation, bricolage and loss of identity (or the production of multiple subjectivities) – can be found in everyday objects and practices.

This chapter begins by briefly outlining what we understand by *modernism*. It then turns to a detailed discussion of post-modernism where we identify three broad characteristics used to interpret postmodernism as:

- a historical moment
- a critical practice
- a stylistic phenomena.

Each of these approaches, though far from exhaustive, furnish us with useful repertoires through which to understand the 'condition of postmodernity' (Harvey, 1995) in which we are said to reside. In following this tripartite schematic, along the way we provide insights into the work of philosophic writers such as Lyotard, Foucault, Derrida and Baudrillard, who have each inspired post-modern practitioners in the discipline. The focus here, though, is more with the development of some useful post-modern concepts related to narrative, discourse, deconstruction and hyper-reality.

Having unravelled some of the mystique surrounding post-modernism we will then investigate how these ideas are re-shaping geographical thought and practice, and engendering new geographical engagements. After discussing these interactions with post-modern theory we will go on to explore how these ideas materialise in space through a focus upon the city and the contemporary practice of urban geography. Finally, we will outline some of the criticisms attributed to post-modern thinking in order to understand why these ideas continue to cause such consternation and heated debate among academics, activists and other political thinkers. But before diving headlong into the pool of post-modernism it can be helpful to outline what is meant by modernism.

Modernism and modernity

Modernity is used to refer to the period of Western Enlightenment which is said to derive from the 18th century onwards where the laws of physics, theories of evolution, scientific discovery and medicine came into ascendance. These modernist ideas undoubtedly shaped academic knowledge and geographical thinking as the 'Foundations' section of this book illustrates (see Chapters 1 and 2). In particular, the Enlightenment period is associated with the birth of 'reason' and 'rationality'. The modern age is then closely linked with technological advancement and the 'white heat' of scientific development. It is an epoch that was to see scientific knowledge supersede ritual and religious belief as the primary conduits of intellectual production. In this sense, we should recall that modernism once appeared radical or *avant-garde* in its own right as it challenged and usurped popular wisdom of the time. By mobilising the power of reason, illusion, myth, superstition and fakery could be consigned to the past in favour of a reliable reality. In this regard the Enlightenment stands apart as an emancipation project that liberated human beings through knowledge, scientific discovery and development.

Despite such significant advancements, the cultural geographer David Clarke (2006: 110) suggests, 'modernity's dream of universal knowledge could very easily turn into a totalitarian nightmare'. Bauman's book, *Modernity and the Holocaust* (1989), is a key reference point for this apocalyptic vision. Here Bauman reminds us how the Holocaust, which involved the persecution of millions of Jews, was not a return to some pre-modern barbarity, but was carried out in the name of modernity as a wholly rational project. Fascism could only work at the scale of the nation state with the agency and beliefs of people. The endless 'will to truth', regarding who was a Jew or Gentile, based on the 'logic' of a supreme, über-modern Aryan race is a stark reminder of the problem of an unreflexive modernity. An inability to see that the purported scientific truth of race was a mirage is a warning against accepting modernist universal truths and failing to place them under scrutiny (see Chapter 8).

As we shall further discuss when we consider the historical dimension and connections to post-modernity, modernism is also equated with the rise of a rational industrial order. The development of factories and industrial machinery has seen a large-scale production of goods for mass market being delivered through a conveyor-belt culture. The growth and expansion of cities, along with socio-economic changes to urban lifestyles, mark out modernity as a period of enormous transformation. Although some differences exist, modernist ways of life are often regarded as relatively homogeneous, where local communities were forged around the pit, shipyard, factory or mill. A strong sense of local culture is further derived from habits and practices that may derive from leisure activity that surrounds labour, including Miner's galas, Sunday worship, pigeon-fancying, leek growing contests and the shared communal activities taking place in working-men's pubs and clubs. The abiding conviction is that modernist life is ordered, regulated and, at least compared with today, offers a fairly unitary existence. It is a world designed around a specific mode of production. If this sounds constraining as we will now go on to see, post-modernism pulls away at these seams. It is seen as the moment in which the certainties of a production-based society become primarily governed by consumption, media signs and new information communication technologies. To develop our argument we will provide three key interpretations of post-modernism and post-modernity by considering it stylistically, historically and as critical practice.

Post-modernity: a historical moment

An initial way of approaching post-modernism is to consider how it signals an interruption – rather than definitive break – with the past. Interpreting post-modernism periodically as the time after modernity might appear to proclaim a new dawn. However, post-modernism has grown out of modernity and, despite some distinguishing markers of difference with the modernist age, it is important not to over-exaggerate this fissure. As Lyotard (1992: 90) reminds us, 'The idea of linear chronology is itself perfectly "modern"'. Modernity and post-modernity then connect and intersect with one another as part of a continuum, as subtly illustrated in Warhol's factory manufactured 'pop art'. For this reason not all critics agree that we even live in a post-modern age and many prefer to speak more cautiously about a period of advanced capitalism, 'high' or 'late modernity'. For Zigmunt Bauman the project of modernity is not just 'unfinished' but 'unfinishable' (see also Doel, 1999). This more subtle reworking acknowledges that large-scale social, economic and cultural transformations *are* going on around us but implicitly draws attention to their connections with the period of modernity.

As we saw previously, the defining characteristics of modernity linked to the Enlightenment can be seen in the way that life rapidly became organised through scientific knowledge, mass industrialisation, urban growth and large-scale manufacturing, which witnessed the transformation of Western societies. In contrast with the functional aspects of modernity, post-modern societies are characterised by fragmentation, uncertainty and the break-up of 'older' modernist ways of life. This is most powerfully conveyed in processes of de-industrialisation and the post-Fordist move away from a conveyor-belt culture of mass production that once produced standardised products for a standardised market. The growth of a service-sector economy coupled with the 'new economies of signs and space' (Lash and Urry, 1999) that surround, are emblematic of a post-modern age. The period of post-modernity is then characterised by consumption, new media technologies and processes of globalisation that are uneven and risk-laden.

This aspect of 'risk' (Beck, 1992) can be seen in the way in which whole ways of life are uprooted with the dismantling of steel industries or the closure of coalmines and shipyards – the modernist infrastructure that defined earlier forms of community. Although the seemingly stalwart certainties of the modern era are disintegrating, feminist geographers and environmentalists have been critical of those who romanticise the industrial heritage and the masculine forms of labour that are occasionally celebrated. For many young women the opportunities of a post-modern age have allowed for new transitions into adulthood, a terrain that was once largely configured around the tight conventions of marriage and motherhood. The changing labour market, which has seen a movement away from a 'masculinised' industrial base of heavy industry towards more flexible, 'feminised' forms of employment, found in the service sector, knowledge economy, arts and creative industries, suggests that 'new femininities' are emergent in the contemporary late-modern period (Nayak and Kehily, 2008). As more women enter into different sectors of the labour market as full- or part-time workers, the patriarchal notion of the masculine 'breadwinner' familiar to modernity is rapidly becoming redundant. Where working-class young men in the manufacturing economy were once 'learning to labour' (Willis, 1977), in a consumer society they are more likely to be 'learning to serve' (McDowell, 2003).

It is noticeable that contemporary societies no longer cohere around the modern symbols of heavy industry such as the pit, mill, shipyard or factory plant but are increasingly seen as 'post-modern' – dispersed, fragmented and intricately diverse in their structure, composition and cultural ties. Advances in technology and communications, coupled with a need for workers to be increasingly mobile, mean that post-modern relations are often more diffuse, complex and elastic than in previous eras. The rise of the information age, including new media technologies and forms of communication, is altering societal relations and everyday practices. While the post-modern age, if there is such a thing, is clearly linked to an extension of modernity, it is also a precarious moment to inhabit. The need to be increasingly flexible, mobile and entrepreneurial to meet the changing demands of late-modernity has meant that older forms of 'community' are being eroded as a greater emphasis is placed upon creative forms of '**individualisation**' (Beck, 1992; Giddens, 1991). Here the onus is upon individuals to market and style their identities in such a way that they can develop a reflexive biography of the 'self' designed around the accumulation of skills, presentation and performativity. The atomisation of traditional notions of work, family, societal and gender relations has seen new opportunities and practices emerge as features of contemporary post-modern living. Somewhat paradoxically then, while the period of post-modernity is often characterised by greater choice, flexibility and individuality, it is also associated with greater 'risk', uncertainty and insecurity. The pace of change has meant that some people feel unsettled by what is happening around them while others seek to capitalise upon the opportunities afforded.

Post-modernism: a critical practice

A second way in which we might approach post-modernism is as a profoundly challenging 'critical practice'. By this we mean that post-modernism can be used to destabilise existing norms that centre around the Enlightenment fascination with order, logic, coherence and rationality. This aspect of a broader post-modernism is often referred to as *post-structuralism*, a movement which developed through cultural studies and literary practice. This post-structuralist turn in literary theory was concerned to 'de-centre' the subject and usher in new meaning by opening up the relationship between author, text and readers. Part of this process of de-centring the subject involves an understanding that individuals are positioned through multiple discourses of gender, ethnicity, class, age and locality. This means that our identities are not unitary, stable points of reference but are dynamic creations always in the making. The multiple identities we inhabit are then fluid, contingent constructions that exist as unfinished projects. To illuminate how post-modernism as a critical practice can unsettle the rubric of conventional thinking, we shall briefly turn to the work of four influential French philosophers – Jean-François Lyotard, Michel Foucault, Jacques Derrida and Jean Baudrillard. Their biographies inform us of the status of philosophy in French academia and the influence of this type of creative thinking. Our synopsis of the post-modern ideas generated from this quartet is as concise as it is selective. Our aim is to extract just one or two key points derived from the radical application of these ideas in order to consider what it might mean for geographical thought and practice. Those wishing to bypass this section and turn to its geographical implications may be inclined to turn to the summary of these critical thinkers found at the end of this section. For those more patient, the concepts we shall now explore are grand narratives; deconstruction; power and discourse; **simulation**; **simulacra** and hyper-reality. Applying these ideas to your own examples is a good way of demonstrating your knowledge.

Jean-François Lyotard – grand narratives

Post-modernists remain thoroughly sceptical of the post-Enlightenment 'will to order' and the continual demand for neat and tidy explanations of social phenomena. For this reason the French philosopher Jean-François Lyotard (1984) in his text *The Postmodern Condition* critically reflects upon what he regards as the 'tyranny of rationalism'. Here, Lyotard riles against the dominance of particular **meta-narratives** – what he calls '*grand narratives*' – that seek to provide an over-arching explanation of the world in its totality. Rather than reflecting a pre-given reality, modernist scientific thought can be understood as a grand narrative, a fiction among other fictions. It is here that the radical impulse of post-modernism can be felt as a critical movement marked by what Lyotard (1984: xxiv) describes as an acute 'incredulity toward metanarratives'. Evolutionary theories concerning the 'survival of the fittest', capitalist ideologies related to wealth creation, the doctrines of Christianity, Marxism, scientific positivism and Islam can all be considered grand narratives of a kind; big stories that self-present as *truth*. It is this act of factual presentation that Lyotard and other post-modern critics are drawn towards as they interrogate the textual construction of worldly relations. Rather than treating these narratives as an unyielding 'reality,' Lyotard regards them as meta-discursive constructs, by which he means they do not reflect the truth but are the outcome of particular knowledge regimes, institutional arrangements and negotiated power struggles.

In order to feel the textual force of narration we might turn to the making of modern religions such as Christianity, which is rightly considered a grand narrative. For many believers the Bible is regarded as the word of God, a totemic text that encompasses the rules we should live by in the present, that even carry through into an afterlife. Post-structuralist critiques of grand narratives seek to take apart their 'unitary coherence' to consider how they are assembled and made to perform as truth yielding devices. In doing so we might consider how what we now know of as the Bible is derived from the Hebrew text of Judaism, which, with the later inclusion of Christian gospels, has come to form the modern Bible. At this point the Hebrew version was re-defined as the 'Old Testament' and the Bible became a new tome. A post-structuralist reading would also focus upon the critical practice of writing. The Bible is not a seamless organic text but an amalgam of perspectives; for example, the story of Christ is told in the gospels according to Mark, Matthew, Luke and John, who were each writing at different periods of time. As the phrase 'according to' designates, this is a subjective and particular narrative account. Indeed feminist post-structuralists may question the historical conditions that have led *men* to be the sole authors of sacred texts and the primary custodians of religious authority. A further aspect of this type of post-structuralist literary criticism is to explore internal contradictions, gaps and fissures in the text itself. By extrapolating the ideological components of grand narratives in this way, post-modernist and post-structuralist critique can be considered a controversial but radical critical practice.

Lyotard's suspicion towards a 'real' underlying truth – an ontological reality – is an attempt to demonstrate how knowledge is produced through language and culture, always carrying with it this social imprint. He refutes the idea that we can adequately represent 'reality' through language. While some discourses are found to be more compelling than others, and must be so in order to achieve the status of 'grand narrative', this in itself does not render them any closer to reality. The construction of grand narratives such as religious texts is then an accomplishment of power, a sleight-of-hand that allows selective ideas to masquerade as an impartial, objective and seemingly coherent truth. In a post-modern age the status of grand narratives is little more than the empty triumph of rationalism; a pomposity that post-modernists argue must be punctured. For Lyotard modernist knowledge is governed through scientific thought and methods of classification that regulate rather than liberate. He asserts that post-modern knowledge 'refines our sensitivity to difference'. In post-modern thinking, the modernist claims to objectivity, logic and factual knowledge are revealed – rather like the symbolic disrobing of the Emperor's new clothes – as just another construction, a floating signifier or 'master logo'. Such grand narratives, it appears, conceal as much as they reveal.

Michel Foucault – discourse and power

So, if grand narratives are constructs rather than self-evident truth, how do they manage to mark out certain ideas as permissible, and others as socially unacceptable? It is here that we can turn to the work of a second French philosopher and social historian, Michel Foucault, whose ideas we discussed in Chapter 7. Although Foucault is not a post-modern thinker in the conventional sense (he tends to be regarded as a post-structuralist) his work has been taken up by numerous writers who are drawn to the way in which he explores the interplay between discourses, bodies and subjectivities. He argues that discourses are not universal but are conceived historically at particular moments in time. In this sense they become ways of knowing and understanding the social world, generating values and particular ways of being.

Unlike those structuralists before him who used semiotic techniques to interpret language, texts and images, Foucault is less interested in relations of meaning and more concerned with relations of *power*. This is reflected in his attention to discourse rather than language, where it is the inter-textual combination of statements, actions, opinions and so forth that are capable of engendering a particular *discursive formation* to spring to life, what he calls an 'episteme'. In this sense discourses are dynamic rather than static, where particular words or phrases ('Chav families', 'loose women', 'broken Britain', 'bogus asylum seekers', etc.) keep alive and give meaning to a host of historically generated ideas about society. Even if many of these taxonomies are fabrications, Foucault's interest lies with the way in which they operate as 'discursive practices'. Here, discourses may lead to particular forms of governance through social policy, law, medical legislation or education.

The emphasis Foucault gives to discourse rather than language reflects a concern with ideas and practice. His argument that meaning cannot exist outside of discourse is not a denial that things have a material existence in the world, but the recognition that discourses produce our objects of knowledge. In other words discourses do things, they are productive and involved in a dynamic process of meaning-making. Discourses about health, medicine, law, education and media incite subject positions and so are formative of our identities. In his meticulous accounts of such wide-ranging topics as crime and punishment, madness and medical practice, sexuality or the disciplinary effects of education, Foucault demonstrates that the ideas and forms of regulation that surround these notions have no meaning outside of their discursive formation. The tropes that identify subjects as 'criminal', 'lunatic', 'prostitute', 'hysterical woman' or 'homosexual' only exist within the discourses that produce them. Importantly, for such labels to become part of everyday discourse they must be widely legitimated by specialists including medical experts, judges, child psychologists, welfare officers, teachers or religious leaders. For example, being gay has been seen as pathological, perverse, a genetic aberration and a practice that has required magistrates to impose a legal penalty outlawing sodomy (see Chapter 7). Despite human rights legislation, a person can still serve a prison sentence for being gay in countries such as Libya, Nigeria, Jamaica, Syria, Kuwait, Bangladesh, Malaysia, Singapore and numerous other states. In some places such as Iraq the act carries the death penalty, demonstrating the unflinching manner in which heterosexuality is constantly legitimated and given credence through legal rights and everyday norms. Here such discourses may come to mark out specific identities and practices as abject or different, as we shall later witness in our discussion of Edward Said's (1995) monumental text *Orientalism* in Chapter 11, in which whole nations could be condensed into a 'geography of sameness' and rendered Other. Over time such taxonomies can come to carry an enormous amount of discursive weight.

This radical overturning of accepted norms has been highly influential in post-modern thinking. It suggests, first, that our ideas of truth and knowledge are socially produced and therefore historically constructed. And, second, that if our ideas of what constitutes truth and the social values we go by are historically specific, these ideas may change over time. The notion of truth as a discursive construct – an accumulation of unstable competing, complementary and sometimes conflicting knowledges – is part of the vitality of post-modern approaches. Here, truth is not seen as eternal but performs as a dynamic and unstable signifier. Knowledge is put to work through 'technologies of power' that are historically specific and driven through the live wires of institutions. Foucault's attention is with the discursive apparatus of the law, education, family, church or medical board (and we could now add global media) that serve to make these discourses intelligible. He argues that knowledge is inscribed with power and, while it may pose as formative of 'the truth', its power lies in its ability to make itself so. The *work* of discourse to produce and sustain selective knowledge has led Foucault to intermesh this into a couplet *power/knowledge*. Our commonsense understandings of work and education, crime and punishment, health or sexual relationships are not an observed reality but the outcome of an on-going discursive struggle for power. Here knowledge is seen to be embedded in a grid of power formed through a nexus of socio-historical relations.

What is interesting is that Foucault regards discourses not only as ways of knowing and understanding the social world, but as generative of values and particular ways of being. They form patterns and function as 'regimes of truth' combining the cumulative power of a host of institutional knowledges and procedures. In his volumes on the 'history of sexuality', Foucault mines what amounts to an archaeology of social production. He points to a 'discursive explosion' in the field of sexuality, as the state attempts to incorporate and regulate the irrational excesses of the sexual in order to make them knowable and identifiable within the rational order of the state. However, we would be mistaken in believing that this process of policing is entirely repressive. Instead, Foucault argues that discourses are *productive* wherein subjects are embedded in the strictly relational character of power relationships. He observes that 'points of resistance are present everywhere in the power network' (1976: 95) and come to be distributed in irregular fashion across a dense capillary of nodes.

In this way, while sexual practices between two people of the same sex have been a commonplace event throughout the centuries and across geographical regions, it was only in the 18th century that the term 'homosexuality' arose to describe, define and subjugate these forms of sexual expression (Weeks, 1981). In time this would lead to state laws banning homosexual activity, which was marked out as unhealthy, immoral and potentially threatening. Interestingly these laws were reserved for men, as lesbians were both inconceivable and largely invisible in the eyes of the nation state. However, such taxonomies are not simply repressive but have been highly productive in the political formation of gay identity today. As these 'spoiled' identities were brought to light, a number of individuals began to define themselves through the new discourse of sexuality available. Political campaigns centring on 'glad to be gay', 'out and proud' and new self-identifications with 'Queer' reveal how a negative epithet can be overturned and transformed into a positive signifier for individual and collective affirmation. These struggles have seen the Commission for Human Rights and Equalities in the UK incorporate legislation designed to avoid discrimination against individuals on the basis of sexual orientation. The Foucaultian impulse that 'power is everywhere' reminds us that where there is power there is also resistance. As gay pride movements and recent legislation have shown, discourses are not simply regulatory devices. They are productive and can incite new subject positions and bring them into being – or out of the closet. The changing ideas concerning sexuality within Western nation states is indicative of a post-modern contention that seemingly established truths are social constructs that may be contested, overturned and re-signified.

Jacques Derrida – deconstruction

So far we have seen how post-modern and post-structuralist ideas shake the foundations of 'truth' or 'reality' and question accepted orthodoxy. In developing the critical practice of interpretation, the French philosopher Jacques Derrida furnishes us with another critical tool for unpacking realism: *deconstruction*. Derrida draws upon the post-structuralist idea that the material world is a textual construction. If we were to think of a particular place – let's say Paris – we might consider how it is composed through an amalgam of 'signs'. These signs might include cinematic images of high fashion, perfume advertisements on billboards, romantic musical soundtracks, or tourist brochures promoting the individuality of Parisian cafes and the uniqueness of French haute-cuisine. Indeed for most tourists a visit to Paris would be incomplete without visiting the Louvre art gallery to see the Mona Lisa, or taking photographs of the Eiffel Tower. It is through these textual experiences that Paris is brought to life and written into our imagination. It is the social construction of Paris as a place, and the way in which we participate in the romantic narration of 'gay Paris' through associations with love, that are of interest here. In this rendering of Paris it is construed as a site of **jouissance**, an ecstatic form of pleasure that always exceeds the mundane.

However, following post-structuralist deconstruction, it is possible to engage with the textual making of Paris and read it 'against the grain'. 'Reading the text against itself', to paraphrase Terry Eagleton (1996), might enable us to explore both how this particular romantic image of Paris is constructed, and also how it comes to hide other narratives. For example, in recent years Paris has also been the site for some of the largest race riots in recent European history, and is equally renowned for serving up poor-quality food at exorbitant prices. This deconstruction suggests the dominant narrative fiction of Paris as romantic is partial, unstable and contingent. The unifying image of Parisian romance then conceals a multitude of deceits. The purpose of deconstruction here is not to get to a 'real', underlying, geographical truth of the city but to point to the fissures, fractures, gaps and absences in the narrative text that comprise what we know of as 'Paris'. Although post-structuralist forms of deconstruction are often seen as 'playful' or 'frivolous', we might also identify an anarchic tendency here, concerned with breaking-open established truths. It is worth briefly clarifying why Derrida's focus on deconstruction is so informative. He suggests that texts, images and other signs carry with them a surplus of meaning, so are open to multiple readings. These texts are not the accomplishment of seamless coherence and unity, but they are riddled with fragmentation, disunity and chaos. This means that they can be 'read against the grain', in the way that we have done with Paris, to explore contradictions, discontinuities and absences. In doing so post-modern criticism is imbued with a radical potential that is sometimes overlooked.

Perhaps the most cited phrase in Jacques Derrida's (1976) extensive philosophical repertoire occurs in his book *Of Grammatology*, in which he states, 'There is nothing outside the text' (p. 163), a statement that has led critics to celebrate and agitate about the future of post-modernism in equal measure. The quote is often misrepresented and has come to be associated with a type of 'linguistic terrorism' intent on turning away from the 'real' world in favour of words, philosophic theory and abstract **relativism**; criticisms we shall address towards the end of this chapter. But if we follow Derrida's argument more closely, we find that much of what is signalled is actually the limits of textual ways of knowing. Words, maps, sounds, images and other signs do not easily represent and communicate what an author or artist may intend. This asymmetry is part of what Derrida (1991: 62) refers to as 'the problematic of the sign and writing'. In this sense a text can be read as saying something quite different from what is intended and may even 'betray' itself. The text then, is not a closed circle for Derrida, where 'pure' truth can be achieved but is a space in which meaning is continually *deferred*. The endless deferral of meaning, where it can be extended, elaborated, reconfigured, contested or overturned, suggests an inherent instability in the text. For Derrida meaning is never complete, but is always in the process of deferral. As we have seen in Chapter 1, the imperial magnitude of geographical knowledge may now have diminished, but its effects live on in contemporary approaches to expeditions, fieldwork and geographical practice: a deferral of meaning that is now the grounds upon which post-colonial geographies are being assembled (Chapter 11).

Derrida deploys deconstruction to expose how words operate to secure meaning, but reminds us that these meanings are unstable and in a constant, slippery process of deferment. To better understand the deferral of meaning we might think about how news is constructed in contemporary media as an exterior reality. News is constantly presented as factual despite being subject to forms of media ownership and distribution, technologies of production, modes of regulation, editorial selection, and different forms of audience consumption and participation. Rather than offering an objective rendering of truth, news is, above all, a *mediated event*. It is composed by a number of human and non-human actors including editors, news readers, foreign correspondents, cameras, print media, journalists, sound engineers, audiences, digital audio equipment and so on. Moreover, different media forms and channels speak to one another – so producing an 'intertextual' narrative – which at first glance appears seamlessly stitched together. When we consider the number of people employed in global media, we perhaps should not be surprised that there is never a day of 'no news'. Rather, news is a thing that must be constructed from an *absence*. Furthermore, while it is impossible for journalists to be present at every event, somehow news coverage creates the fiction of 'being present'. For Derrida, though, such claims to reality are just another detour into deferral, achieved through 'sliding signs'. Although Derrida is speaking philosophically, we might consider how applicable his ideas are to contemporary media signs such as global news production:

> The sign represents the present in its absence. It takes the place of the present. When we cannot grasp or show the thing, state the present, the being-present, when the present cannot be presented, we signify, we go through the detour of the sign. We take or give signs. We signal. The sign, in this sense, is deferred presence. (1991: 62)

For Derrida the power of a sign to manifest as reality is dependent 'entirely upon the possibility of acts of repetition. It is constituted by this possibility. Its "being" is proportionate to the power of repetition' (cited in Derrida, 1991: 12). In this sense the iteration of news stories across various global media formats serves to make a particular event 'newsworthy', whether it is of enduring significance or not. It becomes a compelling means of giving substance to the insubstantial. We might think of deconstruction as an important circuit-breaker that disrupts the flow of endless repetition. Our example of global media is worth considering further as we now turn to a contemporary and contentious post-modern scholar, Jean Baudrillard.

Jean Baudrillard – simulation, simulacra and hyper-reality

In a post-modern world of signs, symbols and endless masquerade the post-modern theorist Jean Baudrillard argues that there is no singular reality to be found. Whether one is in a shopping mall, at the cinema or on Facebook, we seem to be surrounded by multiple realities, or what Baudrillard calls **simulation**. The once solid truths of modernism, he suggests, have 'eaten themselves'. In a time of endless simulation reality appears to have dissolved and all that remains is a vapid layer of surreal representations. The chilling realisation that Baudrillard leaves us with is not so much that reality is under threat, but that it has altogether disintegrated.

To begin to grasp this overwhelming statement it is worth pausing to consider the relationship between post-modernism, consumption and new technology. Both Lyotrad and Baudrillard are interested in the information age and the plethora of new forms of communication that saturate late-modern society, or what Bauman (2000) calls 'liquid modernity'. Today this media overload might include the proliferation of visual signs and aural signals that we receive from film, satellite television, newspapers, email, magazines, social networking sites, advertising, music, BlackBerries, mobile telephones and other advanced communication technologies. Indeed the weight of information has led Baudrillard to remark upon the alienating aspects of post-modern life and the way in which contemporary society is saturated with signs. These arguments were developed in his book *Simulacra and Simulations* translated in 1983. Here Baudrillard argues that the exchange of money for goods is no longer simply about 'use-value', need or social relevance. Instead he suggests that post-modern consumption involves the exchange of signs for other signs. No longer a material practice, post-modern consumption entails the production of economies of signs and space where objects are 'emptied out' of meaning (Lash and Urry, 1999).

To understand the virtual nature of these exchanges we might consider how a plastic credit card can be used to purchase a holiday online as one repertory of signs supplants another. Baudrillard's (1983) concern with the endless exchange of signs is particularly evident in contemporary forms of branding, where he articulates that the commodity has become the sign. Here the value of a product does not lie in its materiality (what it is), but in its signifying properties (what it signals). The way in which the McDonald's 'golden arches', the Nike 'swoosh' symbol and other global signifiers appear to have 'a life of their own' is testimony to a broader 'aestheticization of everyday life' (Featherstone, 1998: 65). We might think of the current fascination with designer accessories where the sign-value of owning a Chanel, Gucci or Prada handbag far exceeds the use-value of the bag itself when compared with other 'ordinary' handbags. The power of the sign only intensifies when it is subject to other forms of replication seen in the rise of global fakes where it is increasingly difficult to prise apart the authentic product from the simulation. For Baudrillard this split between the 'real' and the 'not real' is illusory; each item is as real or hollow as the other. In a sign-enwrapped world it appears that signs are continually slipping in and out of control.

In Baudrillard's writings there is a strong sense of post-modern culture as a site of disappearance. He uses the notion of 'cyberblitz' to suggest the ways in which we are constantly zapped or bombarded with electronic messages, so much so perhaps that our identities are now mediations of particular technologies, networks and electronic sign systems. The popularity of social networking sites is a good example of 'cyberblitz' where our online manifestations of self, signified through the number of friends, hits, motifs or status-ranking indicators, imparts a new reality. For many individuals this assemblage means there is no split between our virtual and embodied identities. Baudrillard suggests that this sensory overload has given way to endless simulation, image hallucination and a distorted **hyper-reality**, a key post-modern phrase we shall return to. The speed-up of media communication, then, engenders an acute acceleration of information and images, but to such an extent that we are now showered with a 'blizzard of signs', a cacophony of white noise which threatens to drown out communication and through which we struggle to be heard. Commenting on the increasing velocity and circulation of signs, Lash and Urry (1999: 1–2) state that, 'With an ever quickening turnover time, objects as well as cultural artefacts become disposable and depleted of meaning.' The feeling is that within the 'eye of the storm' reality and simulation fuse together in a 'depthless' culture where floating signifiers, artifice, affective intensities and simulation pervade. Baudrillard's premise is that signs no longer signify a hidden depth, but only give way to other signs and opaque surfaces. As Featherstone (1998: 83) concedes, 'the end of the deterministic relationship between society and culture heralds the triumph of signifying culture'.

So, Baudrillard confronts us with the proposition that where signs once concealed an underlying reality, in the post-modern world this is no longer the case. The speed-up of social relations through technology effectively means that reality has exceeded itself. Surface and depth, illusion and reality, simulation and authenticity are no longer distinguishable. Baudrillard uses the example of Disneyland to demonstrate this immateriality. His contention is that Disneyland presents itself as an artificial fantasy space to conceal the fact that, 'all of "real" America . . . is Disneyland' (Baudrillard, 1983: 25). For Baudrillard there is no exterior reality to fall back upon, instead places such as Disneyland, Dubai and Las Vegas only serve to remind us that the everyday world is not so much real, but '*hyper-real*'. Similarly, films such as *Memento*, *Inception* and the *Terminator* film and television series offer representations of hyper-reality that may briefly make us forget the hyper-realness of our own daily experience even as we participate in the more-than-real spectacle of cinematic viewing. Our encounters in theatres, supermarkets, museums, social networking sites or nightclubs are then no more or less real than the pseudo-experiences associated with the fantasy world of Disneyland; all are tinged with a hallucinogenic hyper-realism. As we participate in this mediated world, Baudrillard contends, reality implodes: it collapses into a system he terms *simulacra*. Indeed, the flawlessness of Disneyland and other glossy simulations are for Jean Baudrillard an effect of the 'precession of simulacra'. At this point the gap between fantasy and reality is closed, forming a loop through which signs endlessly circulate detached from material objects.

Baudrillard's thesis may make our heads spin, so let us pare down his ideas. He suggests that our encounters in a post-modern consumer society are frequently ephemeral, transient and fleeting. Events and objects become devoid of any real meaning or 'use-value'. The endless simulacra Baudrillard (1983) identifies suggests a world of appearances where 'depthless' reality, 'empty signs', 'floating signifiers' and 'unanchored meanings' abound. This is particular to the information age where the line between reality and representation is eroded. When this blurring happens, signs are no longer tethered to objects and things, but they constitute that 'reality' as they appear to have a life of their own – they are hyper-real.

Baudrillard's most contentious illustration of this is found in his media analysis of the initial occupation of Iraq in which he infamously declared 'the Gulf War did not take place' (1995). In this much cited but often misrepresented quote, Baudrillard is not questioning if the event occurred. Rather his focus is upon the endless simulation used to masque the visceral nature of armed conflict. Media coverage of the Gulf War developed its own synthetic vocabulary, referring to 'smart bombs' and advanced weaponry that could make 'surgical strikes' on Iraqi installations without harming the civilian population. It was only after the 'shock and awe' bombing of Baghdad that some of the 'surgical strikes' alleged to have taken place on ammunition factories were revealed to have destroyed hospitals, public institutions and killed thousands of citizens. Baudrillard's essay goes on to consider the virtual qualities of conflict-at-a-distance that see fighter pilots guiding computer-generated missiles at targets on video screens in an ultimate act of simulation.

Baudrillard's point is that we also participate in this hyper-reality and give it meaning. The media broadcasted images, rolling news and supposedly 'live' recordings of action as if to affirm that these events were taking place after the fashion reported. But for most people far removed from the conflict, these 'media-scapes' became the 'event'. Reality could not then be set apart from representation; it had collapsed allowing simulation and media signs to pervade. In some respects Baudrillard is a handmaiden to the 'post-representational' theories we discuss in Chapter 12. For him and other post-modern thinkers working on the 'cultural turn', this phantasmagoria of parody, pastiche and simulation offers a final realisation: reality is disposable.

You may not be surprised to know that the disposable, 'sign-saturated' society that Baudrillard describes has been subject to fierce critique. It is worth noting here that a number of post-modern writers, including Baudrillard, grew out of the political left. He gradually became vilified by sections on the left as he became increasingly disillusioned with the revolutionary claims of Marxism in an age that had witnessed the global triumph of capitalism with the collapse of the Soviet Union and its allied socialist states. Baudrillard rejects the dogged focus on economic relations proposed by Marxists and is more entranced by the powerful spectacle of global media and its intimate connections to consumption. As he grew increasingly acerbic towards Marxist formulations of class, he in turn became a target for left-wing criticism, notably from war protestors and political activists who were outspoken about the rendering of war into a vacuous hyper-reality. These critics vehemently attest that, in war, bodies are blown apart and people are maimed, while families endure death and grief. The idea that reality can be discarded is, for many of us, despicable.

Baudrillard may be cavalier in discarding the depth of emotional geographies connecting us to events, an issue we unwrap in Chapter 12. He may also underplay the potential for political resistance. And we would further attest that he risks writing out some of the materiality of everyday life through the use of ethereal metaphors, giving recourse to an endless flow of signs and simulation. Despite these observations we remain in little doubt that Baudrillard's dark reflections on contemporary post-modern society are penetrating. The idea that post-modernity is filled with an excess of simulations, simulacra and a hyper-reality is not so far removed from Marxist criticisms of consumerism and the sentiment that 'all that is solid, soon melts into air' (see Chapter 4). However, while Marx believed in a material reality that was being 'concealed' through the spectacle of capitalism and processes of commodity fetishism, contemporary post-modernists such as Baudrillard recoil from any idea of an ontological truth. For Baudrillard we now exist in a high-tech hyper-reality in which there is too much reality, so much so it has undermined itself. Post-modernism is then associated with forms of simulation and parody in which events are seen as constructs rather than stable objects for analysis.

Summary: Reality and its Other

This section has shown how in different ways philosophers are challenging our modernist sense of reality. If Lyotard draws our attention to the prominence of *meta-narratives* in society then Derrida uses the post-structuralist technique of *deconstruction* to expose their 'inner workings' and 'break open' this apparent unity to reveal how meanings are unstable and continually in a process of *deferral*. These writers first reveal, then take apart the master narratives of modernist society. Foucault, on the other hand, is concerned with how these truths come into being and the ideological processes that enable certain ideas to be given more weight than others. He considers the disciplining effects of *discourse* and how *power* operates through repertoires of knowledge that govern and *produce* modern subjects. Finally, Baudrillard explores the contemporary condition of post-modernity, aligning it with a proliferation of *media signs*, the flourishing of technology and a mass consumerism that has given way to a 'hollowing out' of authentic encounters which are displaced by a hyper-reality. Each of these writers inform us of how modernity was a quest for rationalism, objectivity and truth but the point at which this unintentionally takes us to is ultimately the end of reality, at least as we know it. Above all these visions present us with ways of seeing post-modernism as a radical movement that disturbs, disrupts and shakes open 'objectivity' and 'reality' whether we like it or not. The question of what all this means for geographical thinking is discussed on p. 221 of this chapter.

Post-modernism: a stylistic phenomenon

Having explored two understandings of post-modernism – historically and as critical practice – a third way is to see it as a stylistic influence that can be traced in a series of different genres as seen in the composition of music, film, advertisements, literature, comedy, sculpture, paintings and architecture. To illuminate the stylistic practice of post-modernism we shall briefly consider examples from the fields of art and architecture. Through these illustrations, as with Hip-Hop, there is a tendency to be 'referential'; often referring back to the past or to different genres in ways that are knowing and even ironic. This artful practice may be considered a form of *pastiche*. The act of pastiche acknowledges its own pretence and involves mockery, playfulness and simulation. Pastiche is a self-aware practice that may engender a 'hollowing out' of form, content and convention. It can also involve the practice of sticking together different materials in a novel expression of creativity, as we saw in our example of DJs who 'cut-n-mix' vinyl to create new soundscapes. The enactment of pastiche, then, serves to debunk some of the Western Enlightenment ideals of scientific truth and authenticity, where 'reality' is shown to be just another representation in a supermarket of signs. The idea of pastiche also implies a type of re-making in which elements from the past may be mixed into the present. A feature of such pastiche is that it is capable of mocking, uprooting, distorting and dispersing tradition.

Art, commercialism and the cult of celebrity

When Marilyn Monroe, the 1950s matinee idol of the silver screen, committed suicide in 1962, few would have referred to her as a post-modern icon. Yet today her influence can be seen in the world of art, music and film. Following the star's untimely death, artist Andy Warhol produced a series of stylistic portraits commemorating her iconic status. Warhol's images are crisp, bright and bold (see photo). In a recorded interview Warhol explained the process and motives underlying his glossy reproductions.

> In August 62 I started doing silkscreens, I wanted something stronger that gave more of an assembly line effect. With silkscreening you pick a photograph, blow it up, transfer it in glue onto silk, and then roll ink across it so the ink goes through the silk but not through the glue. That way you get the same image, slightly different each time. It was all so quick and chancy. I was thrilled with it.

In many respects Warhol's 'pop art' marks a post-modern intervention into modern life. It does so by holding modernism and post-modernism together in a creative tension. So, how does Warhol's post-modern art challenge the ground rules of modernity?

To begin with, Warhol breaks with many of the rule-bound conventions that were associated with modernist art forms. By engaging with the cult of celebrity, his prints inform us that artistic expression is everywhere and does not have to be sited within the confines of a gallery. In championing the artificial and superficial, Warhol reveals a post-modern interest in the act of representation. It is often said that while Warhol's art makes known the illusion of stardom, it has also preserved the cult of celebrity through mimicking the flawless features of Monroe familiar in film and magazines. There is then a dual aspect to Warhol's post-modern images – they are iconic and ironic. Warhol's famous remark how, 'In the

Andy Warhol's silkscreen print of Marilyn Monroe

Source: Bridgeman Art Library Ltd / Private Collection / Alinari Private Collection / Alinari / © DACS, © The Andy Warhol Foundation for the Visual Arts / Artists Rights Society (ARS), New York / DACS, London 2011

future everyone will be famous for fifteen minutes', today seems a prescient statement given the reach of global media and mass communication.

Second, at a material level Warhol subverts commonly held notions that works of art should use familiar items such as paint, canvas or sculptural materials that can be shaped and moulded accordingly. Instead, he uses silkscreen to offer synthetic representations of subjects, a technique he also deploys in images of other 1950s icons such as Elvis Presley and Jackie Onassis. The non-representational colours of silkscreen – a popular technique for mass production – do not reflect the artist's inner feelings, but gesture instead towards the exterior world of popular culture. The Marilyn Monroe series is designed to generate different moods and sentiments. Indeed, in later images of Monroe, Warhol fades out her visage, rather like a camera would at the end of a film, in what is perhaps a fitting salute to her passage from life into death. It is worth noting here that Warhol cut his teeth in the advertising industry and regularly brings together art with commoditisation. Having been an artist, ad man, record producer, photographer and film-maker, Warhol often uses these practices to create a collage effect that is typical of a post-modernist 'scrap-book' style. In this sense post-modern artefacts are often regarded as ambivalent products that playfully react to the solemnity of modernism by embracing commercialism and the popular.

Third, Warhol challenges the notion of representation itself. The recurring image of Marilyn Monroe is actually taken from a publicity shot by Gene Korman for the film *Niagra* (1953), which was directed by Henry Hathaway. In this sense Warhol forges a copy from another copy, which he seamlessly replicates in a series of montages of Monroe. Here we find a form of inter-textuality – a knowing dialogue between different texts – seen through the filmic image and Warhol's print where the audience can realise how both mediums speak to one other. In setting up this dialogue Warhol transposes the art of celebrity from silver-screen to silkscreen. This is less a case of art imitating life but art imitating art. Both are mediated forms that close the spacing between 'reality' and 'representation' to the extent that we are only left with a series of representations, copies of a copy. A familiar charge aimed at Warhol and other post-modern artists remains, 'is this art'? Indeed, by getting audiences to question the very grounds upon which art and creativity are understood, post-modernism succeeds by raising bigger questions about the worth and reality of things in society. Ultimately his work undermines the very frames of reference it is positioned within.

In later silkscreen prints Warhol extends the myth of Monroe by reducing her iconic image simply to the smile, that small part of celebrity that can come to stand in for the whole. Although Warhol's prints are designed to have a modernist factory-based feel to them, interestingly they are marked by imperfection. By drawing attention to the artificial nature of images, Warhol is less concerned with achieving artistic integrity and originality. Rather than being blemish-free, he reflects how each of the prints are 'slightly different each time'. This difference is heightened when Warhol uses alternative shades, tones, colours and backgrounds to show Monroe, literally, in a different light. Some of his prints contrast vivid colours with faded black and white photographs to reveal a fascination with celebrity and mortality. Warhol openly admits to the ephemeral and fleeting nature of his portraits, which, like fame itself, he depicts as 'quick and chancy'.

So what can we learn from Warhol's homage to Monroe about post-modernism? It is noticeable that Warhol is playfully responding to modernism, while at the same time recognising that his work grows out of this tradition as expressed in the 'assembly-line effect' his prints generate. Warhol is as fascinated with the popular and 'trashy' world of celebrity as he is with the profoundly deeper questions of mortality. At the same time his work is self-parodying, ironic and even schizoid. Even so, in the commercial marketplace Warhol's pop art, in all its gaudy, glossy splendour, sells for millions of dollars. Furthermore, his timeless silk prints appear to signal the end of originality, or at least the myth of what 'truth' and 'authenticity' might come to represent. Instead they superimpose and celebrate the fictional, illusionary and artificial. As we shall go on to find, many of these features – parody, cultural consumption, simulation and so forth – characterise post-modernism as an eclectic movement that has radically influenced the arts, humanities and social sciences. We shall return to a number of these themes later in this chapter, for they offer some of the most profound and destabilising aspects of post-modern theory.

Architecture and the built environment

One of the most visible ways in which the influence of post-modernism can be seen is through its material expression in the built environment, as our case study of Vancouver reveals (see Box 9.1). Here it wilfully attempts to offer a counterpoint to the uniformity of modernist architecture where its functional design is epitomised in towering monuments of steel and concrete. The modernist approach is famously apparent in the sleek designs of Le Corbusier and the 'machine-age aesthetic' (Ley, 1987: 40) he gave to buildings, in his desire to fashion them as usable engines of habitation. Le Corbusier's radiant highways and high-density tower blocks became a blueprint for modern living and are characteristic of many post-socialist cities across the world where there is an emphasis on utility, being economic and functional in presentation. These aesthetics also found favour across many British cities in the 1960s and can be seen in the urban planning of T. Dan Smith in Tyneside in his quest to build high-rise, futuristic 'cities in the sky'. Smith, a prominent Labour politician and head of Newcastle City Council (1960–1965), sought to make Tyneside into the 'Brasilia of the North', through modern high-rise buildings for living and working. This grandiose modernist vision has since been dismantled, along with many of the buildings, as many people felt alienated from these featureless, geometric monoliths. Even so, there are many aspects of modernist architecture that have retained their appeal including large, open public spaces, ideas about the garden city and the use of buildings with large windows to capture natural light.

If parts of modernist architecture can appear somewhat sombre and formulaic, post-modern design is less driven by convention and often eclectic, mimicking styles from previous movements and moments in time. As Cloke *et al.* (1991: 175) surmise,

> the key manoeuvre has been to abolish the homogeneity and the sheer blandness of modernist architecture, and in so doing to replace the mammoth slabs of gleaming white concrete with an accent on variety, colourfulness, attention to detail and the deliberate mixing of building (and ornamentation) styles from any number of sources.

Post-modern architecture is concerned with the social and physical manifestation of style and how it interacts with different subcultures and communities. Where modernist planning is thought to be impersonal, post-modern form has shown greater sensitivity to the local histories and geographies that create our 'sense of place'. The emphasis here is

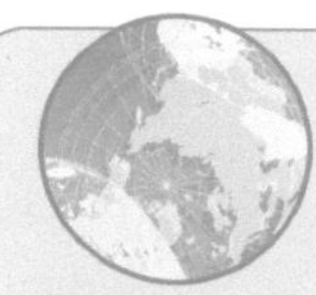

Box 9.1
Modernism and post-modernism: the city as a canvas for ideological struggle

In the inner area of Vancouver Canada, David Ley (1987) records the different attitudes to architecture, planning, design and landscape. He reflects how these styles emanate as the result of broader political and cultural processes influenced by differing ideologies; in this case namely liberal and neo-conservative. Ley discusses how parts of Vancouver's formative modernist planning was overturned by a group known as The Electors' Action Movement (TEAM). They comprise young professionals, teachers, architects, university lecturers and lawyers who mobilised to be elected and then came to form a majority on the planning council. In the area of False Creek Southside, TEAM affect what Ley regards as a 'post-modern strategy of careful place-making' (p. 45) where there is a focus on 'the livable city' and 'quality of life'. This liberal perspective becomes 'expressive of a cross-section of local urban subcultures' (p. 42). In a concerted attempt to foster greater diversity mixed housing initiatives come to include rental apartments, family townhouses, studios, cooperatives, privately owned condominiums and non-profit building for the elderly or handicapped. This attempt to cultivate a post-modern aesthetic through pluralism was to see a mosaic of groups enter into housing enclaves which were each 'differentiated by colour, materials and design' (p. 47). Ley is able to show how the material and symbolic aspects of landscape can be utilised to develop a particular 'sense of place'.

Ley contrasts the imaginative False Creek initiative with a subsequent project begun in 1981 in British Columbia Place, incorporating the Expo 86 site. This later scheme is governed by the ideology of neo-conservatism and designed around high-density living, commercial districts and what he regards as the modernist consolidation of mass society. Where the south shore presents an expressive liberal landscape, the north is given to instrumentalism. There is a market orientation to high-density living, profit, mega-structures and expansive highways. In comparing these aspects of modernist and post-modern development in the north and south side respectively, Ley seeks to peel back the ideological layers and cultural processes that go into the making of place.

False Creek also incorporates Granville Island which exemplifies a post-modern tendency to develop culture in post-industrial landscapes. Granville Island includes a local brewery, a covered farmer's market selling fresh produce, craft stalls displaying 'ethnic' goods, theatres, a hotel and an art college and studios for architecture and design. In the post-modern moment culture is designed into space and is everywhere (see Chapter 5). Parks, marinas and routes for cyclists and pedestrians come to the fore as automobiles are fittingly buried in underground car parks. Although Ley is largely positive about such post-modern cultural regeneration – and the welding of the 'old' into the 'new' – he questions if this fascade serves only to enhance the middle-class elitism of creative spaces.

Modernist architecture is seen to be homogenerous and overbearing, where post-modern initiatives allow for eclecticism, experimentation, local variation and difference. It is these qualities that form the signature for celebratory accounts of post-modernism and hold open the potential for more pluralistic ways of city living. Ley's discussion of Vancouver's landscape 1968–1986 informs us how modernist (neo-conservative) and post-modernist (liberalist) visions are struggled out – in what Don Mitchell (2000: 3) calls 'culture wars' – and made manifest in the built environment. Evidently modernism and post-modernism are not mutually exclusive phenomenon but co-exist in architectural and ideological tension. Post-modern style has emerged out of modernism and the metaphoric, symbolic and material aspects of each moment can unfold through different geographies of power.

upon 'people places' and a return to the embodied use and affective attachments different user-groups might have with places. At this level of design post-modern architecture may use renovated or recycled materials and in terms of structure re-shape our perceptions of what a book shop, art gallery, office block or residential home might look like. Selfridges department store in central Birmingham, with its smooth curves and shiny chrome disks, appears at once like a sculptural work of art, a flying saucer or a humorous, if wilful,

Selfridges Department Store, located in the Bull Ring, UK
Source: iStockphoto / Chris Hepburn

resistance of modernist iconoclastic austerity. Furthermore, the shape invites the spectator to be drawn into the sensual and tactile properties of the building, perhaps encouraging connection rather than alienation from the built environment. As with our illustration of Hip-Hop culture, such buildings defy immediate classification and appear to turn the unspoken rules of hard modernism upside-down. The stylised post-modern challenge then subtly uses pastiche, *bricolage* and melange to disrupt tradition, hollow out reality and render it disposable.

Geographical engagements: theory, method and practice

Post-modern geographies draw upon the ideas and type of critical thinking described above. The influence of post-modernism in geography can be traced at the level of philosophical theory, method and practice.

Theory

Theories of post-modernism set out to disrupt the flow of geographical knowledge that defines, classifies and tabulates information and packages it as 'fact'. Geographers working in a post-modern vein unwrap this packaging to consider how particular meanings come to be accepted, and the silent cartographies that are hidden in the process. Here we might reflect on previous chapters in the book which recall how women's experience is 'written out' of geography's traditions (Chapter 6) or the colonial assemblage of knowledge that rides roughshod over native people's beliefs and defines them as inferior, uncivilised, Other (Chapter 1). Bringing these alternative geographies to bear upon assumed 'geographical facts' is then a formative part of a reflexive post-modern practice. For these writers, 'Geography is cracked, fissured and fractal' (Doel, 1999: 103). As Steven Flusty (2005: 170) surmises, this manoeuvre towards multiplicity meant 'geography was pluralised: it had become geographies'.

Where much of human geography has relied upon the production of distinct categories for analysis, post-modernists suggest that this modernist way of thinking is centred upon Western scientific ideals and encourages simplistic, 'flattened out' explanations of social processes. In contrast to the totalising visions that have gone before, post-modern geographies appreciate the 'messiness' of geographical research and do not seek to write out ambiguities, contradiction or ruptures in the text. Rather than claiming scientific objectivity, these accounts openly profess to be partial, forever incomplete renderings of social life marked by 'difference and disjuncture' (Appadurai, 1990). This doesn't mean that post-modernist explanations are weak or lack rigour, but that they recognise the socially constituted production of knowledge and are open to counter-explanations and future elaboration. Post-modern writing remains, then, theoretically open to the possibility of multiple and ambivalent meanings in research. In considering knowledge production in this way, most contemporary human geographers now ask – who makes 'knowledge', how, why and under what conditions – resulting in deeper engagements with power and social context.

At a theoretical level, a number of geographers has turned to post-modern and posttructuralist philosophers in order to develop their spatial accounts. In a prescient essay, Lawrence Berg (1993) remarks how the insights generated by post-modern theory are both 'positive' (p. 498) and 'valuable' (p. 497), allowing geographers opportunities to work between the lines of modernism and post-modernism. This, he claims, has led to a greater awareness of positionality and the situatedness of knowledge fields. Berg warns against seeing modernism and post-modernism as opposing forces, and usefully points to the connections and overlaps between these movements. Moreover, he contends that post-modern debates in the discipline 'increased sensitivity to space and place' (p. 490). In a similar vein, Marcus Doel (1999), in his book *Poststructuralist Geographies*, discusses how thinkers such as Baudrillard, Derrida, Lyotard, as well as Deleuze and Guittari, can enable us to develop a 'splayed out' geography that better appreciates the infinite plasticity of space. The use of diverse theories displays 'an openness both to a wide range of theoretical sources and to the very definition of theory' (Barnes, 2001b: 547) in the creation of an 'unhinged geography' (Doel, 1999).

The theoretical influences of post-modernism are noticeably apparent in the works of some feminist geographers (see Chapter 6), as seen in Gillian Rose's (1993) work on feminist theory and identity. Rose usefully draws upon post-modern ideas to demonstrate how feminist politics based around a shared category of 'woman' cannot be homogeneous but must account for different positionalities with regard to class, nationality, ethnicity or sexuality. A similar orientation is also apparent in the writing of Jane Gibson and Katherine Graham who seek to offer alternative perspectives from which to bring into being the numerous 'diverse economies'. This includes exploring work practices which traditionally lie outside the formal economy such as housework, care work, reciprocal labour, voluntary and community activities, and ethical or illegal markets. By drawing on a post-structuralist notion of economic geography the traditional (and often masculine) focus of work as tied to capitalist monetary exchange can be displaced in favour of a pluralised account. Interestingly the collaborators use the moniker J.K. Gibson-Graham (2005) in their work in what amounts to a post-modernist blurring of Self and Other through the formation of an indeterminate identity, a connection found in our opening example of Hip-Hop artists. The joint name attempts to honour their collaboration and politically undermine the established hierarchy of academic convention that gives prominence to first-named authors. This playful gesture shows how post-modern 'language games' can inject an element of chaos into the modernist tradition of academic inquiry. Such manoeuvres fit with Doel's (1999: 103) appeal for a 'scrumpled geography' that avoids the geometrical linearity of modernist, precision logic.

A number of feminist geographers in particular are drawn to the politics of 'deconstruction' as a means to move beyond gender binaries and dismantle established categories. Above all, post-modern readings emphasise the *plurality* of subjectivities – masculinities and femininities – and in so doing pay attention to similarity as well as difference. In doing so such work points towards the multiple and fragmented habitation of gender within society. A post-modernist intervention has been to regard identity as mutable rather than a meaningful unit of analysis. Instead of seeing 'masculinity' and 'femininity' as stable identities from which other actions follow, we might consider the habitual ways in which gender is performed through work, consumption and leisure (McDowell, 1997; Gregson *et al.*, 2000; Nayak and Kehily, 2008). The idea of performance suggests that gender is an incomplete project, forever in the making. Rather than being fixed, gender is only tenuously secured in relations, processes and power-laden projects of self-identification. This recent work reflects an important shift away from regarding sex and gender as 'objects', towards a more active understanding of gender as a 'doing' (Butler, 1990), an issue addressed in Chapter 6.

Method

The influence of post-modernism can also be traced in human geography methods and approaches. Since the 1990s there has been a spiralling number of qualitative techniques, methods and approaches to the study of geographical phenomena. In eschewing notions of 'scientific objectivity', qualitative approaches tend to offer contingent and context-bound readings of geographical processes. Such has been the proliferation of qualitative methods within the discipline that Mike Crang (2002) has questioned whether these approaches now comprise a 'new orthodoxy'. The plurality of methods, approaches, techniques and philosophies is indicative of a broader post-modern sensibility that may incorporate multiple methods and styles of inquiry. In the process this may present us with different forms of data and even contradictory accounts. Rather than silence these ambiguities, post-modern geographers are willing to discuss and direct our gaze to anomalies in their writing, and that of other scholars. Through these 'textual interruptions' post-modern geographers can draw attention to the text itself as a social construct. Since post-modernists generally believe that all knowledge is the product of a given context and situation, this does not invalidate the responses but is felt to lead to a more open form of interpretation that allows multiple and occasionally dissonant voices to speak.

At the level of method, post-modernism exerts an enduring influence upon geographical practice. Michael Dear established a forum for critical debates in post-modern theory and method when he became the initial editor of the journal *Environment and Planning D: Society and Space.* He remained entranced by the possibilities of post-modern ideas but, in suitably modernist fashion, seemed fearful of its extremist deconstructive tendencies to anarchy and reletivism (Peet, 2001 [1998]). Today few human geography practitioners now claim their accounts reflect universal truths, are securely objective or remain closed-off to other readings/meanings. This suggests that geographical papers should not be treated as coherent, finished products beyond reproach. This open-ended nature of inquiry is a feature of post-modern understandings. However, post-modernists remain sceptical of social science conventions and are wary of the formation of an established canon where deference is paid to particular bodies of work and the 'founding fathers' of a discipline. This critical reflection does not aim to disregard this work as irrelevant, but rather point to the social construction of disciplinary knowledge and the manner in which these traditions or literary canons come to shape future ideas.

Post-modernists further suggest that if the world is a place where chaos, disorder and ambivalence abound, then 'pure' or 'hygienic' explanations will not suffice. Rather than iron out difference, overlook complex social relations or present homogenising accounts, post-modern approaches lean towards 'heterogeneity, particularity and uniqueness' (Gregory, 1989: 70). This appreciation of 'difference' explores how relations of gender, ethnicity and sexuality might make us rethink geographical practice and subvert the 'master narratives' of modernism. Post-modern researchers pay close attention to their own positionality and how our gender, class, sexuality, age or ethnicity may shape the research process. For example, the white feminist geographer Gillian Rose (1993: 15) remarks on the multiple subject positions she holds: 'my own position is empowered by my whiteness. I may feel marginalised in geography as a woman, but my whiteness has enabled my critique of geographical discourses by allowing me to get close enough to them to have a good look.' The belief here is that as researchers we do not impart a 'gaze from nowhere' (Haraway, 1991: 188), but are always already positioned within networks of power. For Donna Haraway, objectivity is an illusory position that hides and protects the interests of those privileged to benefit from this stance. Post-modernism, then, offers a way of imploding this power-base of objectivity by alerting us to the fact that all knowledge is subjective and constituted through and by relations of power. In order to gain deeper insight into the spatial practice of post-modernism we shall now turn to examples of the post-modern city.

Practice

One of the more noticeable areas where post-modernism is practised is in the field of urban geography. To examine how different understandings of post-modernism can be used to interpret cityscapes, we can consider the city of Los Angeles (LA) in the US, which is an exemplar of post-modern processes and has been subjected to a refined geographical analysis (see, for example, Davis, 1990; Soja, 1995, 2000; Dear and Flusty, 1998). In the voluminous work completed on LA, it is worth drawing attention to some notable post-modern characteristics. LA has been restructured from its former industrial base into a site for high technology, large craft-based industries, finance, insurance and real estate. It has become part of a select pool of 'world cities' that includes New York, London, Tokyo, Hong Kong, Paris, Sydney and Sao Paulo. It is a city that has dramatically expanded, whose spokes radiate outwards forming new distinct areas and trans-urban connections. Increased social polarisation has been accompanied by more complex social class formations that include corporate managers, yuppies, IT whiz-kids and ethnic entrepreneurs, as well as low-paid Latino workers and female employees, domestic servants, a stagnant urban underclass and an expanding homeless population. In this sense LA reflects the historical shift to post- or late-modern society where new jobs and inequalities abound.

A feature of post-modern cities is their inherent diversity. Like many world cities, LA is richly multicultural, with around a third of its nine million residents foreign born, engendering various ethnic enclaves and mixed-areas to preside. Some of these multicultural spaces are generative of what Jencks (1993) terms a type of 'hetero-architecture' where pluralism, hybridity, border transgression and the formation of new languages, habits and practices teem forth. While some post-modernists celebrate difference, such diversity has also seen the 'social mosaic of the post-modern city become more kaleidoscopic, with . . . new forms of inequality and geographically uneven development' (Soja, 1995: 133). Mike Davis's expansive book *City of Quartz* (1990) goes on to document how LA has been transformed into a fortress, or 'carceral city' (Soja, 1995, 2000) in response to a political economy of drugs. Here, gated communities policed by armed guards, high-security shopping centres and high-tech spatial surveillance now form a part of a panoptical, 'postmetropolis experience' (Soja, 2000). Filtering into the post-modern dream can then be found dystopian shadows of apocalyptic nightmares. The beating of the innocent black teenager Rodney King by the Los Angeles Police Department, and the ensuing race riots in 1992, bring to the fore the deeply racialised cartographies of power that splinter post-modern cities. Ed Soja captures this tension:

> The new topography of race, class, gender, age, income, and ethnicity has produced an incendiary urban geography in Los Angeles, a landscape filled with violent edges, colliding turfs, unstable boundaries, peculiarly juxtaposed lifespaces, and enclaves of outrageous wealth and despair. (1995: 134)

What is noticeable in such post-modern readings of cityscapes is that they avoid rendering people or place as 'objects' that have fixed qualities and are internally coherent. Rather these approaches look at the complex social relations that define places and communities, regarding them as embroiled in a continual process of becoming in which new and sometimes surprising exchanges and encounters may occur.

A further aspect of the post-modern city is an unnerving sense that so much of what it comprises is hyper-real, as the city is composed through arbitrary signs, symbols, images, artifice, advertising, myth and ephemeral experiences. The 'unreal' qualities of LA are epitomised by Hollywood, Disneyland and the glamorous lifestyles paraded in Beverley Hills. As the title of the soundtrack *California Dreaming* intimates, much of the region is steeped in the unreal – sex, surfing, hallucinogenic drugs, hippies, beat-nicks and endless blue skies. This hyper-real quality is seen where the California State is now fittingly governed by one-time 'action hero' Arnold Schwarzenegger who has stepped out from the Hollywood cinematic screen into 'real' life. Baudrillard's radical claim that in a post-modern world reality is disposable appears prophetic – 'California dreamscapes become simulacra' (Dear and Flusty, 1998: 56). Have a nice day.

In the moment of simulacrum, marketing directors, self-help gurus, porn stars and surfer dudes preside amid 'sound bites and spin doctors, virtual reality, cyberspace, cyberpunk' (Soja, 1995: 135). In a place famous for escapism, F.J. Turner's 'frontier thesis', which recalls how the American frontier remains a defining presence in US life through its associations with infinite space and spiritual 'freedom', carries forward into everyday life. In Southern California the frontier myth can be applied to the economic horizons fantasised by Mexican immigrants hoping to cross the state border, aspiring wannabe movie stars, junkies on a trip, surfers looking for the next big wave, gang warlords and countless others mesmerised by the spectacle of the American Dream. Everyone is trying to become someone else, or so it seems, as image supersedes reality and hollows it out. This post-modern 'project of the self' (Giddens, 1991) is worked out not only in imaginary form, but is embodied in the corporeal transformative processes familiar to a region where sun-tans, dietary regimes, body-building, teeth whitening, breast enhancement, 'make-overs' and a cornucopia of surgical procedures are commonplace enough to have become a norm. In the intense space of simulation and simulacra what is 'plastic' and what is 'real' is indistinguishable; once again there is no 'real', only copies modelled on other copies of pseudo-perfection. People and places in our case account of Southern California appear composed of 'multiple layers of abstracted representations and surface reflections that can only be played with endlessly' (Flusty, 2005: 171). Confronted by the spectacle of simulacra it is impossible not to participate in the fiction and lend life to the fantasy spaces of the contemporary post-modern city.

Post-modern criticisms

So far we have seen how post-modernist ideas can be used to enrich our 'geographical imaginings' but their impact can be traced in various sub-disciplinary areas. For example, social and cultural geographers have made wide use of these approaches through theoretical observation exploring the activity of representation as opposed to subscribing to notions of 'reality'. The influence of post-modern perspectives can also be seen in post-structuralist feminist geography with its insights on power, diversity and multiplicity (Gibson-Graham, 2005; Rose, 1993); in political geography with a view to developing a project along the lines of a discursive popular geopolitics of media signs (Dittmer, 2005; Dodds, 2006); and in attempts at 'retheorizing economic geography' (Barnes, 2001b) from the ground up through considering the multifarious cultural inflections that make up the economy. Human geographers have also made use of these ideas through methodological application – considering positionality, reflexivity, and the social production of knowledge in research. Although the take-up of post-modern ideas has been somewhat piece-meal in the discipline, many aspects have been embraced with little reference to post-modernism itself. This perhaps informs us of the success of at least some post-modern interventions where certain ideas have become the norm.

Despite this renewed awareness to the 'doing' of geography, a number of geographers remain sceptical of the post-modern turn and are deeply critical of post-modernist practice. These include those working in the fields of public policy (Hamnett, 2003), quantification (Fotheringham, 2001) and Marxist radical geography (Peet, 2001; Harvey, 1995) who collectively have turned post-modernism into something of a 'whipping boy'. In a consciously provocative commentary Chris Hamnett (2003) shows concern that post-modernism might render geography obsolete through its lack of relevance and policy use. Echoing a view shared by some quantitative geographers such as Fotheringham (2001), Hamnett (2003) goes on to argue that new generations of geographers risk turning their backs on quantitative methods in favour of what he sees as an 'ad hoc qualitative impressionism' (p. 2) derived from various interviewing methods allied to discourse and textual interpretation. Both writers worry about the future of the discipline, the rise of qualitative methods generally and the tenacity with which philosophic approaches critique realism. Although there is something defensive about some of these reactions, a number of interesting critiques about post-modern geography have erupted that engage more closely with its philosophic tenants.

A familiar criticism launched at post-modern ideas is that they lack the capacity to 'change the world'. In this sense post-modernism has been seen as an empty gesture that is ultimately apolitical. The geographer Richard Peet remarks on the debilitating characteristics of post-modernism: 'While suggestive of a politics of difference, post-modernism is inherently and deliberately incapable of producing the necessary alliance or network of organizational ideas or institutions' (Peet, 2001: 216). If we follow Peet's line of argument it seems that post-modernists, for all their posturing, are incapable of translating radical ideas into radical action. Hamnett (2003: 2) makes a similar claim when he suggests that 'in reality' (note the thoroughly modernist phrasing) 'there has been a retreat from substantive political engagement and social analysis in favour of superficial academic radicalism'. His point is that much of what comprises post-modern geography is merely cosmetic radicalism articulated in academic writing rather than social activity. The Marxist geographer David Harvey is even more direct. In a vitriolic attack upon the political limits of post-modernism, Harvey declares:

> The rhetoric of post-modernism is dangerous for it avoids confronting the realities of political economy and the circumstances of global power. (1995: 117)

Harvey's point is that post-modernist philosophy with its emphasis on pastiche, illusion and the discursive dissemination of signs is blind to the material 'realities' of global economic inequality.

An important entrée into this debate was offered in a paper by the Marxist literary critic Frederic Jameson (1984) in the journal *New Left Review*. Here, Jameson attempts to dismantle the idea of post-modernism as 'left-wing', 'radical' or a breach with the past. Instead, Jameson argued that post-modernism is the 'cultural logic of late capitalism'. In considering post-modernism as a historical moment, Jameson reminds us that while it may be a reaction to the perceived rigidity of modernity, post-modernism has nevertheless emerged from it. Rather than offering an alternative vision of society, this more cynical reading suggests that post-modernism can be incorporated within the Marxist superstructure (see Chapter 4). In other words, post-modern characteristics are revealing of the complexity of capitalism and are not an alternative to it. Moreover, capitalism works through 'difference'; the multiplicity, fractures and ruptures of post-modern society do not necessarily signal that capitalism is 'in crisis' but could equally suggest that more complex, diverse and 'uneven' forms of exploitation are in ascendancy.

A second critique frequently ventured in the direction of post-modern writing is its tendency to be overly theoretical and even opaque. The language may use self-referential 'jokes' to show the 'play' of words, or write in ways that are open to double meanings or designed to mystify and alienate. While we would caution against 'dumbing down' and any lingering strands of anti-intellectualism – good ideas often require close, critical re-readings – much post-modernist writing requires a wide knowledge base often rooted through the discipline of philosophy. Hamnett (2003) fears that the rise of post-modern human geography is creating a 'theoretical playground', manufacturing theory for theory's sake. Furthermore, the experimental aspects of post-modernist theory mean fixed definitions and categorical statements are regarded with scepticism as part of the modernist quest for coherence. There is a post-modern will to be deliberately slippery, a refusal to be held down. At the same time, there can be an aesthetic property to post-modern writing. Consider, for example, the tone, pitch and mood Marcus Doel (1993: 389) exerts in the following extract:

> Pull threads. Release laugh catastrophes. Engage in theoretical terrorism. Stitch pick. Immerse your rigidities in flows. Some have called this strategy post-modernism . . .

Doel's writing is at once playful and rather too self-satisfyingly aware. Many sentences are crammed with an excessive vocabulary which we have consciously omitted here. For Peet, who scrutinises a number of these passages, the writing is marked by a 'terminological vagueness', perhaps a consequence of a post-modern tendency to create open, polysemic texts (i.e. writing that is open to a plurality of meanings). Peet's frustration is with the rarefied air of exclusion that surrounds much post-modern writing by geographers. Subsequently he is unrestrained in his criticism of post-modern writers such as Doel, who he claims offer an:

> invitation to personal indulgence reminiscent of those days of privilege when academics had triple-barrelled names, and were upper-class dilettantes researching quaint topics of pure esoterica. (2001: 242)

Blistering criticism indeed! No doubt Doel would form part of the 'new intellectual dilettanti' Hamnett (2003: 3) also refers to. But if Doel's intention is to provoke and agitate in the guise of a post-modern linguistic terrorist, Peet's response intimates he has succeeded. Doel delights in the play of language; a pretentiousness that Peet regards as constitutive of an elitist 'amoral geography' (2001: 241–242). But if we look between the lines of each of their arguments we might discern the unfolding of a more familiar story. Peet's polemic remarks, above, reveal his Marxist leftist traditions and concern with developing a critical geography that is politically meaningful. This leads him to be suspicious of Doel's self-conscious flourishes, restless 'language games' and post-modern stylistic excess. In the open-ended spirit of post-modern readership, you might consider which viewpoint you agree with and why. As with Ley's account of architecture discussed previously, the battle over presentation, style and choice of materials (or words) is profoundly ideological.

In addition, post-modernism has been seen to be destructive rather than constructive in its approach. If post-modernism is against the 'regimes of truth' which it constantly seeks to deconstruct, we might equally ask what is it actually *for*? For Marxist, feminist, post-colonial or most other radical geographers, their manifesto and what they stand for is self-evident. In contrast, post-modern geography could be seen by sceptics as just a hollow, soulless project; a vacant sign on the road to nowhere. For Marxist geographers such as Harvey, post-modernism 'lends itself to nothingness' (1995: 116) as it appears caught in an endless act of self-annihilation. He regards post-modernist attempts to consistently implode narrative truth as a self-defeating project. Harvey's preference is for working with grounded notions of realism in order to displace the fictions of capitalism and any lingering notions of 'false consciousness'. In this way he operates with foundational notions of truth that are familiar to modernism and would be rejected out of hand by post-modern thinkers.

Finally, post-modern geographers have been accused of being relativist. If a secure ontological truth is forever beyond our grasp then the realisation is that we can only produce fictions, signs and the semblance of reality. This argument has been used to suggest that, for post-modernists, one truth, narrative or representation is as valid as another. In wide-ranging discussions, David Harvey contends that a Marxist focus upon commodity chains can identify the fingerprints of exploitation upon a bunch of supermarket grapes;

or reveal the truth of why African-American workers in the Deep South may perish in the event of a fire in a dilapidated chicken factory. Harvey's point is that the truth of capitalism as a fundamentally exploitative system goes 'hiding in the light' and needs to be exposed. This truth can be revealed through Marxist critical praxis that explores the production and reproduction of social inequality across time and space. He therefore riles against relativism:

> Obsessed with deconstructing and deligitimating every form of argument they [i.e. post-modernists] encounter, they can end only in condemning their own validity claims to the point where nothing remains of any basis for reasoned action. (1995: 116)

Where Harvey seeks to impose the 'truth' of Marxist geography to implode the 'lies' of neoliberalism, this may appear something of a 'black and white' approach. Post-modernists often tend to work in the grey areas of uncertainty, although at times they may generate more heat than light. There are indeed moments where some post-modernists may revel in the infinite possibilities of relativism. In this vein Marcus Doel urges geographical practitioners to 'Pull the thread which chokes, binds or fascinates you the most, even if it is a factious, factitious and fictional one' (1993: 389). The extreme type of terroristic post-modernism triggered here aims at shooting down any notion of ontological truth. On one hand this experimental approach has enabled marginalised and neglected aspects of the discipline to prosper, including important work on consumption, sexuality and difference. On the other, it has meant just about any topic – from dog-walking to sleeping – is worthy of academic publication in geographical journals. In practice most geographers with post-modernist leanings are likely to avoid relativism, by using modernist techniques such as interviewing, to develop some form of 'empirical truth' while being open to the knowledge of the partiality of their accounts. A more pronounced post-modernist reading would regard interviews as mere discursive representations, signs that conceal as much as they reveal. Such accounts point to the tensions in working between the lines of modernism and post-modernism. What is apparent in the variety of criticisms that have sprung forth is that post-modernism, in its contemporary offering, is a contested and negotiated terrain for most human geographers.

Conclusions

This chapter has shown the heterogeneity of post-modernism and how it can be defined in multiple ways. While it remains an elusive and 'exceptionally slippery idea' (Mitchell, 2000: 58) this in turn means that the criticisms often lodged against it are in danger of fixing and holding it up to scrutiny as a discernible 'object'. To avert this tendency we have provided some different ways of understanding post-modernism: as a historical moment in time known as post-modernity; as a critical practice evident in post-structuralism; and stylistically as an artistic form. To develop these ideas we have drawn upon the generative work of famous French philosophers including Lyotard, Derrida, Foucault and Baudrillard. At the same time, we have stressed that geographers have an opportunity to creatively use these insights to understand the contemporary aspects of cities as places that are fashioned out of a wide pallet of symbolic registers including fantasy shopping complexes, mediated images of place, heritage-scapes, commercial branding signs and touristic encounters with the uncanny. Here we saw what post-modern approaches can do for geography in terms of theory, method and approach.

Finally we considered some of the more cogent criticisms of post-modernism, including its abstraction from the material world, the obsession with language and deconstruction, and a creeping sense of relativism. As an open and eclectic cluster of ideas, post-modern geography papers can range from those with audacious flashes of 'sheer brilliance' (Peet, 2001: 226) to pedestrian wanderings into the work of others that regurgitates dynamic ideas in a tired, self-indulgent manner. Notwithstanding these critiques there is little doubt that post-modernism has expanded the horizons of geographical thought and practice. A willingness to experiment, to encounter difference, and make a mockery of convention has enabled post-modernism to persevere as an unruly child occasionally given to genius. Although many critics may regard post-modernism as 'lost in words' the questions it poses are enduring. To what extent they are 'radical' may ultimately depend upon one's political convictions, knowledge and interests. In large part the significance of post-modernism rests upon how we come to define and utilise this concept across time and space.

Summary

- As a product of the Enlightenment, modernism values scientific rationalism.
- Post-modernism emerges through modernism and can be explored historically, as a critical practice or stylistically.
- Post-modern ideas counter modernist tendencies by questioning the very basis of truth and reality.
- Geographical engagement with post-modern ideas can be traced at the level of theory, method and practice.
- Post-modern geography has been critiqued for resorting to self-indulgent language games without fully engaging with politics, policy and economic inequality.
- The influence of post-modernism in geography can be seen where a number of ideas, methods and approaches are now integrated in the discipline.

Further reading

Clarke, D. (2006) Post-modern Geographies, in S. Aitken and G. Valentine (eds), *Approaches to Human Geography*, London: Sage, pp. 107–121.

Dear, M. and Flusty, S. (1998) Post-modern Urbanism. *Annals of the Association of American Geographers*, 88(1): 50–72.

Flusty, S. (2005) Post-modernism, in D. Atkinson, P. Jackson, D. Sibley and N. Washbourne (eds), *Cultural Geography: A Critical Dictionary of Key Concepts*, London: I.B. Tauris, pp. 169–174.

Gibson-Graham, J.K. (2003) Poststructuralist Interventions, in E. Sheppard and T. Barnes (eds), *A Companion to Economic Geography*, Oxford: Blackwell Science, pp. 95–110.

Soja, E.W. (1995) Post-modern Urbanization: The six restructurings of Los Angeles, in S. Watson and K. Gibson (eds), *Post-modern Cities and Spaces*, Oxford: Blackwell, pp. 125–137.

Critical geo-politics

Learning objectives

In this chapter we will:

- Explore the origins of geo-politics in the study of human geography and the environment in the late 19th and early 20th centuries.
- Examine the emergence of 'critical geo-politics', a body of work that seeks to challenge dominant understanding of geo-politics using social theory.
- Identify the different forms of geo-politics, as identified by scholars of critical geo-politics: formal, practical and popular.
- Introduce feminist and pacifist work that has sought to critique the methods and moral commitments of critical geo-politics.

Introduction

On 26 November 2008 the Indian city of Mumbai experienced a series of coordinated terrorist attacks across a number of sites clustered in the southern parts of the city (see photo). The targets included high-profile hotels, such as the Taj Mahal Palace and the Oberoi Trident, and the main railway station the Chhatrapati Shivaji Terminus. Various media reports suggested that the attackers were singling out American, British and Israeli hotel guests to hold as hostage or kill. The unrest lasted for around three days before Indian Army troops were able to regain control of the hotels. Details of the attacks are, at the time of writing, relatively scarce, although the total death toll

Mumbai attacks in 2008
Source: Getty Images/India Today Group

is considered to be around 188 people, with over 290 injured. Responsibility for the violence was initially claimed by an unknown organisation calling itself the Deccan Mujahadeen, although the only terrorist captured alive claimed that he represented Lashkar-e-Taiba, an Islamist group fighting Indian control in the disputed territory of Kashmir at the Indian/Pakistan border. The response to the attacks by both politicians and media agencies was one of outrage as the violence was presented as an assault on Indian democracy. These responses often drew on the language of geo-politics to explain, first, the prosecution of the violence and, second, its likely impact. For example, in terms of the violence, Shashi Thadoor writing in the *Guardian* newspaper suggested that 'by singling out Britons, Americans and Israelis, they demonstrated that their brand of Islamist fanaticism is anchored less in the absolutism of pure faith than in the geopolitics of hate' (Tharoor, 2008). In these terms Tharoor inserts the attack into the wider international politics of the US-led 'war on terror'. Tharoor uses the phrase 'geopolitics of hate' to bring attention to the way in which victims were targeted on account of the state of which they were citizens. Geo-politics, in these terms, indicates the imagined geography of the attackers, as they divide the world into allies and enemies as a response to US foreign policy. The 'geopolitics of hate' is therefore as much about identity as geography: the sites of perceived enemy states were (Tharoor argues) embodied by the citizenship of target victims.

The idea of geo-politics has also been drawn upon to analyse the wider impact of the violence in Mumbai. In particular, political and media commentators have been keen to reflect on the 'geo-political implications' of the attacks. For example, Devangshu Datta writing in the Indian newspaper *Business Standard* on 8 December 2008 suggested that there is 'the possible impact of geopolitical aftershocks from the Mumbai attacks or of follow-on attacks. This is frankly incalculable since events could unfold in many directions.' Datta's comments refer to imagined 'geopolitical aftershocks', in particular alluding to the possibility of further tension between India and her neighbour Pakistan. In these terms, 'geo-political' is taken to signify the relations between states. Rather than just 'political' the 'geo' prefix serves to locate possible repercussions in the international arena. These two examples illustrate, in brief, the challenge of attempting to provide a precise definition of 'geo-politics'; people use the term to mean different things. In this chapter we will explore the power of geo-political language, its roots in 19th-century geography and the ways in which geographers are critiquing its use today.

The diversity of uses of the term 'geo-politics' indicates that, like many terms we have covered in this book, the definition of geo-politics is contested. To some, geo-politics is simply another word for international relations. For others, geo-politics relates to a specific struggle over space by competing states. It would be satisfying to open the chapter with a universally agreed definition of geo-politics, but this is not possible. Instead we need to trace the historically and geographically situated nature of the idea of geo-politics, from its roots in 19th century Swedish geography through to its current use as a means of gesturing at the geographical aspects of foreign policy. But to understand the current position of geo-politics within geographical thought we need to go further than simply providing a historical account.

In particular we want to explore the ways in which the study of geo-politics has incorporated the perspectives of post-modernism to produce a distinctive theoretical and methodological approach entitled *critical geo-politics*. Rather than understanding geo-politics as a specific body of knowledge, this approach has drawn on the work of French theorist Michel Foucault to examine geo-politics as *discourse*. As we saw in previous chapters, and following Ó Tuathail and Agnew (1992), discourse can be understood as a set of socio-cultural resources through which individuals and groups make meaning about their world and activities. Specifically, a discursive approach to geo-politics allows us to explore three interconnected processes. First, analysing geo-politics as discourse draws our attention to the uneven power relations among those who produce geo-political knowledge. Some individuals and organisations are extremely influential in producing knowledge, creating norms that we naturalise and take as 'common sense'. Second, a discursive approach highlights the specific historical and geographical contexts in which geo-political knowledge is produced. These are not timeless truths about the world, but rather are situated perspectives that suit particular powerful actors at specific times. Third, and perhaps most significantly, exploring geo-politics as discourse highlights the close connection between ideas about the world and subsequent political interventions. The production of geo-political knowledge shapes subsequent political outcomes, either through explicit international interventions or through more long-term and less tangible international relations. Discourses make meaning, and these meanings allow certain political actions while ruling out others.

This chapter is divided into four further sections. In the first we explore the origins of the term 'geo-politics' rooted as it is in 19th-century European geographical scholarship. This section goes on to illustrate the shared traits of much early work that was either conducted under the banner of geo-politics or has, after the event, been identified as geo-political by more recent commentators. This body of early work is defined as 'classical geo-politics'. The second section examines the shift from classical geo-political thinking to more recent studies that have drawn on the wealth of theoretical and philosophical developments in geography, particularly after the cultural turn (see Chapter 5), to develop a body of scholarship called 'critical geo-politics'. We go on to explore how this work has drawn on the theorisations of Michel Foucault and Edward Said in order to critique classical geo-political reasoning. In the third section we develop this discussion of critical geo-politics by identifying the three locations of geo-political reasoning: formal, practical and popular arenas. In the final section we look at emerging geographical scholarship that has been prompted by, but often looks to move beyond, critical geo-politics. This work draws on feminist and activist approaches in order to explicitly re-politicise critical geo-political inquiry and think about some of the moral questions that have been prompted by this work.

Origins of geo-politics

In order to understand the origins of the term 'geo-politics' we need to delve back into the history of geographical thought at the end of the 19th century. It would be worthwhile reading the discussion of Ratzel's work in Chapter 1 alongside this section as there are some shared theoretical approaches.

As we have seen in the Introduction, geo-politics means different things to different people. One starting point for understanding the contested nature of geo-politics is to explore the origins of the term 'geo-politics'. In Chapter 1 we identified the central role of geographical knowledge in supporting the colonial ambitions of imperial states in the 19th and early 20th century. It is within this work that we can see the first uses of the term *geo-politics.* In its early iterations this term related to the geographical dimensions of state power, in particular the resources, territory and environments of individual states. The term was coined by the Swedish geographer Rudolf Kjellén in 1899 (see Box 10.1). Kjellén's approach identified the natural and geographical attributes of the state as the key factors for consideration, relegating societal and governmental variables as relatively insignificant. For Kjellén, the ultimate aim of the state was to become 'a geographical individual – to unite within an organic area' (Holdar, 1992: 312). Kjellén therefore suggested coastlines, mountain ranges or relatively uninhabited areas such as deserts or swaps as the ideal boundaries of states.

The inspiration for Kjellén's geo-politics was the work of German geographer Friedrich Ratzel in the late 19th century. Ratzel was an early pioneer of assessing geographical phenomena through the lens of the biological sciences. Ratzel drew on the ideas of Charles Darwin to suggest that the state is a living organism that strives against others to ensure its survival and expansion. For Ratzel, the territory of the state corresponds to the body of an organism. Within this framework the capabilities of a state are not influenced by politics or history, but rather by its natural processes determined by its geographical environment. This biological understanding of the state proved extremely influential – it seemed to provide a scientific basis through which to study the comparative power of individual

Box 10.1
Rudolf Kjellén 1864–1922

Rudolf Kjellén was a Swedish political scientist who gained his PhD from Uppsala University in 1861 and taught at Gothenburg University from 1891, later becoming a professor of political science and geography. Kjellén's work was committed to discovering scientifically derived laws of the state, and to this end he coined the phrase *Geopolitik* to refer to the study of the territory of states. It is important to note that this was not the only focus of Kjellen's expansive work: he also expounded theories on the population of states (*demopolitik*), the economy of states (*ekopolitik*), societal politics of states (*sociopolitik*) and governmental politics (*kratopolitik*). Kjellén's scholarship provides a useful illustration of the nature of early work in geo-politics: he saw this area of research as an empirically grounded area of study that could shape political action. In empirical terms, Kjellén saw the geography of states as an alternative to the abstract and overly theoretical approaches to statehood that had dominated political science. His approach sought to connect humans with their environment through grounded research. Sven Holdar (1992: 308) talks of Kjellén's summers spent bicycling around Sweden studying various geographical and geological aspects of the landscape. This approach drew on the prevailing environmental determinism within geographical thought (see Chapter 1), where the climate, topography and resources of a given state were considered factors that influenced the political life and cultural characteristics of its people. In terms of policy action, Kjellén sought to present geo-politics as a 'practical guide to political action' (Murphy, 1997: 6). With this in mind, Kjellen sought to confront key policy issues facing the Swedish state, such as its uneasy union with Norway. By drawing on geo-politics, Kjellén sought to demonstrate the scientific grounds against Norway's independence. This work began a long-term engagement with the writings of Friedrich Ratzel, most notably his conception of the state as an organism. Kjellén argued that the natural boundary of Sweden was the coastline; any alteration to this is an unnatural human denigration of geography. For not the first time, we see a scholar's ideological position (in this case Norway's union with Sweden) justified in terms of natural or scientific observations. Kjellén's work has been extremely influential, not simply on account of the appropriation of the term 'geo-politics' by geographers and politicians, but in addition his work directly influenced scholars such as Mackinder and Haushofer.

Further reading

Holdar, S. (1992) The Ideal State and the Power of Geography: The life-work of Rudolf Kjellén, *Political Geography*, 11(3): 307–323.

Kjellén, R. (1899) *Staten som lifsfom* [The State as an Organism], Stockholm: Hugo Gebers Forlag.

Kjellén, R. (1899) Studierofver Sveriges Polistic granser, *Ymer*, 19: 283–331.

Murphy, D.T. (1997) *The Heroic Earth: Geo-political Thought in Weimar Germany, 1918–1933*, Kent, OH: The Kent State University Press.

states. But underlying Ratzel's thesis were two assumptions. First, that states are striving in competition with each other, rather than suggesting the possibility of a cooperative equilibrium. In this model, interstate conflict and antagonism are assumed. Second, and related, Ratzel's reliance on evolutionary biology would suggest that such conflict can be naturally assumed. Interstate aggression or expansion is classified as a struggle for survival rather than a calculated political project to advance the territorial or economic ambitions of an individual state. In this way Ratzel's work exhibits an attachment to the political philosophy of *realism*, where interstate relations are considered to be governed by anarchy with each state struggling against others to ensure its security and status in the global political system.

While early scholars in geography did not necessarily identify their work as 'geo-politics' (as we will see, some actively tried to distance themselves from this term) the writings of Ratzel and Kjellén proved influential in the institutionalisation of the geographical discipline in both the UK and US over the late 19th and early 20th centuries. This emerging

Box 10.2
Four traits of classical geo-politics

1 *Emphasis on practical knowledge*: the most renowned geographers working in Western Europe and North America in the late 19th and early 20th century were striving to produce knowledge that would, in their eyes, advance the standings or territorial ambitions of the states in which they worked. As we saw in Chapter 1, geography was, at this time, a tool of statecraft, a means through which state interests could be justified as a 'natural' fact.

2 *Interest in the state*: the prima ry focus for research in this early work in geo-politics was the state. The influence of Ratzel's work meant that scholars felt there were universal laws of statehood that could be discerned through careful observation.

3 *Scientific basis of knowledge claims*: this early phase of geography sought to identify natural laws of states; geographers drew on theories and phrases from evolutionary biology through which to establish a scientific basis for their observations. In this case the geographical aspects of the state shaped its global status. As Gearoid Ó Tuathail (1994: 325) states, although none of this intellectual work was outside politics and ideology, 'all claimed the mantle of science and objectivity'.

4 *Shaped by contemporary political context*: as with other areas of geographical thought, the emergence of geo-politics as a distinct field of knowledge cannot be disconnected from the prevailing political context. The early part of the 20th century was a key moment in interstate competition, culminating in World War I, while Britain was also beginning to confront the waning of its imperial power. In addition, this period saw the emergence of the US as a world power and the antecedents of its Cold War stand off against the Soviet Union. As we will see, these political contexts shaped the emergence of geo-politics as a scholarly pursuit.

work is often presented as 'classical' geo-politics, as distinct from the more recent move to 'critical' geo-politics. Indeed, as we will see, strenuous effort has been made by recent political geographers to distance themselves from the theoretical and political underpinnings of much of this early work. By way of an overview of what follows, we can identify four traits shared by these key scholars working in the era of 'classical geo-politics' (see Box 10.2).

Mackinder

Ratzel's idea of the state as an organism in competition with others provided a framework for a wealth of scholarship into the geographical aspects of the state. In particular, geographers were interested in how the natural environment or resource-base shaped the capabilities of individual states and imperial projects. Perhaps the most famous example of this work originates from the British geographer, parliamentarian and entrepreneur Halford Mackinder. As we saw in Chapter 1, Mackinder was a great defender of the British Empire and his geographical scholarship was directed towards outlining the future threats to Britain's global pre-eminence. Mackinder was particularly interested in the arrangement of the globe's landmasses, and he felt that control of the major Eurasian landmass would become the most significant factor in establishing global dominance. In his 1904 paper, 'The Geographical Pivot of History', Mackinder divides the world into zones radiating from a central 'pivot' or 'heartland' on the Eurasian landmass (see Figure 10.1). For Mackinder it was geography that served as the key determinant of human history (and not the other way around). Reflecting Mackinder's practical stance, this approach was designed

Figure 10.1 **The natural seats of power by Halford Mackinder**
Source: Royal Geographical Society, with IBG

to serve as a call to the British Empire to intervene to prevent the pivot area falling into the hands of competing states. Through Mackinder's work we see the four traits of classical geo-politics: an interest in practical knowledge, a focus on the geographical dimensions of the state, a belief in the rational scientific basis of his approach and a close connection between his work and the prevailing political context of threats to the pre-eminence of the British Empire. This work highlights an important point: that, in practice, these four traits contradict each other. Mackinder attempts to present his thesis as a 'natural' and 'detached' scientific observation, but simultaneously he is striving to support a particular ideology: that of the continued dominance of the British Empire (see Agnew, 2002: 68). This contradiction draws our attention to the significance of exploring the production of geo-political knowledge in its historical and geographical context.

Haushofer

Following Mackinder, the term *geo-politics* gained notoriety through its adoption by German political geographer Karl Haushofer. As outlined in Chapter 1, Haushofer founded a German school of *Geopolitik* in the 1920s and 1930s. Haushofer had also been influenced by Ratzel and promoted in particular his idea of *lebensraum* or 'living space'. This theory suggested that people were most influenced by their geographical location and would naturally expand to fill the space available. This idea of a state's entitlement to greater territory was particularly influential in Germany in the 1920s and 1930s, where the Treaty of Versailles had restricted German territory and economic prospects. Most infamously, the idea of *lebensraum* entered the lexicon of German leader Adolf Hitler, who drew on these ideas in his text *Mein Kampf* (see Box 10.3). The 'natural' basis of *Geopolitik*, where states were perceived best served by 'organic' boundaries, provided a pretext upon which German territorial expansion during World War II could be justified.

Box 10.3
Nazi geographies

What role did geography play in the expansionist policies of Nazi Germany over the 1930s and 1940s? This question has been the subject of great debate and, although the direct connections between geographical thinkers and the Nazi policy formation remain unclear, it is evident that the terminology of the emerging geographical discipline was enrolled by Nazi leaders to justify European military campaigns. Like much of the early work in human geography, this connection can be traced back to the work of Friedrich Ratzel. As we have seen, Ratzel incorporated Darwin's ideas into geography to identify the state as a living organism that struggled against other states to establish its entitled *lebensraum* (or living space). This interpretation of Darwin's ideas has been criticised in the intervening years, not least by the geographer David Stoddart (1966) who felt it was based on an incomplete understanding of Darwin's theory of evolution. However this did not reduce the political impact of evolutionary ideas within German political thought. The Versailles Peace Treaty (1919) significantly reduced Germany's territory, both in terms of the confiscation of its overseas colonies and the reorganisation of its national boundaries, for example granting Alsace-Lorraine to France and Upper Silesia to Poland. German scholars of political geography looked to Ratzel's concepts to explain how Germany needed greater *lebensraum* and subsequently justify expanding its borders into surrounding smaller and less populous states. A key figure within such debates was Karl Haushofer (1869–1948). Haushofer had been an officer in the German army but following World War I he took a lectureship in political geography at Munich University, rising to a professor in 1933. Haushofer's incorporation of Ratzel's theorisations with Mackinder's ideas of state territorial strategy was crucial to the formulation of what was termed a German school of *Geopolitik* over the 1920s and 1930s. This may have remained an obscure scholastic endeavour, were it not for the fact that it tapped into a popular imagination of Germany's territorial loss and that one of Haushofer's students was Rudolf Hess, the future deputy to Adolf Hitler. Historians have noted that Hess served just over seven months in Landsberg prison at the same time as Hitler, another inmate, was writing *Mein Kampf* (My Struggle). Through this link, and Hess's later Nazi party connections, Haushofer's ideas and concepts were incorporated into Nazi party strategy. But as political geographer John Agnew (2002) cautions, we must be careful that we do not overstate the role of Haushofer, or geography more broadly, in the violent crimes enacted by the Nazi regime. We cannot easily conflate Haushofer's *Geopolitik* with the pernicious combination of anti-Semitism and racial purity that the Nazi party used to pursue violent action against domestic and, later, neighbouring populations. What we can trace quite clearly is the implications of geography's entanglement in the violence of Nazi philosophies: that the discipline retreated away from normative theorisations (how the world *should* be) following World War II towards more rational and scientific approaches that used positivist approaches (how the world *is*).

Bowman

Understandably, the enrolment of scientific justifications for Nazi policies was widely critiqued. A particularly prominent critic was the US geographer Isaiah Bowman (1978–1950). Bowman is a fascinating figure who in many ways embodies the shifting intellectual and political position of American geography over the late 19th and early 20th centuries. His life has been vividly analysed by the geographer Neil Smith across a range of academic articles and then in the magisterial work *American Empire: Roosevelt's Geographer and the Prelude to Globalisation* (2003). These works explores the rise of US-led globalisation through Bowman's life course, exploring how Bowman's ideas and career shaped, and were shaped by, changing global political contexts. While we will cover some details below, as a general introduction Smith's research portrays Bowman as something of a contradictory figure (see Smith 2003: 27). Bowman was keen to portray himself as an internationalist who travelled and saw the importance of Americans understanding the world around them. At the same time he was a staunch nationalist who worked tirelessly for US interests and often fell back on understandings of the world that were structured around a natural hierarchy with America at the top. These contradictions reflect wider changes in global international relations that Bowman seemed to well understand: this was a period where the struggle for empire building changed, from claiming territory to trying to extend economic influence, while World Wars I and II presented opportunities for the US to reconfigure international relations to its own advantage. These were profoundly geo-political issues and, while Bowman often strove to distance himself from the label 'geopolitician', in particular on account of its Nazi connotations (see Bowman, 1942), he is often seen as a key figure in the history of classical geo-political thought.

Bowman rose from humble beginnings in rural Michigan to become a key figure in US geography and academia, serving as President of the American Geographical Society and Johns Hopkins University. Like Mackinder and Haushofer, Bowman was interested in shaping policy; he viewed scholarly pursuits as advancing the interests of the US state. This position was reflected in the key role he played within the US delegation to the negotiations for the Versailles Peace Treaty at the end of World War I. Bowman was also a firm believer in the scientific grounding of geography. His background was in physical geography; he spent his early career under the tuition of the renowned geologist William Morris Davis. This specialisation prompted an interest in cartography and, in particular, mapping river erosion in South America. Through these fieldtrips Bowman began to foster an interest in human development, exploring the potential benefits of US economic expansion. His best-known work in this field was *The New World: Problems in Political Geography* (first published 1928). In this wide-ranging text Bowman attempts to address the political, social and economic consequences of World War I. This text betrays Bowman's support for colonial (and particularly British colonial) forms of rule, discounting a return to what he terms 'disorder and cruelties of native rule' (Bowman, 1928: 22). Despite his admiration for British imperialism, Bowman felt that the US could follow a different path. He advocated what he termed *economic lebensraum* for the US, whereby there was no need for a US strategy to expand territorially into neighbouring states but instead the resources of neighbouring states needed to be opened up for US businesses.

Despite Bowman's use of Ratzel's terminology, he studiously distanced himself from the German school of *Geopolitik*. This should come as no surprise, since any scholar would hardly seek to implicate his or her work in Nazi expansionism in the 1930s and 1940s. But Bowman was particularly vociferous in distinguishing his own 'geography' from German 'geo-politics', arguing that:

> [G]eopolitics presents a distorted view of the historical, political, and geographical relations of the world and its parts. . . . Its arguments as developed in Germany are made up to suit the case for German aggression. It contains, therefore, a poisonous self-destroying principle: when international interests conflict or overlap *might* alone shall decide the issue.
>
> (Bowman, 1942: 646)

In contrast to what he perceives to be the aggressive and ideology laden nature of geo-politics, Bowman presents his own work as scientific and of universal value. For example, in the same paper he explains that one of his major works, the 1935 *Pan-American Atlas*, would be of universal benefit to trade. The atlas, he argues:

> was not a proposal to learn how to use science to 'conquer' Latin America after the fashion of the German *geopolitikers*, but how to work together for common ends, and specifically how to do so through cultural exchange, trade and general economic improvement.
>
> (Bowman, 1942: 649)

The association between geo-politics and the expansionist German state of the 1930s and early 1940s was to have significant implications for the geographical discipline. The aura of Nazi policies led to geographers rejecting the idea of geo-politics as a arena of inquiry, and focusing instead on more quantitative subject matter with less directly political ramifications.

Decline of geo-politics

Following World War II, the priorities of geographers moved away from examining the nature of states and interstate relations towards more quantitative and behaviour-based modelling. The pre-war close connection between geographers and those in power had strengthened the discipline by institutionalising geography within the curricula of schools and universities. But the relationship between these early prominent geographers and imperial forms of rule was seen as an embarrassment in an era of decolonisation of the British Empire. Over this period the language of geo-politics was a rarity in Western academic institutions; it became (according to a military officer at a meeting of the Association of American Geographers) a 'dirty word' (see Ó Tuathail, 1986: 73). It must be noted that certain scholars, particularly in international relations, continued to draw on classical geo-political ideas and vocabulary in order to generate 'policy relevant' research (see, for example, Gray, 1977). Political geography as a sub-discipline fell into decline, with a shift away from the founding concepts of the relationship between environments and territorial units (states). This move away from the naturalistic approach that had characterised earlier research stimulated an identity crisis within political geography. John Agnew suggests that over this period 'leading figures in political geography focused more on policing the boundaries of the field, declaring what was and what was not Geography, rather than engaging in much innovative or novel research' (Agnew, 2002: 96).

Critical geo-politics

In the mid-1980s there was a resurgent interest in geo-politics among political geographers. This does not mean these scholars returned to the theories of Ratzel and Kjellén in order to explain the world. Rather, geographers began to take an interest in how geographical ideas within world politics were constructed, circulated and reproduced. In particular this work focused on the way in which foreign policymakers used geography to naturalise particular representations of the world. In this way, the new study of geo-politics was almost a complete inverse of the classical geo-political tradition – as Simon Dalby explains: this new approach argued that it was *not* the task of geographers to 'provide state policy makers with rationales for foreign policies that promote imperial power or coercion'. Instead the analytical gaze is 'turned precisely on these activities, and in the process becomes an explicitly critical activity' (Dalby, 2008: 417). This new approach consequently was given the title *critical geo-politics* (in particular following Ó Tuathail's 1996 book of the same name) and its early key protagonists were political geographers Gearoid Ó Tuathail, David Campbell, John Agnew and Simon Dalby. Crucially, this new approach did not appear from nowhere, but was rather a reflection of the broader academic and political contexts in which these geographers were working. We will examine each of these contexts in turn.

Academic context

First, in academic terms, political geographers began to adopt the social and cultural theories that had been popularised in other areas of geography, most notably feminist geography, and in other disciplines, specifically post-structural international relations. These theoretical inspirations had two consequences. First, the adoption of post-structural approaches fostered scepticism of essential truths about the world. Instead, all knowledge is situated and shaped by the context in which it is produced. Second, feminist approaches helped illustrate the importance of *positionality*, that aspects of identity (in particular gender) shape the production of geo-political knowledge. In short, ideas about the world are not neutral reflections of an external reality, but are instead value-laden and reflect the interests and perspectives of the person or institution that created them. In order to better understand geo-politics, a critical perspective explores geo-political pronouncements as practices in ordering the world and rendering it meaningful for a particular audience. Therefore we can understand geo-political ideas and images as tools through which powerful actors legitimise their own positions and make certain interventions in the world possible.

The critical geo-politics perspective was summarised by Ó Tuathail and Agnew as a form of critique that views geo-politics as discourse. Specifically, they argue that geo-politics is a 'discursive practice, by which intellectuals of statecraft "spatialise" international politics and represent it as a "world" characterised by particular types of places, peoples and dramas' (Ó Tuathail and Agnew, 1992: 190). This approach is indebted to the French theorist Michel Foucault. As we have seen in previous chapters, Foucault has had a profound impact on geographical thought, in particular in helping understand the mechanisms through which power is communicated and reproduced in modern societies. One of the key instruments that Foucault draws our attention to is *discourse*. This is a necessarily difficult concept to define and it is often mistakenly thought to refer simply to words, phrases and speech, stemming from its popular meaning as written communication and debate.

But Foucault was trying to illustrate something more subtle and profound – he uses the idea of discourse to illustrate the ways in which particular ideas are made meaningful and are consequently assumed to be appropriate or 'common sense'. Discourse consequently draws our attention to the ways in which particular ideas become dominant and simultaneously encourages reflection on the ways in which particular discourses are reproduced. We can best illustrate this point with a series of examples. In Foucault's own work he drew on the example of the construction of norms of sexuality in Western societies and the way in which heterosexuality was conveyed as a 'biological norm' and alternative forms of sexuality were therefore rendered strange and even illegal. The geographer Philip Howell (2007: 293) explains that Foucault wanted to illustrate that ideas of sexuality are *produced*. 'Sexuality is not,' Howell continues, 'a universal biological fact, something that stands outside individuals and societies to a greater or lesser extent directing or determining them. It is, instead, a historical product – a historically and culturally specific through which a new and insidious form of power, the "truth" of sex, assumes a locus in the body and its pleasures' (Howell, 2007: 292). Foucault's work forces us to question categories that are assumed to be natural.

Perhaps the best known example of an attempt to attempt to explore Foucault's idea of discourse in practice is found in the work of Edward Said, in particular his 1978 text *Orientalism* (reprinted 1995). In this book Said critiqued the representations of the Orient in Western 18th- and 19th-century art and literature. Said's point is that the Orient is not a place that is simply 'there' waiting discovery and representation; rather, the Orient was produced through a Western discourse that he labels **Orientalism**. Said presents Orientalism as a 'way of coming to terms with the Orient that is based on the Orient's special place in European Western experience' (p. 1). Said's contention is that 'without examining Orientalism as a discourse one cannot possibly understand the enormously systematic discipline by which European culture was able to manage – even produce – the Orient politically, sociologically, militarily, ideologically, scientifically, and imaginatively during the post-Enlightenment period' (p. 3). Said therefore suggests that Orientalism produces the Orient as an imagined geography. We would like to make three points about this work. First, while Said's analysis clearly engages with an 'imaginary' realm (texts, art and ideas) he is clear that the effects of Orientalism in terms of legitimising colonial encounters and Western domination could not be more real. 'Orientalism', Said argues, 'is not an airy European fantasy about the Orient, but a created body of theory and practice in which, for many generations, there has been considerable material investment' (p. 6). Second, Orientalism is not just about a distant 'Other', but is equally a discourse that reproduces self-congratulatory ideas of the Western 'Self' as progressive and modern in comparison with a backward and barbaric Oriental 'Other'. Said's critique of Orientalism therefore draws our attention to the dual process of identity formation. Third, Said's writings have proved extremely influential within geography, and in particular within critical geo-politics. Recently the geographer Derek Gregory used Said's critique of Orientalism to investigate the production of geographical knowledge in the US-led 'war on terror' since 2001. In his book *The Colonial Present* (2004) Gregory traces the imaginative geographies that connect military and symbolic violence in Afghanistan, Palestine and Iraq following the 9/11 attacks on New York. He argues that the policies enacted in the name of 9/11 demonstrate that Orientalism is 'abroad again . . . hideously emboldened' (p. 18), as policies emerging from the US have divided the world into a series of dualisms such as 'us' against 'them', 'civilisation' and 'barbarism', 'darkness' and 'light'.

While Said's study is of a specific historical and geographical frame, his analytical technique of using Foucault's idea of *discourse* to assess the production of knowledge concerning places and identity has had a wide appeal. A discursive understanding therefore helps us to illustrate the historically and geographically situated nature of social, political and cultural phenomena. The work of Arturo Escobar has helped illustrate this point in relation to international development. Escobar (1995) argues that development is a discourse, which he defines as a 'domain of thought and action' (p. 10). Escobar's aim is to explore the mechanisms through which development knowledge is produced and the effects of this knowledge production on the outcomes for poor countries. Escobar sees development as a form of knowledge that has reproduced the domination of wealthy former colonisers while excluding or ignoring the expertise of those living in 'developing' countries. Therefore 'development', Escobar argues, is a form of knowledge that produces 'permissible modes of being and thinking while disqualifying and even ma king others impossible' (Escobar, 1995: 5). This understanding allows Escobar to explore three interrelated aspects of a discursive understanding of development: first 'the forms of knowledge that refer to it and through which it comes into being and is elaborated into objects, concepts, theories'; second 'the system of power that regulates its practice'; and finally 'the forms of subjectivity fostered by this discourse, those through which people come to recognise themselves as developed or underdeveloped' (1995: 10). Therefore Escobar's approach echoes that of Said: although we are exploring forms of knowledge production, these practices are considered to have 'real' material effects.

By the later 1980s there was a growing body of scholars in both geography and related disciplines (such as international relations and development studies) drawing on Foucault's idea of discourse to enrich understanding of complex social, geographical and historical phenomena. By exploring foreign policy as *discourse*, political geographers were able to explore how powerful actors on the world stage validate particular understandings of the world through geo-political ideas. This was a critical geo-politics because it sought to understand how ideas of space and place were inserted into foreign policy pronouncements in order to naturalise or normalise particular political claims (Kuus, 2007: 7). But as with the classical geo-politics explored earlier in the chapter, this new theoretical dimension did not appear from nowhere; it has also been shaped by the political context in which it is embedded.

Political context

Throughout this book we have stressed the interconnections between shifts in geographical thought and the prevailing political context. The emergence of critical geo-politics as an area of inquiry is no different. The 1980s were a key moment in the Cold War, the name given to the era of diplomatic and military confrontations between the US and the Soviet Union from 1947 to 1991. This narrative of tension and conflict between these two power blocs dominated foreign policy since World War II. But the role of the US was changing in the 1980s, with a relative decline in its military strength in relation to emerging powers such as Germany and Japan. These events prompted political geographers, most notably Gearoid Ó Tuathail, to become interested in the 're-militarisation' of US foreign policy (see Dalby, 2008: 414). In a 1986 paper, Ó Tuathail explores the geo-political language used by the US in relation to El Salvador in Central America. Over the early 1980s the US markedly increased its economic and military aid to El Salvador directed to supporting the right-wing government in a civil conflict with a number of paramilitary groups. Ó Tuathail explains how the country became 'a symbol of the new US administration's strong commitment to a tough foreign policy in the Caribbean basin' (1986: 75). In order to illustrate this new policy position Ó Tuathail traces the shift in rhetoric between the Carter and Reagan administrations. This dynamic focus is important, since it illustrates the importance of context to the production of geo-political ideas. Ó Tuathail is clear: he insists that geo-political language 'must not be abstracted from the concrete historical and ideological contexts within which it originates' (p. 74).

It is valuable to explore this early work in critical geo-politics since it identifies a number of enduring aspects of this style of academic analysis. The methodology Ó Tuathail employs is an analysis of formal political documents (in particular US congress hearings relating to El Salvador and presidential documents) and newspaper reports of press conferences and policy speeches by members of the Reagan administration. As we will outline later in the chapter, this reliance on textual data has been critiqued in recent years as the number of methodologies employed by scholars of critical geo-politics has increased, in particular adopting qualitative methodologies such as ethnography and participant observation. But Ó Tuathail's approach allows the identification of three particular geo-political devices used by the Reagan administration to reassert American military strength in Latin America:

1 *Soviet expansionism*: in studying the speeches of President Reagan and his administration, Ó Tuathail notes a reliance on the idea of the Soviet Union as an aggressive expansionist power. Rather than exploring the local or regional reasons for the conflict in El Salvador, the violence is narrated within US foreign policy documents as a 'deliberate attempt by the Soviet Union and her "proxies" (Cuba and Nicaragua) to export revolution and destabilize pro-US governments' (Ó Tuathail, 1986: 75). The impulse of the Soviet Union to seek to expand beyond its territories and convert states to Communism is presented as an unquestioned truth. As we have seen, it is the objective of scholars of critical geo-politics to question such essential narratives and focus instead on uncovering (and often resisting) the political motivations that underpin such geo-political reasoning. That the Reagan administration viewed the attempt by the Soviet Union in the plural (destabilising 'pro-US governments') is important, since a second aspect of the foreign policy documentation studied by Ó Tuathail centred on the potential knock-on effects of El Salvador turning to communism: the so-called domino theory.

2 *The domino theory*: US foreign policy documents throughout the Cold War allude to or directly reference a fear of a 'domino effect' of countries bordering one another 'going communist' one after the other (like falling dominoes). The imaginary of countries 'falling' to communism had been an important justification by the US government in the military intervention in Vietnam in the 1960s and early 1970s. If we take a moment to consider this idea it seems somewhat absurd: why would countries naturally follow neighbouring countries in adopting their political system? Fundamentally this idea relies on a belief that Communist states are expansionist or spread (like viruses) beyond their borders. The domino effect is inherently geographical in its foregrounding of proximity between states as an influential factor in their make-up, but simultaneously it denies geography by overlooking specificity and difference between states and the reasons that motivate revolution in the first place (that is, poverty and perceived disenfranchisement). As Ó Tuathail notes: the Reagan administration only gave 'tacit recognition of class conflict, income disparities and the poverty of the vast majority of the people of El Salvador' (p. 78).

3 *Geographic proximity and strategic interests*: Ó Tuathail identifies a final geo-political device employed by the Reagan administration to justify increased US intervention in El Salvador: geographical proximity. He asserts that the Reagan administration 'argued that both the geographical proximity of Central America and the US strategic interests

in the Panama Canal and the Caribbean sealanes make the US involvement in the region a matter of "national priority"' (1986: 76). This proximity, coupled with the centrality of sea trade to US economic health, rendered the fate of El Salvador a crucial part of US strategic thinking.

The early work by Ó Tuathail, and later in collaboration with John Agnew, charted a new direction in the study of geo-politics. Rather than seeking to present policy relevant material that would bolster the interests of powerful states, as had been the case in the classical era of geo-politics, this work attempted to question and critique the geo-political imaginaries of powerful state actors. Consequently the classical presentation of geo-politics as a natural science of statecraft that linked the prowess of states to their geographical resources was rejected as an unquestionable fact. Instead the focus shifted to how such geo-political discourses served powerful states in order to advance particular strategic agendas. In order to study the production and circulation of geo-political discourse scholars of critical geo-politics have identified three types of geo-political discourse: formal, practical and popular. Understanding these varieties of discourse helps us not only to explore the power of geo-politics but also strategies of resistance.

Formal, practical and popular geo-politics

Ó Tuathail's early statement of a 'new' geo-politics proved influential and inspired a series of further reflections on the nature of US foreign policy during the Cold War. Rather than reproducing the idea of a binary division of the world between a capitalist West and a Communist East, this work explored the Cold War as an object of knowledge that 'naturalised' certain foreign policy interventions. As Ó Tuathail noted in a later paper, the discourse of the Cold War had material effects since it gave 'a distinct organizational structure to global politics and helped bring into existence not only two distinct geoeconomic and geo-political orders but two permanent military-industrial complexes in West and East' (Ó Tuathail, 1993). We must be careful, however, how we relate discourses to foreign policy outcomes. As Merje Kuus has illustrated through her recent research into the geo-politics of Estonia's incorporation into NATO, discourses do not 'cause' particular policy responses. Rather, she notes, 'they frame political debate in such a way as to make certain policies appear reasonable and feasible while marginalizing other policy options as unreasonable or unfeasible' (Kuus, 2007: 10). Consequently, scholars of critical geo-politics have talked of discourses as 'frames' of foreign policy, and this seems like a helpful metaphor. Discourses of the Cold War as a binary struggle between East and West helped frame policy responses by setting a boundary around what responses were deemed logical or acceptable. But these frames did not emerge simply through foreign policy iterations. Rather they were reproduced through the activities of intellectuals and mainstream cultural outlets (such as films, books, comics and music). In order to understand how geo-political ideas circulate within these diverse arenas, scholars have suggested that we explore geo-political discourse in formal, practical and popular spheres (see Table 10.1). It should be noted that there is considerable blurring between these arenas of geo-political discourses, but they provide a helpful starting point for understanding the analytical process of critical geo-politics.

Table 10.1 **Formal, practical and popular geo-political discourses**

Geo-political discourse	Producers	Examples
Formal geo-politics	• Academics • Intellectuals • Think tanks	Refers to the geo-political models and imaginaries produced by what are termed intellectuals and academics in the service of the state. The classical works of geo-politics, such as Mackinder's 'The Geographical Pivot of History' (1904), are often considered by scholars of critical geo-politics as key examples of formal geo-political discourse. We find a more up-to-date example in Thomas Barnett's work, in particular *The Pentagon's New Map* (2004). This text presents a geo-political vision based on a division of the world between two zones: the core (North America, Europe, Australasia and India) and the 'non-integrated gap' (everywhere else).
Practical geo-politics	• Politicians	Practical geo-political discourse is found in the speeches and policy statements of politicians and policymakers. The study of political speeches has been a fertile ground for scholars of critical geo-politics, particularly those made by the Bush administration during the 'War on Terror' (see Bialasiewicz *et al.*, 2007; Jeffrey, 2009). Recently, scholars have used this methodology to explore a variety of geo-political events and trends, such as the expansion of NATO into Central and Eastern Europe (Kuus, 2007) and the multiple explanations of the Beslan school siege in North Ossetia, Russia (Ó Tuathail, 2009).
Popular geo-politics	• Film makers • Cartoonists • Artists • Musicians	The label given to geo-political images, expressions and texts found in popular culture, such as films, music, literature, artwork and cartoons (Dodds, 1996). As Ó Tuathail reminds us (2006: 9), this 'informal geo-politics circulates geo-political understandings beyond the political class to ordinary people.' There has been a wealth of research in this area, as scholars have sought to deconstruct the images and forms of representation in popular culture to expose the often hidden communication of geo-political ideas. A path-breaking example can be found in Sharp's (2000) exploration of the geo-politics of the *Reader's Digest* magazine. More recently, Dodds (2003, 2006) has provided a series of commentaries on the geo-political knowledge conveyed in James Bond films, and Jason Dittmer (2005) has explored the geo-politics of the cartoon *Captain America*.

Over recent years scholars in political geography, international relations and cultural geography have provided numerous examples of critiques of formal, practical and popular geo-politics. It should be noted from the outset that these forms of discourse are never distinct; they always blur into one another in practice. For example, a political speech is usually communicated through a mediating source, whether a newspaper or television news coverage. Likewise, popular culture does not stand apart from practical geo-political reasoning, since politicians often have an influence on the production of films either directly (as in the case of the US film *Top Gun*, to which the US military lent hardware and provided film locations) or indirectly through regulation (denying work permits or preventing the use of film locations). Tracing these connections of political influence, and the subsequent implications for geo-political discourse, is a key task for geographers and other scholars.

In order to illustrate the practices of critiquing formal, practical and popular geo-politics, we will provide examples of each. In the case of formal geo-politics, scholars have sought to reconsider the geo-political imaginations of classical scholars of geo-politics, such as the work of Sir Halford Mackinder. In 2004 a special issue of *The Geographical Journal* included a reprinted version of Mackinder's 'The Geographical Pivot of History' paper (1904), 100 years on from its original publication in the same journal (see Dodds and Sidaway, 2004). A number of scholars provide commentaries on the paper and reflections on the legacy of Mackinder. These scholars (both explicitly and implicitly) adopt the methodologies of critical geo-politics in order to deconstruct Mackinder's work and place it within the political context from which it emerged. While Mackinder presented his arguments as models of human history and geography, the recent accounts have analysed his work as geo-political discourse: situated in the particular political context of British imperial anxiety and the emergence of the geographical discipline. The contributions point to the legacy of Mackinder's thought in the geo-political discourses of both the Cold War and 'war on terror'. In recent years critiques of formal geo-politics have extended beyond the work of classical scholars such as Kjellén, Mackinder or Haushofer. In particular, attention has been paid to individuals and institutions that formulate geo-political knowledge in close proximity to political elites. For example, recently the geographers Klaus Dodds and Stuart Elden (2008) have examined the links between the Cambridge-based policy institute the Henry Jackson Society and the UK Conservative party. Their paper explores the geo-politics of Conservative party foreign policy, and its links to the interventionalist politics of the Henry Jackson Society. Their paper urges us to take seriously the links between policy institutes and political elites, since these groups play a key role in formulating and communicating geo-political discourse.

There has been a wealth of scholars exploring practical geo-politics, from studies of political rhetoric of the Cold War (Ó Tuathail and Agnew, 1992), to analysis of the speeches of former US President George W. Bush (Elden, 2009). We can see a particularly well-worked example of a critical study of practical geo-politics in the work of geographer Merje Kuus. Over the past decade, Kuus has examined the expansion of NATO and, to a lesser extent, the EU into Central and Eastern Europe. In her text *Geo-politics Reframed* (2007), Kuus explores the security discourses that have circulated the struggles for NATO membership in former Soviet Republics, in particular Estonia. Kuus's narrative draws attention to the entanglement of discourses of geo-politics (the inscription of difference onto space) and security (the mechanisms necessary to ensure survival and prosperity).

To this end Kuus highlights the popularity of Samuel Huntington's *Clash of Civilizations* (1996) thesis in Estonian politics, since it bolsters a state security discourse of impending threat from the Russian east. Huntington's work 'nicely complements Central Europe's own versions of the "Return to Europe" narrative' (Kuus, 2007: 44). Within these geo-political discourses Kuus identifies a broad shift in concepts of security: from a purely militaristic understanding based on the threat of violence, through to a cultural understanding of security based on the threat to the cultural integrity of a given state. While the perceived threat of Russia remains, Kuus identifies a shift in how it has been presented within geo-political discourses of the Estonian state from a military adversary to a culturally distinct and non-European other. The key point in Kuus's work for our discussion of critical geo-politics relates to the implications of these discourses of threat. In a number of places in her text Kuus is careful to point out that these discourses do not prescribe policy responses. For example, the description of Russia in certain terms does not instrumentally lead to a particular foreign policy towards Russia from Estonia. Rather, these discourses *enable* certain responses in the future by framing the relationship between Estonia and Russia in particular terms.

The turn in political geography towards the study of cultural practice and forms of representation is reflected in the interest scholars have developed in popular geo-political discourses over the last decade. Rather than simply studying the self-identified scribes of geo-political thought (as in the case of formal geo-politics) or the iterations of state elites (as in practical geo-politics), popular geo-political discourse is found in myriad sites across popular culture from art and literature, to film and music. This has provided a fertile ground for exploring how geo-political divisions and identities are communicated in cultural artefacts. We could spend some time outlining this body of work, from Klaus Dodds's exploration of the geo-politics of James Bond films (2006) and Steve Bell cartoons (1996), to Joanne Sharp's (2000) examination of the geo-political discourses of *Reader's Digest* magazine. But we will focus on a body of work by the political geographer Jason Dittmer (2005) on the geo-politics of Captain America cartoons. Dittmer grounds his work in a study of the role of popular culture in shaping people's geo-political imagination. 'Popular culture,' Dittmer (2005: 626) argues, 'is one of the ways in which people come to understand their position both within a larger collective identity and within an even broader geo-political narrative, or script.' By studying the plots, imagery and characterisation of Captain America, Dittmer argues that the cartoon – and most significantly the character of Captain America – play an important role in formulating American geo-political imaginations. Specifically, Dittmer (2005: 627) argues that Captain America cartoons are able to:

> connect the political projects of American nationalism, internal order, and foreign policy (all formulated at the national or global scale) with the scale of the individual, or the body. The character of Captain America connects these scales by literally embodying American identity, presenting for readers a hero both of, and for, the nation.

This work has therefore helped illustrate the role popular culture plays in normalising particular understandings of place and identity within particular states. Challenging this work, we could ask the question of how popular culture is interpreted; that is, how do we know that the interpretation granted by the scholar of geo-politics is shared by other members of a community? Work in this field is still at its early stages, but Dittmer and Dodds (2008) attempt to assess this point in a study of audience reception and 'fandom'. The authors point out that 'scholars of popular geo-politics should complement an interest in the discursive analysis of representations with a concern for audiences and the meanings that they construct out of popular culture and related texts' (Dittmer and Dodds, 2008: 453). As we will see below, this approach is part of a broader critique of the primacy of discourse analysis within critical geo-politics and an attendant call to incorporate other research methodologies, such as participant observation and ethnography. For example, we can only start to understand how popular culture is received and interpreted by asking cultural consumers how particular aspects of popular culture shape their geo-political imagination.

Beyond critical geo-politics

The critical geo-politics approach has been widely adopted over the past 15 years, as scholars have sought to illuminate the spatial narratives and identities that are communicated in world politics. As we see above, this has led to a proliferation of work exploring the geographical assumptions and ideas encapsulated in the output of think-tanks, politicians, journalists and creative artists. This was a significant shift for scholars of political geography, who saw their sub-discipline adopt social and cultural theory in order to explain the processes of knowledge production within international politics. But in recent years geographers, among other scholars, have attempted to move beyond critical geo-politics and, in doing so, have identified a number of limitations in this style of critique. We will explore three pathways beyond critical geo-politics in this section: feminist geo-politics, anti-geo-politics and normative geo-politics. In doing so we will point readers in the direction of emerging areas of research that have sought to expand the theoretical and methodological concerns of critical geo-politics.

Feminist geo-politics

As we have seen, scholars of critical geo-politics sought to contextualise geo-political discourses in the intellectual and political contexts in which they were produced. This ever-expanding body of scholarship has sought to deconstruct texts and images in order to illustrate geo-political narratives and assumptions. Although this has widened the object of inquiry for political geographers, incorporating theoretical styles from cultural and social geography, this has also led to critique by scholars concerned at the silences of critical geo-politics. Dowler and Sharp (2001), among other feminist scholars (see Hyndman, 2006), have pointed to a number of these absences and suggest that attention to feminist geography can act as a corrective. We can draw out three specific criticisms. First, Dowler and Sharp point to the efforts that have been made to *situate* the geo-political narratives within the work of critical geo-politics. Rather than a 'view from nowhere', scholarship in critical geo-politics has involved a deft critique of the production of geo-political knowledge in formal political spheres (such as speeches and policy statements) and illuminated the power relations from which they emerge. But this approach has failed to simultaneously situate the pronouncements of scholars of critical geo-politics themselves.

Scholarship within critical geo-politics has emerged primarily from men within elite Western academic institutions, and can therefore be criticised as paying insufficient attention to the marginalised voices of women and scholars from the global south. Second, Dowler and Sharp (2001) suggest that the focus on formal political arenas has led to an absence of studies of the production of geo-political knowledge in everyday sites and spaces, which are more likely to involve women. In particular, they suggest increased intellectual scrutiny of the home as a site of geo-political discourse, as Hyndman (2006: 276) states, feminists 'have long argued that private-public distinctions serve to depoliticise the private domestic spaces of "home" compared to more public domains'. Third, they argue that the somewhat disembodied nature of critical geo-political texts leads to a failure to suggest alternatives. For scholars of critical geo-politics, Dowler and Sharp (2001: 167) argue, resistance is 'a textual intervention, a subversion of a sign or a displacement of meaning'. The preoccupation with deconstruction of texts may prove intellectually satisfying, but it fails to table transformative political alternatives. These critiques of critical geo-politics have fed into an alternative focus that prioritises the geo-political narratives of oppressed and marginalised individuals and groups, while simultaneously broadening the methodologies used to expose and explore the production of geo-political knowledge. These disparate attempts have been brought under the label 'anti-geo-politics'.

Anti-geo-politics

Anti-geo-politics seeks to draw attention to the forms of resistance to the production of geo-political knowledge within formal political arenas. In the words of geographer Paul Routledge (2006: 233), a key proponent of this field, anti-geo-politics draws attention to 'alternative stories' emanating from '**subaltern** (i.e. dominated) positions within society that challenge the military, political and economic and cultural hegemony of the state and its elites'. Therefore, anti-geo-politics draws our attention to the production of geo-political knowledge outside the state, through actors within civil society who are often seeking to resist the geo-political imaginaries of state actors. One key example of the production of anti-geo-political imaginaries is provided by Gearoid Ó Tuathail's (1996b) account of the reporting of journalist Maggie O'Kane during the Bosnian war (1992–1995). Ó Tuathail highlights the neutral and abstract geo-political explanations given by Western state elites for the conflict in Bosnia. These discourses described the war as a consequence of 'ancient ethnic hatreds', where the primordial ethnic differences between members of the Bosnian population led naturally to violent conflict. In contrast to these accounts, Ó Tuathail (1996b: 175) describes Maggie O'Kane's reports in the UK's *Guardian* newspaper as 'grounded and situated within the war ravaged sites of that country: within its starved cities, overflowing hospitals, blockaded roads, teeming refugee centers and vile concentration camps'. Drawing on the work of feminist scholar Donna Haraway, Ó Tuathail argues that O'Kane deploys an 'anti-geo-political eye' which severs the 'distancing scripts' that enframed Bosnia within dominant state discourses. The anti-geo-political eye provides a more grounded, situated and messy account of the conflict that illustrates the violence and complexity of the war, rather than reducing it to simple geo-political narratives that absolve any responsibility for the violence and reproduce a notion that it could not occur outside the borders of Bosnia.

As Routledge (2006: 233) states, anti-geo-politics 'can take myriad forms, from the oppositional discourses of dissident intellectuals, the strategies and tactics of social movements to armed insurrection and terrorism'. Through this definition we see that forms of anti-geo-political knowledge are not necessarily progressive (advancing goals of social justice and equality) but are rather defined by their *resistance* to dominant geo-political narratives. As an example, in the second edition of the *Geopolitics Reader*, Ó Tuathail *et al.* (2006) include Osama Bin Laden's (2002) *Letter to America*, a call to arms by the al-Qaeda leader that posits an alternative geo-politics to mainstream accounts of the 'war on terror' offered by the US presidential administration of George W. Bush from 2001 to 2008. For example, Bin Laden strives to connect the ongoing plight of Palestinians in Israel/Palestine with the emergence of Islamist terrorism towards the West, a connection that had been repeatedly rejected by political actors within the US and UK. Adopting a layered metaphor of society, anti-geo-politics is therefore often perceived as 'geo-politics from below', a form of knowledge production that seeks to displace and unsettle the geo-political stories of powerful agents of the state (see Box 10.4).

Box 10.4
Anti-geo-politics in practice: the *Narmada Bachao Andolan*

We can draw an example of anti-geo-politics from Paul Routledge's 2003a paper in the journal *Political Geography* exploring resistance to the Narmada Dam project in India. The paper explores the economic, political and social implications of the Narmada Valley Development programme (NVDP) which comprises a series of 30 mega-dams and over 3,000 smaller dams throughout the Narmada river valley area in the Indian states of Madya Pradesh and Gujurat (Routledge, 2003a: 244). Routledge examines the forms of resistance that have been mobilised against the NVDP, in particular focusing on the social movement *Narmada Bachao Andolan* (save the Narmada movement or NBA), a network of activists and non-governmental organisations (NGOs) that have demanded the halting of the dam project. Through an exploration of the NBA's structure and activities, Routledge charts the disconnect between official state sanctioned discourses that present the NVDP as part of India's post-colonial development and the NBA, which seeks to expose the human, social, economic and political costs of the dam project. In doing so, Routledge (2003a: 260) highlights the *discursive resistance* of the NBA to the NVDP, and explains:

> Discursive resistance, like its material counterpart, acts as a political disruption in the unanimity implied by state discourses regarding development. In India, development is coded within a moral idiom; being equated with a prosperous, civilised, advanced society. However, through various discourses, the NBA critiques state and corporate sponsored development, and articulates alternative forms of development.

Routledge draws particular attention to the use of *adivasi* (tribal people) and peasant testimonials to highlight the impact of the dam programmes on the everyday lives of people living in the area. These testimonies refute the geo-political narratives of progress and development emerging from the Indian State and the World Bank, and instead draw attention to the local consequences of the engineering projects in displacing individuals and families from their homes and livelihoods. Routledge (2003a: 263) explains:

> The testimonials offer visceral and passionate denunciations of state practices, and underpin the moral legitimacy of the NBA's struggle. They are entered as grievances in the public record, and together with other testimonials form part of the overarching narrative of *adivasi* experience in the Narmada valley.

The NBA has therefore provided an arena through which civil society can articulate resistance to hegemonic geo-political narratives. But more than simply resisting the proposed plans, these anti-geo-political discourses serve to present alternative ways of understanding the landscape of the Narmada Valley region. Routledge (2003b: 243) highlights the spiritual connection to place among the *adivasi* of the Namada Valley which 'intimately informs their customs and practices of everyday life'. While these narratives emphasise a local connection to place we must not assume that anti-geo-political discourses do not resonate beyond the local scale. The NBA's success is grounded in its ability to communicate the *adivasi* testimony beyond the Narmada Valley region and utilise information technologies in order to publicise its campaign and forge links with other activist groups across the world. We would therefore emphasise Routledge's (2003a: 260) point that the contestation between hegemonic geo-politics and anti-geo-politics takes place 'within different, yet entangled spatial registers'.

Whether this anti-geo-political focus attends to the criticisms of feminist scholars of critical geo-politics (see above) is still open to question. The geographer and solidarity activist Sara Koopman has argued that the term 'anti-geo-politics' is limited, since it focuses solely on resistance rather than the more productive concept of building an alternative to hegemonic geo-politics. Koopman (forthcoming) is clear. 'I am interested not only in pushing back against what I *don't* want, but I do that in order to make space for nurturing what I do want in the world.' Echoing feminist scholars, Koopman argues that critical geo-politics has been preoccupied by questions of discourse and representation, without taking serious political practice. In her work on the NGO Fellowship of Reconciliation (FOR) in Columbia, Koopman argues for the concept of 'alter-geo-politics', which she defines as grassroots struggles to build peace through solidarity. In her

Box 10.4 continued

doctoral research Koopman conducted long-term fieldwork with FOR, a group that undertakes protective accompaniment – the placement of US citizens (or other citizens of the global north) alongside communities in remote regions of Colombia threatened by violent displacement by armed gangs. The strategy to make spaces of peace through the placement of international accompaniers, Koopman argues, posits a form of alter-geo-politics, alternative geopolitical discourses that focus on the security of the body and community, rather than hegemonic discourses of state security. But perhaps just as importantly, Koopman's work draws our attention to the alternative methodologies adopted by scholars and activists interested in the legacy of critical geo-politics. The use of methods of ethnography and participant observation point to the need to collect situated and embodied accounts of geo-political meaning-making, rather than relying on abstract texts and images.

Normative geo-politics

As we have seen, criticism is central to critical geo-politics. This may seem like a self-evident point, but it creates some complex issues. In particular, it raises the question of to what extent is the practice of geo-politics a normative undertaking: going beyond a descriptive practice and instead presenting what *should be done*. It is worth reflecting on the philosophical question that undergirds this question: is act of criticism an implicit statement that there is an alternative way of proceeding that will lead to a better outcome (however that is understood)? The very language of anti-geo-politics and feminist geo-politics is suggestive of alternative ways of acting in the world that will lead to a more just and equitable world. Having such a vision for the world is one of the primary motivators for political action.

These may seem like very abstract considerations, but it is to these normative questions that a number of political geographers have begun to turn. The political geographer Nick Megoran (2008) has sparked debate by examining the normative language used by scholars of critical geo-politics and in particular Ó Tuathail's writings on Iraq and Bosnia. Megoran's central point is that scholars of critical geo-politics have been reticent to discuss questions of normativity, and this has led to inconstant moral commitments. He suggests that within scholarship on critical geo-politics, vocabularies of normative moral judgement 'tend to be almost throw-away remarks, rhetorical gestures that are neither elaborated nor critically grounded in any form of justification or explication' (Megoran, 2008: 474). There is an argument, advanced by the political geographer Simon Dalby (1996), that it is not the place of scholars of critical geo-politics to offer political alternatives, and it is instead their task to explore the political implications of knowing the world in particular ways. But Megoran challenges this position, suggesting that scholars of critical geo-politics *do* intervene in political debates, in particular concerning decisions of whether a particular military intervention should (or more usually should not) have happened.

Megoran explores the moral reasoning of critical geo-politics through Ó Tuathail's writings on Iraq (specifically the US invasions in 1991 and 2003) and Bosnia (where he focuses on the US response to the 1991–1995 war in Bosnia). Megoran's analysis exposes an interesting variation in Ó Tuathail's political position: he is opposed to the US interventions in Iraq, although he seems to lament the lack of military intervention in Bosnia. This divergence is explained by Megoran as a product of an implicit attachment to just war theory: a perspective that suggests there are certain times that war is necessary in order to avoid a greater harm that is possible in its absence. While Megoran advocates an alternative pacifist agenda, it is not so much the moral alignment that is of concern, more the lack of explicit discussion of moral issues:

> [I]n what circumstances, if at all, should a state be considered right in making or joining a war? The argument in this paper is simply that critical geo-politics has not properly grappled with this question in a systematic and consistent way. By virtue of opposition to certain wars but advocacy of others, by implicit use of just war categories and language in moral reasoning, it is de facto operating within the parameters of a version of just war theory. However, because this appropriation is not made explicit – indeed, because just war theory is at times summarily dismissed – its appropriation is partial. (Megoran, 2008: 493)

The growing body of research adopting a feminist or anti-geo-political perspective is suggestive of an increased need to consider questions of normativity, since both of these areas of work combine activism with intellectual inquiry. This is particularly appropriate in situations that deviate from the conventional geo-political concerns of interstate competition and conflict and perhaps bring in more localised scales of political action: for example, the city, the neighbourhood and the home.

Conclusions

This chapter has sought to introduce the concept of geo-politics and trace its emergence within geographical scholarship over the past century. As we stated at the outset, it is worth reading this chapter in conjunction with other chapters in this book, in particular those examining imperialism and feminism, as we see some important shared intellectual and political contexts. The central narrative of this discussion is a shift from classical to critical geo-political reasoning – marked by a move from scholars producing knowledge in the service of the state to a more critical perspective that sought to illuminate the assumptions that underpin the foreign policy pronouncements. As we have made clear, this classical to critical transition does not mean classical geo-political thinking has simply disappeared; rather it is no longer the primary approach of geographers.

As with every intellectual narrative, this story has not ended. The study and practice of critical geo-politics is continually evolving, as evidenced by the popularity of the international journal *Geopolitics* and the wealth of workshops and conference sessions dedicated to this theme. But just as we have argued the context within which these ideas emerged matters, so too does the ability of the academic to speak back to political contexts. In the final part of the chapter we saw how scholars of critical geo-politics are beginning to examine the moral and political implications of critique.

Summary

- Geo-politics is a contested term that is used in different ways in different situations. A particular distinction can be drawn between its usage in mainstream media as a synonym for international relations and its use within geographical scholarship as a term relating to the relationship between political divisions and space.
- Geo-politics emerged in geography as a term that related the political boundaries of states with questions of natural resources and environments.
- In the 1980s a practice of intellectual inquiry known as *critical geo-politics* emerged that used the ideas of Michel Foucault to study the geographical assumptions of foreign policy pronouncements.
- Studies of critical geo-politics have sought to illuminate the spatial assumptions across a range of formal (scholarly), practical (governmental) and popular (films, cartoons, music) arenas.
- Recent scholarship has looked to develop new areas of research beyond critical geo-politics, incorporating feminist concerns and advocating more activist approaches that do go beyond simply critiquing power relations and seek to transform them instead.

Further reading

Bowman, I. (1942). Geography Vs. Geo-politics. *Geographical Review*, 32(4): 646–658.

Dodds, K. and Ingram, A. (eds) (2009) *Spaces of Security and Insecurity: Geographies of the War on Terror*, Aldershot: Ashgate.

Hyndman, J. (2003). Beyond Either/Or: A feminist analysis of September 11th. *Acme*, 2: 1–13.

Kuus, M. (2007). *Geo-politics Reframed: Security and Identity in Europe's Eastern Enlargement*, New York and Basingstoke: Palgrave Macmillan.

Mackinder, H. (1904) The Geographical Pivot of History. *The Geographical Journal*, 23(4): 421–444.

Megoran, N. (2008). Militarism, Realism, Just War, or Nonviolence Critical Geopolitics and the Problem of Normativity. *Geopolitics*, 13(3): 473–497.

Ó Tuathail, G. (1996) *Critical Geopolitics*, London: Routledge.

Ó Tuathail, G. (2002) Theorizing Practical Geopolitical Reasoning: The case of the United States' response to the war in Bosnia. *Political Geography*, 21(5): 601–628.

Ó Tuathail, G. and Agnew, J. (1992) Geo-politics and Discourse: Practical geo-political reasoning in American foreign policy. *Political Geography*, 11: 190–204.

Smith, N. (2003). *American Empire: Roosevelt's Geographer and the Prelude to Globalisation*, Berkeley, CA: University of California Press.

Post-colonial geographies and the colonial present

Learning objectives

In this chapter we will:

- Introduce the concepts of post-colonialism and post-colonial geographies.
- Discuss the contributions of key post-colonial writers.
- Demonstrate how post-colonial theory can be put into practice through the doing of post-colonial geographies.

Introduction

The 20th-century novelist Somerset Maugham once described Raffles Hotel in Singapore as 'all the fables of the exotic East'. Venturing into the Long Bar in Raffles one is immediately transported to a different time and space, far away from the humid streets and buzzing traffic outside. It is here that the 'gin sling' cocktail was said to have been invented, to quench the thirst of Western travellers and colonial administrators. Today, Raffles is a space where tourists, ex-pats and businessmen accumulate to savour the lingering atmosphere of a passing colonialism. Such tourist attractions can be said to represent what Derek Gregory (2001) describes as 'the fatal attractions of colonial nostalgia . . . inscribed in contemporary forms of travel'. In a nostalgic recreation of the past, Raffles guests can take afternoon tea out on the veranda, visit the billiards room or lazily loll around in the wicker-cane chairs enjoying a refreshing, if extortionately priced, drink in the Long Bar. In here time stands still.

Staging colonialism: the Long Bar in Raffles Hotel, Singapore
Source: Alamy Images/Slick Shots

The bar is named after Sir Thomas Stamford Bingley Raffles, a British statesman known for founding the city of Singapore. In many senses, Raffles is a space for the staging of colonialism as the past is selectively preserved, celebrated and transformed into built heritage. Litter-dropping is not permitted in Singapore but visitors to the Long Bar are encouraged to throw their peanut shells onto the floor in an aristocratic, colonial gesture that mimics defiance. Above, wooden ceiling fans serenely flap like bird wings and it is easy to forget that the environment is thoroughly air-conditioned as the markers of modernity are hidden from view. Raffles is a site for manufactured nostalgia and 'post-colonial melancholia' (Gilroy, 2004). This is produced through selected acts of remembering and forgetting in which the Raffles experience has also been recreated in other cities in different countries. The artificiality of Raffles and its claims to colonial authenticity reflect what Brenda Yeoh (2003: 371) calls the 'not-so-hidden histories and not-so-absent geographies of imperialism'. It is this intersection between the past and the present, and the real and imagined that this chapter seeks to locate, observe and open out for inquiry.

In many ways this chapter is a direct response to geography's imperial past, as discussed in Chapter 1, as it attempts to expose and critique this master narrative. Rather than taking these practices for granted, post-colonialism seeks to demonstrate the enduring power of the colonial legacy in contemporary geographical practice and show how these power lines can be overturned. We begin by defining what is meant by the term post-colonialism and exploring how this might shape our view the world. We then go on to consider how these ideas are being taken up by a section of human geographers implicated in the production and practice of what can be termed *post-colonial geographies*. The core of this chapter centres upon a critical discussion of three key post-colonial scholars; namely, Edward Said, Frantz Fanon and Homi Bhabha. These writers have been chosen not only because they are prominent post-colonial theorists, but also because their work continues to influence and exercise the geographical imagination in many different ways. In this chapter we aim to draw out these connections between geography and post-colonial theory to develop a richer understanding of what a post-colonial geography might look like. In order to empirically ground these ideas the final section discusses the art of 'doing' post-colonial geographies to illustrate the multiple ways in which post-colonial theory can be put into practice.

Understanding post-colonialism

While the term post-colonialism is undoubtedly 'contested and diverse' (Blunt and Wills, 2000: 167), it may be helpful to begin to think about the idea in at least two ways. First, as a *historical concept* post-colonialism can be said to refer to the moment after colonial rule. This is the point at which colonial nations relinquish their authority over the territories, resources and the governance of people in another nation state. On the surface this may suggest a clean break with what has gone before, a decisive split between the colonial past and post-colonial present. However, as many post-colonial writers appreciate, the term is as much about recognising our historical relations with the past as opposed to writing these connections out of existence. Work done in this vein is undertaken 'after', 'beyond' and 'in the knowledge of' (Gregory, 2000) the colonial aftermath. As Jane Jacobs (1996: 25) asserts, 'post colonialism may be better conceptualised as an historically dispersed set of formations which negotiate the ideological, social and material structures of power established under colonialism.' Connecting the past to the present is integral, especially as older colonial powers may still maintain imperial control over certain territories and many newly independent states find it difficult to escape the shadow of the colonial past as they negotiate forms of neo-imperialism (see Chapter 1). This more cautious reflection of the post-colonial condition has led Anne McClintock (1995) to suggest that the term 'post-colonial' was perhaps prematurely celebratory. Here the past and the present are not held up as separate, tidy cubicles into which we can place and neatly separate out objects, relations, people and place. Instead they are the overlapping coordinates of time and space that fold into one another and indicate that the colonial present is very much alive.

Post-colonial geographers are keen to pluralise our understandings of geographical histories and traditions. Those such as James Sidaway (2000: 592) are invested in acknowledging the varied and complicated 'array of post-colonialisms'. The Dutch, British, Spanish, French and Portuguese empires have differently enacted colonial authority over the globe to differing effect. This generates a renewed attention to the imperial past and a critical rethinking of the place of the West in the world (Clayton, 2003). The promise of this revisionist approach opens up the possibility to understand post-colonial studies as a radical project from which to 'address the ways in which colonial power was exercised, legitimated, resisted and over-turned over time and space' (Blunt and Wills, 2000: 170). In this way, we suggest it may be more productive to think of the post-colonial as an ongoing process: a folding into and back upon a geographically situated and historically positioned set of power relations. For example, after the bombing of cities and the loss of lives following World War II, Britain turned to migrant labour from Commonwealth countries to help rebuild the nation and fill employment vacancies in transport, factories, textile production, health and other services.

Migrants from former British colonies including parts of Africa, India, the Caribbean and Ireland were once again encouraged to labour constituting what Avtar Brah (1996) has termed 'cartographies of **diaspora**' – those population movements choreographed through post-colonial linkages. Here, the past and present concertina in the folds of time and space.

A second way in which we might also consider post-colonialism is as a *critical concept*, sharpened with an anti-colonial edge. In this reading, a post-colonial perspective is one that challenges disciplinary norms and established categories. As James Sidaway (2000: 594) suggests, 'post-colonial approaches are committed to critique, expose, deconstruct, counter and (in some claims) to transcend, the cultural and broader ideological legacies and presence of imperialism'. It is less a temporal and more a critical point of departure. In the words of Anthony Appiah (1992) the prefix 'post' signals a type of 'space-clearing gesture', a movement beyond colonial orthodoxy. In viewing the post-colonial as a diverse, fractured but incomplete historical process we can gain new insights into such geographically pertinent issues as migrant labour, 'third world' debt and world development. As Homi Bhabha (1994: 111), whose work we later discuss, expresses, 'colonial domination is achieved through a process of disavowal that denies the chaos of its intervention'. Placing these relations in a post-colonial framework then acknowledges the role of Western powers in perpetuating global inequalities.

Seen in this light, the post-colonial, in its many guises, can open up new areas of inquiry and offer an 'alternative narrative' (Hall, 1996) to the classical story of Western modernity. The notion of **alternative modernities** suggests that modernity always unfurls across societies and that different starting points on that transition lead to different outcomes. For Dipesh Chakrabarty (2000), situating modernity within Europe has not only had an enormous impact upon what constitutes social scientific knowledge but it has meant that non-Western ideas from the 'periphery' must always be translated back towards the 'core'. For post-colonial scholars this act of representation can involve an element of loss in which some ideas and meanings are inevitably 'lost in translation'. Chakrabarty reminds us that for those writing outside of the West, Europe remains an implicit 'silent referent' that cannot be ignored, while Western ignorance of the global south is unapologetic. The suggestion here is that we need to engage with a more global 'cosmopolitan theorising' (Robinson, 2003: 276) that seeks to make provincial the enclaves of Western knowledge.

An important dimension of post-colonial work is the attempt to include the hitherto silenced voices of non-Western subjects and provide new forms of knowledge and experience. The value in 'recovering' and representing at least some of these experiences can be seen today in the increasing interest ascribed to Australian Aboriginal art, Chinese herbal medicine, African music or the complex algebra of Indian mathematicians. When we consider each of these examples it is interesting to note how they tend to be excluded from mainstream genres, literary and artistic canons, although these cultural forms carry with them a dense accumulation of knowledge, history and expertise. Despite the incorporation of Indian yoga, Buddhist meditation or Japanese massage into contemporary Western lifestyles, these practices seldom displace the widely held idea of modernity as a European construct. While these illustrations are suggestive of alternative modernities, the radical impulse to implode Western frameworks rarely occurs. Rather than involving a 'return' to the post-colonial, invariably these examples indicate the power of Western classification and its tendency to marginalise or incorporate these knowledges without altering its foundations. In this respect modernity is itself seen as a white Western construct that has sought to define culture, science and the arts in highly particular ways. The challenge of post-colonial theory is that it can open up new spaces for

Box 11.1
The limits of post-colonial practice – insights from Gayatri Spivak

Before advancing anything approaching a post-colonial project we might consider Gayatri Spivak's poignant question raised in her essay, 'Can the Subaltern Speak?' (1993). Spivak writes as a feminist post-colonial theorist and in using the term **subaltern** she is referring to the subordinated position of colonised people and the way in which their histories, experiences and voices are invariably hidden from view. Spivak reflects that the recovery of the past in its entirety is inevitably impossible. This task is made all the harder as the histories of this period, at least the majority that survive, are written by Western hand. Within this framework and in the context of colonial production, the subaltern has no history and cannot speak. If the subaltern is to speak they must adopt the language, forms of expression and Western systems of logic to make themselves heard. The muted voices of colonised people lead Spivak to argue that if subaltern peoples cannot speak on their own terms, are displaced and have no history outside of the West, then we should aim for alternative dialogues that listen to and work alongside the silenced subaltern subject. As one geographer has ventured, 'Much work that passes as post-colonial within geography hinges on the trials and tribulations of colonizers in other spaces rather than on the intersubjective contours of the contact zone in question' (Clayton, 2003: 364). All this suggests that post-colonialism is inherently a limited and incomplete project of recovery that seeks a reconciliation and reparation with the past.

thinking, understanding and representing the world differently, but this potential is continually contested and located across a grid of power, as illustrated in the work of Gayatri Spivak (see Box 11.1). Even so, if post-colonial imaginings are put into operation as critical ideas to be prosecuted across the discipline there certainly are possibilities for new and exciting geographies to emerge – as we shall now discover.

Post-colonial geographies

The colonial relationship between geography and empire was seen to be both intimate and intricate (see Chapter 1). What is most noticeable in work of this period is that the relationship was rarely questioned: it was simply what geographers did. And while geography's institutional heritage (Johnston and Sidaway, 1979) and imperial past (Driver, 1998; Livingstone, 1992) has been meticulously traced, political critiques of these dynamics are more recent (Godlewska and Smith, 1994). More lately, this has given way to its opposite: the emergence of a conscious and diverse set of practices in which authors explicitly identify their work as contributing to a broader project equated with the production and proliferation of *post-colonial geographies.*

> The aim of a post-colonial geography might be defined as: the unveiling of geographical complicity in colonial dominion over space; the character of geographical representation in colonial discourse; the de-linking of local geographical enterprise from metropolitan theory and its totalizing systems of representation; and the recovery of those hidden spaces occupied, and invested with their own meaning, by the colonial underclass. (Crush, 1994: 336–337)

Inspired by debates that raged in the late 1980s in literature, history, cultural studies and anthropology, geographers have begun the gradual and painstaking task of rethinking the discipline in the light of post-colonial transformations. Within this new framework 'Critical geographers look for signs of ambivalence, contradiction and the assertion of power in the geographical archive, and are showing that empire was deeply inscribed within the discipline of geography' (Clayton, 2003: 359). Another feature of these interdisciplinary engagements, epitomised in the phrase 'the empire writes back' (Ashcroft *et al.*, 1989), is the post-colonial possibility that the 'margins' could speak back to the colonial 'centre'. The effect of this has been to begin the complicated task to 're-orientate' the globe, in what amounts to a geographical and cultural repositioning of East–West relations.

Following this post-colonial reorientation, two prominent human geographers, Trevor Barnes and Derek Gregory, have gone as far to suggest that it 'is now unacceptable to write geography in such a way that the West is always at the centre of its imperial geography' (1997: 14). While Barnes and Gregory provide a righteous and impassioned remark on what *should be* unacceptable, we would contend that much human and physical geography continues relatively untouched by the impulse of the post-colonial challenge. Despite Blunt and McEwan's (2002: 1) earnest contention that the post-colonial challenge has seen a flowering of 'new ideas, a new language, and a new theoretical inflection for a wide range of teaching and research in human geography', we remain cautious about the extent to which these new vocabularies are shaking the discipline. There is little doubt that the empirical and symbolic de-centring of Western knowledge through post-colonial ideas *can* enable new geographical imaginings to emerge, but the extent to which they *have* done is, we suggest, more debatable. To begin with it is probably too early to make an adequate assessment of the influence of post-colonial ideas on human geography, especially given the longstanding 'geographical traditions' that have gone before. But to consider this debate in more depth budding geographers may be tempted to reflect upon their own degree and its modules to assess whether post-colonial ideas are mainstream, marginal or absent from the bulk of teaching and learning.

For those geographers – including ourselves – who are stirred by the theoretical and political insights post-colonial theory affords, it is worth reflecting upon the small but valuable contributions we can make in the development of post-colonial practice. Colonial practices are frequently recognised to be visible in discourse, text and representational form. We contend that post-colonial geographies can strengthen the 'literary-base' of post-colonial theory through providing empirical and spatial accounts of these processes. This enables us to root the 'flows' of post-colonial objects, subjects, cultures, finance and knowledge through geographical interpretations of space and place. In this way geography provides us with important tools through which we can make concrete the occasionally abstract textual referents of post-colonial literary theory and embed this through an analysis of place (McEwan, 2003). This material approach suggests that colonial relations are not simply discursive constructs but are invariably inscribed in architecture, urban form, built environments and numerous spatial engagements. As Anthony King (2003: 398) opines, it is integral that we engage with 'the materialities, built forms and physical spaces of the city' precisely because they 'affect and help to produce and reproduce social relations, identities, memories and subjectivities'. This can be seen in the city of Kolkatta (Calcutta), a headquarters for British colonial rule in India until 1912, where the imprint of the colonial past is designed into the landscape as seen through the numerous buildings, open

The colonial legacy inscribed in urban form: Victoria Memorial Building, Calcutta, India
Source: Alamy Images/World Pictures

parks, pathways and commemorative statues to British soldiers, politicians and colonial ambassadors (see photo). However, developments in sewerage, road surfaces and building infrastructures such as schools or hospitals suggest that colonialism may not be seen as entirely negative from the perspective of Indian citizens. With its meticulous radial planning, wide circular pathways and white-washed civic buildings, other major cities such as Delhi are testimony to the material power of colonialism in practice. Despite the violent histories and geographies buried within this architecture, the colonial splendour of these buildings is a reminder that post-colonial relations are imbued with *ambivalence* that continues to persist in the present despite Indian independence in 1947.

The development of post-colonial geographies has enabled a more critical and reflexive disciplinary turn to transpire often reflected in work in the 'new development studies', critical approaches to globalisation and so forth (McEwan, 2003). It has led to a direct concern with culture and power, and how they continue to impact upon post-colonial relations. This more tentative way of 'doing geography' means that 'cross-cultural understanding depends not on the mastery of meaning but on openness to difference, to developing an ear for the other, and on relations of translation' (Barnett, 2006: 158). It also means recognising the limits of what is possible within this particular geographical imagining.

Having explained the meaning of post-colonialism and gestured towards the potential of post-colonial geographies we will now turn to an analysis of the work of three landmark post-colonial writers: Edward Said, Frantz Fanon and Homi Bhabha. In each of these vignettes we aim to focus upon the key ideas of these individuals and show why their insights might be generative for post-colonial geographers today.

Imaginative geographies: the work of Edward Said

> [J]ust as none of us is outside or beyond geography, none of us is completely free from the struggle over geography. (Said, 1993: 6)

First published in 1978, Edward Said's classic opus *Orientalism* is heralded as, 'the single most important reference point for the emergence of post-colonial theory' (Barnett, 2006: 149). Said's work examines the complex interactions between power, knowledge and textual forms of representation and he remains an inspiration to critical and post-colonial geographers alike. His biography below captures the geographical displacement he experienced first-hand (see Box 11.2).

Said's focus of inquiry is upon the work of Western scholars and experts writing about the Middle East over the past two centuries. These writings formed a specialist area of scholarship known as **Orientalism.** At a rudimentary level, Orientalism is 'a field of learned study' (Said, 1995: 49) simply put, 'an academic discipline' (p. 50). But by elaborating upon the insights developed by the French philosopher Michel Foucault (see Chapter 9), Said convincingly argues that the practice of Orientalism is less a dispassionate agglomeration of facts and more a **discourse** replete with particular ideological meanings. Said likens Orientalism to a 'vast treasure-house of learning' (1995: 51) in which words, images, metaphors and tropes concerning the East become imbued with discursive meaning. This extensive repertoire of knowledge achieves legitimacy when intellectuals, poets, politicians, artists, novelists, military officials and a constellation of different actors contribute to and share in these ideas. This enables them to function, in a Foucaultian sense, as a 'regime of truth'. Said explains how Orientalism as a field of knowledge is then transformed into a highly exacting ideology:

> In a sense Orientalism was a library or archive of information commonly and, in some respects, unanimously held. What bound the archive together was a family of ideas and a unifying set of values proven in various ways to be effective. (Said, 1995: 42)

Box 11.2
Edward Said: a biography of displacement

Edward Said was the son of a wealthy middle-class family living in Jerusalem. He came from a Christian-Arab background, was musically literate and versed in numerous languages including Arabic, English and French. As he entered his teenage years, political conflict led to his exile from Palestine and he completed his secondary schooling in the US. This experience of exile left a fundamental mark upon Said, although it could not still his intellectual ardour. In 1957 Said graduated from Princeton and seven years later he had a Master's and Doctorate from Harvard University. There is no doubt that Said's education, class privilege and diasporic experience were formative of his ideas and sense of being. Indeed, along with the publication of some 20 books and several journal articles, Said was also regularly featured in national newspapers and television media. The breakout of the Arab–Israeli war in 1967 had meant that the question of Palestine would always be central to Said's thinking and writing and he was frequently deployed as a commentator. As a member of the academic elite, living and working in the pro-Israeli environment of US universities, Said was critical of the tendency to cast Arabs as 'terrorists' and 'fundamentalists'. His outspoken critiques of Western foreign policy in the Middle East continued to be politically insightful and deeply controversial, right until his death from leukaemia in 2003.

Orientalism then, comprises an assemblage of information, what Said terms 'a family of ideas', that are woven through and alongside one another in an inter-textual fashion. The interlacing of these different knowledges reveals the 'sheer knitted together strength of Orientalist discourse' (1995: 6). Said uses the power of discourse and textual analysis to reinterpret the selected works of Orientalist scholars. Through these meticulous engagements he concludes that Orientalism is, above all, 'an exercise in cultural strength' (1995: 40). If we consider discourse to be the unstable resolution of an ideological struggle for meaning, Orientalism can be rendered 'a Western style for dominating, restructuring and having authority over the Orient' (Said, 1995: 3).

Having established Orientalism as an ideological device, Said contends that it is also the product and purveyor of an 'imaginative geography' (1995: 49), a theme pursued throughout *Orientalism* and in his later book *Culture and Imperialism* (1993). He remarks how for a number of Orientalist scholars the East was revealed by way of a 'textual attitude' gleaned from a mishmash of secondary sources, fictional accounts and collective imaginings. This is not all that surprising given that with the exception of sailors, bureaucrats, soldiers and business elites most people would not have encountered the Orient directly. Said argues that the Orientalist scholar, posing as an expert, does little to dispel these myths but invariably confirms these geographical imaginings and presents them as 'bare reality'. This imaginative geography means that 'the Orient was something more than what was empirically known about it' (1995: 55) as it was embroidered from the cloth of colonial fantasy. Here, the Orient is frequently imagined as a timeless landscape, a space to be filled with romance, adventure, fear and desire. Seen in this light the Orient is the product of a deep and enduring set of power relations elicited through the 'collection of dreams, images, and vocabularies available to anyone who has tried to talk about what lies east of the dividing line' (1995: 73).

In speaking directly about geography, territories and spaces Said has been credited with nurturing 'a spatial turn in post-colonial studies' (Clayton, 2003: 357). This has inspired geographers to consider the spatiality of colonialism and empire and the ways in which territories are conceived through complex configurations of power. What is also evident is Said's fixation with power and the mutually constitutive relationships between the East and West. His text *Orientalism: Western Conceptions of the Orient* (1995) is, as the subtitle reminds us, 'a book not about non-Western cultures, but about the Western representations of these cultures' (Loomba, 1998: 43). In other words, we would suggest that the 'fabulous' narratives of Orientalists may in fact tell us a great deal more about Western civilisation than they can ever impart upon the 'inscrutable' East. It is not too far-fetched to venture the idea that Europe came to know itself through imperial encounters. This construction of self and Other occurs through an interplay of opposite signs, what we might call a binary, that is used to define one against the other.

Looking at the racial dualisms identified above, it is vital to remember that these are 'fictional representations' masquerading as facts. In this way, Said regards Orientalism as Europe's surrogate and underground self, a distorting mirror which holds up a reflection of the West unto itself. Orientalism is then part of the internal dialogue that arises when European nations believe they have little to learn from seemingly 'barbaric' states elsewhere, as explained in Paul Cloke and Ron Johnston's (2005) clear distillation of the Self–Other concept. In Table 11.1 it is the West that has the power to name and define itself in

Table 11.1 Imagined geographies of self and other: constructing race binaries

The West	The East
• Rational	• Irrational
• Civilised	• Barbaric
• Safe	• Dangerous
• Sexually restrained	• Sexually licentious
• Familiar and knowable	• Strange and unknowable
• Masculine	• Feminine

relation to a subordinated and imaginary East. It appears as the 'master signifier' while the East is constituted as deficit or 'lack'. Furthermore, these projections can be seen as a means through which the West handles its own fears and insecurities by displacing these paranoid fantasies onto the silent and seemingly blank screen of the Orient. It is unlikely, therefore, that inhabitants in the Middle East would have understood themselves in these terms. Nevertheless, the power of these myths over time may lead colonised subjects to begin to view themselves as 'unfamiliar strangers', an issue we shall return to in our discussion of the post-colonial theorist Frantz Fanon.

What is also interesting if we look at the dichotomy above is that it is more ambivalent and troubling than it may at first appear. It is possible to rework these dualisms where the idea that the West is 'safe'/boring lends itself to the exhilarating possibilities that the East is infinitely more 'dangerous'/exciting. The master logos of Eurocentricism are then undermined by their own internal logic and the incapacity to fully represent the Other. These engagements reveal the anxiety embedded in all colonial stereotypes, a theme we will later address in an explication of the work of Homi Bhabha.

Notwithstanding its totemic significance, Edward Said's *Orientalism* has been subject to debate, critique and theoretical elaboration. It has been suggested that his focus has been overly textual, failing to account for the material and social forces of colonialism. While a tendency towards an 'excessive textualism' is evident in much post-colonial theory, Said's focus on text, discourse and representation has been largely generative for human geographers. Indeed, Clive Barnett (2006) provides a strong case for regarding the textual dimension of Said's approach as a radical intervention in geographical thinking. Post-colonial scholars would also suggest that discourses have material effects. If we consider travel brochures and tourist magazines, how we speak about, write, map, paint or photograph places has a physical bearing upon those spaces, which come to be understood in these acts of representation. Textual and material relations are then not so easily separated. Said himself asserts that the Orient is not merely an imaginative construct, but 'an integral part of European material civilization and culture' (1995: 2).

Critiquing *Orientalism*

As *Orientalism* was very much a product of its time, it inevitably has some shortcomings. An interesting absence in Said's thesis is the dearth of non-Western voices, representations and discourses. These silences that mark much Western writing on the East tend to ignore the ways in which colonial authority is contested, resisted and occasionally overturned.

Critics beyond the West have argued that Orientalised subjects cannot be understood as simply complicit in their subordination, 'passive, inarticulate and lacking self-determination' (Morin, 2004: 241), but often challenge many of the imaginative geographies ascribed to them. And while Said is not of the West, his biography imparts that neither was he born into an Islamic family. As Morin further remarks, Said 'was a political exile but a wealthy, privileged, "cosmopolitan" one' who ultimately occupies a 'paradoxical location' (2004: 238). Although this does not detract from his penetrating insights on Islam and the West, it alerts us to the ways in which Muslim commentators and spokespeople from the Middle East are regularly omitted from intellectual and media debates on power and conflict. We could even argue that BBC foreign correspondents among many others are modern-day 'Orientalists'; experts whose voice and opinions become the master narrative in politics and foreign policy.

Second, Said's book has been criticised by feminist scholars for the relative lack of attention paid to issues of gender. More recently these writers have provided rich, textual elaborations of Said's thesis. In her book *Gendering Orientalism*, Lewis (1996) argues that women are barely represented in Said's account and that where gender is spoken it is deployed as a metaphor for the negative characterisation of the Orientalised Other as feminine – the 'lack' we identified previously. She further remarks that Said's focus upon the writings of male Orientalists absolves white women from their involvement in colonialism. As we saw in Chapter 6, Said's writings have inspired a number of feminist geographers to examine the complex and contradictory experiences of women travellers and their ambivalent relationship to empire.

Third, some critics have further questioned Said's textual methodologies and his choice of Orientalist writing. A debate has hinged regarding Said's rather eclectic patchwork selection of Orientalist texts. To what extent, we may wonder, are these arbitrary narratives or actually representative of a broader European Orientalist tradition? Furthermore, Orientalists also used different methods when it came to describing the Middle East. In particular, cultural geographer Clive Barnett (2006) suggests that Said fails to distinguish between those who encountered the Orient textually and those who experienced it first-hand. The implication is that direct encounters with the East are likely to challenge textual imaginings of the Orient or at least produce other ways of 'knowing' that require more detailed interrogation.

A fourth criticism is the tendency in Said's early landmark account to homogenise and occasionally essentialise the West, what he terms the 'Occident'. This only serves to reinforce the binary between the Orient and the Occident. In this vein Morin (2004: 241) appreciates how 'texts, discourses and representations about the Orient are considerably more ambivalent, heterogeneous and dynamic' than they appear in Said's initial deconstruction of Orientalism. At the time when his first book was published Said reflected how, unlike Orientalism, 'no one is likely to imagine a field symmetrical to it called **Occidentalism**' (1995: 50), that is the study of the West. This statement ignores the fact that non-Western scholars produced their own accounts of the Occident, although until more recently these have evaded the Western gaze. Following Said's intervention, several writers have risen to the challenge to produce accounts on Occidentalism. In his book *The Idea of the West: Culture, Politics and History*, the social geographer Alastair Bonnett (2004) draws upon early historical portraits and representations of the West from Russia, Japan, China and Iran. He reveals how non-Western imaginings of the Occident were certainly in existence and, moreover, that these representations cast the West in a deeply ambivalent light. Where the West is often portrayed as free, democratic and utopian, Bonnett is one among a number of scholars who show how it is simultaneously constituted as 'soulless', 'slack' (i.e. work-shy) and, more recently, caught in a paranoid neoliberal dystopia of its own making. Such Occidentalist narratives inform us that the West is neither homogeneous, nor necessarily the glowing beacon of modernity that other nation states strive to follow. This more complex and contingent relation implies that the East can 'write back' and that Orientalist discourses are more contested and less stable than may be imagined. The performance of Occidentalism also reveals the mutually constitutive imagined geographies of East and West.

Summary

- Post-colonialism reflects the period after colonial rule and may also refer to an anti-colonial approach in thinking and writing.
- Said's notion of 'imagined geographies' enables us to see how East and West exist as mutually constitutive relations.
- The term Orientalism implies that it is Western powers that have taken centre-stage in representations of self and Other.

Splitting race objects: the work of Frantz Fanon

Frantz Fanon is one of the freshest and most original voices to have emerged from the fault-line of post-colonial transition. His prose is sharply analytic, occasionally sprinkled with humour but at once deeply unsettling. In his first and most celebrated volume *Black Skin, White Masks* (1978), Fanon adopts an incisive psycho-social approach to the thorny question of race, that entails psychological and social understandings of racism. This focus on the interior territory of the mind is particularly instructive for geographers who have tended to concentrate their accounts on physical rather than mental spaces. Fanon argues that 'what is often called the black soul is a white man's artifact' (1978: 12). The revelation that what is thought of as blackness is in actuality the product of white fantasy becomes a means through which race difference is produced. For Fanon, 'blackness' and 'whiteness' are arbitrary signs rather than 'proper objects'. In analysing the dynamics of colonialism in French-occupied Algiers, Fanon uncouples race signs from their respective bodies to show how signifiers such as 'black', 'white', 'mulatto', 'negro', 'coloured' and so forth are relational and mutually constitutive. However, although whiteness and blackness are produced through one another, there is little doubt that they are marked with different configurations of power and carry unique historical meanings. In colonial arrangements whiteness is a dominating 'master logo', leading Fanon to reflect, 'There will always be a world – a white world – between you and us . . .' (1978: 87). At one level Fanon recognises the material realities of race – 'the fact of blackness' (1978: 77) – that leads to the assemblage of an impassable race binary wherein, 'The white man is sealed in his whiteness. The black man in his blackness' (1978: 9). But, as we shall go on to discover, to read Fanon purely through this simplistic set of dualisms cannot account for the deeper ambiguities he strives towards.

Box 11.3

Frantz Fanon's Biography of Displacement

Born in 1925 on the French-owned West Indian island of Martinique, Frantz Fanon left the island as it came under Nazi control. Escaping to Dominica, Fanon volunteered for military service in the French Army in 1944 and fought in North Africa and Europe. After World War II Fanon returned to Martinique before taking up a French scholarship in Lyon, which was awarded to him as a veteran of war. It was here that Fanon trained as a medical psychiatrist, soaking up the heady atmosphere of student radicalism while becoming intensely aware of the surrounding racism. During these turbulent times Fanon became editor of *Tam Tam*, a magazine aimed at black students. In 1952 Fanon published his remarkable volume *Black Skin, White Masks* before he was on the move again, taking up a position as Head of Psychiatric Department in Algeria a year later. Here he witnessed the outbreak of the Algerian Revolution, as Algeria had been a French colony from the mid-19th century, but for two years Fanon aided Algerian rebels before he was forced to resign in 1956. By 1959 he had published *A Dying Colonialism* and two years on his book *The Wretched of the Earth* signalled the political and psychic consequences of the struggle for Algerian independence. Married to a white Frenchwoman, Fanon continued to work and move around North Africa before contracting leukaemia. After seeking treatment in the Soviet Union it was with deep reluctance that he was forced to seek medical treatment in the US where he died at the end of 1961, aged 36. Algeria officially became independent in 1962 and fittingly Fanon's body was buried in the Algerian battlefields shortly after the North African rebels had triumphed.

Blackness, whiteness and psychoanalysis

For Fanon, race signs are deeply internalised psychic processes, neatly illustrated in his book's title, *Black Skin, White Masks.* In carefully side-stepping the more obvious strap-line – Black Skin, White Skin – we witness instead a conscious and effective rupturing of the racial binary. The appendage speaks, then, of the *splitting* of race 'objects' and a psychoanalytic understanding that race is not tied to bodies. Even so the negative feelings associated with colonial disempowerment are often embodied through an 'epidermalisation' of racial inferiority and the taking up of 'white masks' by black subjects. Fanon's own blackness and racialised persona frequently take centre-stage in these writings as he plays with the oscillating nature of race signs. Hence, Fanon discloses, 'There is no help for it: I am a white man. For unconsciously I distrust what is black in me, that is, the whole of my being' (1978: 136). Thus, we learn how, 'The Antilles Negro who wants to be white will be the whiter as he gains greater mastery of the tool that language is' (1978: 29). In attempting to 'whiten' their identities Fanon suggests that the Negro is trapped within an impossible dilemma. The tension in this situation is unbearable, 'I burst apart. Now the fragments have been put together by another self' (1978: 77). Within this schematic, Fanon is 'Sealed into that crushing objecthood' of being seen as a Negro. His blackness is a prefix that prefigures all that he is and all that he does. The binding quality of the gaze leads Fanon to recall how 'the glances of the others fixed me there, in the sense in which a chemical solution is fixed by a dye' (1978: 77). Such scientific metaphors liberally pepper *Black Skin, White Masks* and are not incidental. Fanon was a trained medical psychiatrist and explicitly declares, 'This book is a clinical study' (1978: 11). In another text, *The Wretched of the Earth* (1968), Fanon adopts a similar clinical approach to Freud, providing in-depth and harrowing psychoanalytic profiles of black male patients who have been traumatised by the brutality of colonialism, for example enduring torture, seeing their wives raped or their children killed. The traces of Fanon's medical expertise can be seen in his forensic analysis of black/white relations, where he notes how gestures, glances and mere utterances serve to Other, and make palpable the idea of the Negro.

> And already I am being dissected under white eyes, the only real eyes. I am fixed. Having adjusted their microtomes, they objectively cut away slices of my reality. I am laid bare. I feel I see in those white faces that it is not a new man who has come in, but a new kind of man, a new genus. Why, it's a Negro! (Fanon, 1968: 82)

The emergence of these immanent race markers is captured in countless biographical and autobiographical episodes which are the product of historical and colonial relations.

Fanon's encounters with whiteness are as tortured as they are inevitable, symbolised in his anguished utterance, 'All this whiteness that burns me . . .' (1968: 81). Fanon's mobile, splintered biography is depicted in Box 11.3. According to Caute (1970: 8–9) 'Throughout his life, Fanon was plagued and embittered by his encounters with racism'. In aiding Algerian rebels to secure independence from French colonial rule, he witnessed first-hand the brutal effects of torture, rape and death on the North African population. He also personally experienced racism, being forever the repository of racial stereotypes. 'I was battered down by tom-toms, cannibalism, intellectual deficiency, fetishism, racial defects, slave ships' (1968: 79), he recalls in listing the multiple typologies which served to produce him as Other. Moreover, this racialised Other – the white man's black man – presents Fanon with an unrecognisable image of himself, 'My body was given back to me sprawled out, distorted, re-coloured, clad in mourning in the white winter day' (1968: 80). So, Fanon is fixed, objectified and 're-coloured' in his interactions with whites. The enactment of white masks can barely conceal the psychic trauma, rupture and splitting these acts of displacement entail. 'What else could it be for me but an amputation, an excision, a haemorrhage that spattered my whole body with black blood?' (1968: 79).

Undoing race

Fanon's mobile, diasaporic biography, outlined above, also bespeaks a type of 'splitting' that has become a hallmark of his writing. The geographer Steve Pile (2000: 268) has remarked that Fanon is 'both coloniser and colonised', having served with the French Army in Algeria: an uneasy relationship to reconcile with his political and academic activity. His biography captures 'a life on the move', and the way in which he is frequently exiled and displaced from his surroundings. It is worth comparing Fanon's biography, above, with that of Edward Said to consider how their experiences may shape and give force to their geographical imaginings. It is a good idea at this point to write down the similarities they share, and the differences they inhabit. How might Said's and Fanon's different experiences shape their ideas?

Fanon has certainly had his critics, but his use of psychoanalytic methods and his concern with the multiple identities individuals inhabit have led to a renewed interest in his work. One of the most radical impulses running through Fanon's work is his commitment to undoing race 'objects'. In numerous statements he alerts us to the absurdity of race – 'The Negro is not. Anymore than the white man is', is just one jarring illustration. His point is that race difference is not real. Writing prior to the rise of black separatist movements in the late 1960s Fanon remarks on the geographical impossibility of presuming common ties or goals between black people across the globe, a core theme that runs throughout *The Wretched of the Earth* (1968). Fanon's rejection of essentialism – the mistaken belief that black people share distinct characteristics – also countered the celebratory affirmation of African culture that many black nationalists and anti-racist activists advocated throughout the Civil Rights Movements in the 1960s. Ever sceptical, Fanon points to the folly of subscribing to these compelling race tropes. He writes of the implacable race binaries this logic enforces:

> . . . the idea of an old Europe to a young Africa, tiresome reasoning to lyricism, oppressive logic to high-stepping nature, and on one side stiffness, ceremony etiquette and scepticism, while on the other frankness, liveliness, liberty and – why not? – luxuriance: but also irresponsibility.
> (Fanon, 1968: 171)

Fanon appreciates the seductive appeal of race binaries but in deciding to resist them he chooses to write 'against race'. It is worth pausing here to look again at Table 11.1 to think how racial binaries can be re-worked. Even where these dichotomies may ascribe 'positive' qualities to blackness and re-signify the traditional race signs that impart, 'Sin is Negro as virtue is white' (1978: 98) Fanon remains unconvinced. In looking beyond racial dualisms he is critical of a nationalism assembled on the bedrock of essentialism and a flawed, 'unconditional affirmation of African culture' (1968: 171).

Just as Stuart Hall (1993) was later to argue for the end of 'innocent notions of a black subject' Fanon also contends that black subjectivity cannot be reified in this abstract and romantic fashion, something akin to early anthropological myths of the 'noble savage'. Rather, as Fanon alludes in *Black Skin, White Masks*, it is because black masculinity is a source of 'fear/desire' within white imaginings that it can be all of these things and yet none at all. Fear and desire are simultaneously projected onto black bodies, as can be seen today in the hyper-real images of black athletes or 'muggers', rap stars and pimps. The circulation of these race signs through global media is now commonplace, where fear and desire can be used to connote black masculinity simultaneously as dangerous and desirable. Ultimately for Fanon the 'fact of blackness' (1978: 77) rests with the knowledge that there is no comfortable essentialist race 'facts' that set black bodies apart and upon which we can anchor meaning.

Geographical contribution

Fanon's geographical contribution may appear more subtle than Said's, although it is equally significant. In many ways Fanon opens up new spaces in geographical thinking by alerting us to the interior landscapes of the mind and how this shapes perceptions of Self and Others. Significantly he shows us how colonialism works at the most intimate scale as it is lived in and through the body. Here we find that some bodies are more mobile than others and can pass through space unimpeded. Others are racially marked and come to be stuck in place as Fanon finds when a frightened child points to him on a bus and screams, 'Look Mama, a Negro!' (Fanon, 1970). The geographical implications of this process are today present when it comes to the enactment of state security, police stop-and-search activities, the control of state borders, school expulsion or other practices in which black bodies are a source of recurrent surveillance and moral panic.

Much of Fanon's writing provides fascinating discussion of borders, territories and communities. In *Wretched of the Earth* Fanon draws attention to the uneven geographies that distinguish the 'developed' from the 'undeveloped' world. Steve Pile (2000) adopts an innovative approach by considering what difference geographical location makes to interpreting Fanon and is drawn to his anti-essentialist understandings of space and place.

> From Martinique, through France, to Algeria, Fanon attempted to understand the political possibilities implicit in these troubled spaces. He invoked a politics of identity suspicious of claims to authenticity, whether located in identity or place. It is here that it is possible to re-imagine a politics of 'where you are', but about the ways in which 'where you are' is bound up in other spaces, other places, other people. (Pile, 2000: 275)

Other geographers have identified Fanon as 'a key precursor to post-colonial theory' (Kipfer, 2007: 703). Kipfer applies Fanon's work to recent debates in France which have outlawed the wearing of religious iconography such as the *hijab*, the Islamic headscarf, in schools. For Kipfer, Fanon's writing is particularly potent as it articulates the spatial organisation of colonialism and illuminates the way in which popular racism operates as an alienating spatial relation. With this regard Kipfer ventures that, 'Fanon's perspective on racism was thus thoroughly historical and geographical at the same time' (*ibid.*). The productive nature of Fanon's writing is also seen in Anoop Nayak's (2003a) ethnographic research on race, place and youth cultures. Here the concept of fear/desire is deployed to elaborate upon the complex social and psychic interplay that has seen many white youth attracted to the 'bad', 'dangerous' and down-right 'dirty' aspects ascribed to black culture which are commodified and materialised through global consumption.

Summary

- Fanon suggests that race has no reality but only exists as a social and psychic phenomenon.
- The metaphor of *Black Skin/White Masks* implies that as a consequence of colonialism many black people are filled with self-loathing and wish to 'whiten' their identities.
- Fanon's writings are geographically important to our understanding of how identities and communities are constructed.

Hybridity and the third space: the work of Homi Bhabha

Homi K. Bhabha is a post-colonial literary scholar whose complex but thought-provoking ideas have enriched the study of colonial discourses. His work is a lynch-pin of post-colonial theory and provides fertile ground for rethinking the geographical constitution of culture. In many of his essays Bhabha draws upon the insights generated by Fanon and Said to develop a psychoanalytic and discursive approach to colonial relations. His contribution is neatly summed up by Jane Jacobs (1996: 26) who declares, 'Bhabha is concerned with colonialism's own vulnerability to itself'. It is this vulnerability – the ambivalence at the heart of empire – that shall be interrogated further. This shall be done in two ways, first by considering the role of 'colonial stereotypes' and second by discussing what Bhabha has called the third space of '**cultural hybridity**'. Bhabha's own hybrid belonging is illuminated in his personal biography (see Box 11.4).

Box 11.4
Homi K. Bhabha's biography of displacement

Homi Bhabha was born in Bombay (Mumbai) in 1949, two years after India secured independence from British colonial rule. In this respect his biography captures the post-colonial *zeitgeist*. Bhabha descended from a Parsi family, a Persian minority, and grew up speaking English and Gujarati. After completing an English degree in Bombay, Bhabha enrolled as a postgraduate student in Oxford where he proceeded to read English at Master's, then PhD level. In 1978 Bhabha was appointed as a lecturer in English at Sussex University where he became especially interested in French philosophy and psychoanalysis. He was outspoken in challenging the negative representations of Islam in the media that were widely circulated after the publication of Salman Rushdie's *Satanic Verses* and the imposition of a *fatwa* (death sentence) by the Iranian Ayatollah Khomeini against the author. Bhabha's edited collection *Nation and Narration* (1990) reveals his broad interests in how nation states are constructed through discourse. Many of his ideas are synthesised in his 1994 collection *The Location of Culture* (1994) which brings together a number of previous essays he has published. In that year Bhabha moved to the University of Chicago, US, and is currently Professor of English and American Literature and Language at Harvard. These migrant, scattered belongings are suggestive of the different geographies Bhabha has moved through and his location as a culturally 'translated' man.

Colonial stereotypes

Following Said (1995) Bhabha argues that colonial discourse proceeds through a familiar rendition of words, images and symbols. He suggests that colonial stereotypes have a 'fixing' quality, holding in place complex social relations that are instantly reduced and then reproduced in miniature form. However, Bhabha takes this further by arguing that underlying this is a deep ambivalence. For Bhabha (1994: 66) 'the stereotype must always be in *excess* of what can be empirically proved or logically construed'. It is because the stereotype is an exaggeration that its excesses seem to leak into everyday life. Bhabha maintains that these excessive words and ideas are fearfully repeated and that the colonial stereotype is 'as anxious as it is assertive' (1994: 70). Bhabha's psychoanalytic interpretation of colonial powers as anxious, uncertain and fearful departs from the cool mastery Said (1995) imputes to colonial rulers. In Bhabha's reading, the coloniser is dependent on these stereotypes and even requires them to dehumanise the colonial subject. Here, stereotypes are much more than simple depictions of a given reality, but are grotesque contortions of enormous imaginative proportions. These stereotypes are brought to life in a colonial system of representation that transforms these dark imaginings into something approaching 'reality'.

Such anxious colonial stereotypes are repeatedly found in Victorian adventure fiction, and the poems of authors such as Rider Haggard, Aphra Benn or Rudyard Kipling, popular texts ripe for post-colonial geographical analysis. They are also seen in contemporary films such as *The Mummy* series, the *Raiders of the Last Ark* trilogy with Harrison Ford, or James Bond films such as *Live and Let Die*. In this latter film, set on an imaginary Caribbean island, the excessive construction of the colonial stereotype is represented through the tropes of human sacrifice, snakes, bones, chanting, voodoo black magic and the spectacle of reincarnation. These all too familiar tropes serve to establish a discourse through which blackness is 'made strange' and Other. For Bhabha (1994: 72) these excessive symbols reflect the ambivalence of the coloniser, which is central if we are 'to calculate the traumatic impact of the return of the oppressed – those terrifying stereotypes of savagery, cannibalism and alienation, scenes of fear and desire, in colonial texts'.

Colonial stereotypes then, are anxious registers. They create a *fetish* through which cultural differences, real or imagined, are reified. In other words, customs are stripped from their social context and endowed with a totemic almost magical quality. As Fanon has shown, black skin is one of the most prominent fetishes in modern society and consistently used to mark out difference, a drama that is spectacularly played out in the white imaginary in numerous anxiously repeated scenarios. The psychoanalytic notion of the fetish is associated with the perverse and the sexually compulsive. In this way Bhabha regards the exertion of colonial power as 'racial and sexual'; it 'is a form of discourse crucial to the binding of a range of differences and discriminations that inform the discursive and political practices of racial and cultural hierarchization' (1994: 67).

If we were to interpret films such as *Indiana Jones and the Temple of Doom*, modern re-makes of *The Mummy*, or *Live and Let Die* through Bhabha's insights on the colonial stereotype, we could consider the enormous effort undertaken to mark out the colonised subject as Other; a difference that is entirely visible and somehow knowable. This can be seen through these films in the familiar creation of Egypt, India and the Caribbean as mysterious, erotic but dangerous places. We might also think about the ways in which these filmic images seamlessly conjoin with earlier poems, novels, paintings and imaginings of 'foreign' places in a modern reworking of Orientalism. In each of these films particular characters come to symbolise and embody the excessive, fictional characteristics ascribed to race. In doing so, they comprise 'a form of narrative whereby the productivity and circulation of subjects and signs are bound in a reformed and recognizable totality' (1994: 71). Here, stereotypes have a 'fixing' quality and are dependent upon other 'types' to form a system of representation through which meaning can be derived and Western anxiety made rational. Such colonial stereotypes of racialised Others silently inscribe what it is to be 'white', 'Western' or 'British'. In this way Bond, played by Roger Moore, is then emblematic of a different national imagining that produces Britishness as suave, sophisticated and shaken-but-not-stirred when under pressure. And although films, advertisements and other texts of this kind are fictional, Bhabha is quick to point out how they present as 'realist' in a format that suggests pseudo-authenticity, very much in the way that Raffles bar does.

Bhabha's interpretation of an anxious colonialism complicates the idea of power and control as total. Unlike Said he provides a detailed interrogation of how colonial authority is taken up, resisted and transformed through its daily enactment. Rather than a colonial power imposing its will upon colonised subjects, Bhabha suggests that many encounters were the product of an intense, if sometimes brutal, negotiation. At the same time, other encounters could be more convivial and more easily accommodated by subjects within the colonial situation. What transpires in many contexts is a more partial, open and contested set of relations. The encounters between the Western 'core' and the non-Western 'periphery' involve a diverse set of transcultural 'flows' in which neither coloniser nor colonised remain untouched. For Bhabha this can create new spaces for negotiation and resistance that are hybrid or mixed. In deploying the term 'cultural hybridity', Bhabha breaks open the seal that purveys culture as authentic, homogeneous and securely bounded. Instead, the 'location of culture' for Homi K. Bhabha is a space of inbetweenness that cannot be pinned down. It is always in process, incomplete and forever in the making.

Cultural hybridity

Bhabha's use of the term hybridity borrows loosely from genetics in which two species of plant may be cross-fertilised to form a new, distinct species. Similarly, the mixing of two (already mixed) cultures inevitably produces new hybrid forms. Forms of cultural hybridity can be found in music, food, drink, fashion or art. For example, chicken tikka massalla is neither South Asian nor British but a 'hybrid' mixture that has developed through a series of Anglo-Asian encounters. This best-selling dish is one of many that have transformed British culinary habits and attitudes to 'foreign' food. In a similar vein, South Asian Bhangra music combines traditional Punjabi folk music with dance, house and Hip-Hop styles to give rise to something new and unrecognisable. It appeals to new generations of South Asian youth and has become a site for the celebration and performance of disasporic identity in the British context. These new spaces create alternative practices and points of identification. They involve an opening out of social relations that is creative and productive. As Bhabha (1994: 12) reflects, 'If the effect of colonial power is seen to be the *production* of hybridization rather than the noisy command of colonialist authority or the silent repression of native traditions, then an important change of perspective occurs.'

To this extent the imposition of colonial cultures upon colonised peoples rarely results in the production of sameness. For example, Bhabha considers the way Christian missionaries tried to impose their beliefs upon Indian society only to find that their ideas would be taken up and transformed through customs, rituals and new articulations of religiosity. This act of translation creates new interpretations and understandings that dissolve the bar that separates the 'strange' from the 'familiar', East from West, or 'Us' from 'Other'. For Bhabha such hybrid, bastardised, practices speak back to colonial authority in ways that appear grotesque and unrecognisable. He regards the hybrid as *'less than one and double'* (emphasis in original, 1994: 116) precisely because it fails to measure up exactly to the imagined 'original' and also because it partly reflects the original with something else besides. It is this 'something else', surplus, or excess that continually evades the binary order.

What is interesting about Bhabha's reading of hybridity is that he treats it not as an act of assimilation or 'selling out' but as a potential site for *resistance*, that can 'turn the gaze back upon the eye of power' (1994: 112). He explains:

> The paranoid threat from the hybrid is finally uncontainable because it breaks down the symmetry and duality of self/other, inside/outside. In the productivity of power, the boundaries of authority – its reality effects – are always besieged by 'the other scene' of fixations and phantoms. (Bhabha, 1994: 116).

Cultural hybridity may then involve strategic and symbolic reversals. Rather than an act of complicity, colonial mimicry can involve the subversion of authority. It can incite, 'new forms of knowledge, new modes of differentiation, new sites of power' (Bhabha, 1994: 120). For Bhabha the linguistic subtle slippage between 'mimicry' and 'mockery' captures the precarious qualities of mimetic, hybrid form.

Critiquing hybridity

Bhabha's approach to post-colonial studies is widely recognised to be sophisticated, generative and enduring, although it is not beyond critique. Robert Young (1990) has drawn attention to the biological origins of the term hybridity and the ways in which it carries negative colonial ascriptions related to miscegenation, that is the threat of 'racial mixing'. Although writers such as Stuart Hall (1993) make evident that all cultures are impure, the lingering history of colonial discourse produces bounded notions of culture and race. The longstanding colonial focus on 'hybrids', 'mongrels' and those deemed 'mixed race' remind us that the powerful inscriptions of race and their connections to the body are not easily crossed over and written out. For Young (1990) these racial typologies are formative of the 'white mythologies' of race and empire.

Although cultural hybridity may hold radical potential, the recuperation of these aspects is all too apparent within neoliberal, capitalist societies. As McEwan remarks, 'Hybridity and difference sell, but in the meantime the market remains intact, power relations remain unequal and marginalized peoples remain marginalized' (2005: 279). In encounters with a global market culture, difference and hybridity are transformed into marketable commodities. An example of this can be in the case of black cultural forms such as Hip-Hop music. The production and circulation of race signs is proliferated through music, image, film, MTV and other technologies where black bodies and imaginings of blackness are paraded through their associations with being 'urban', 'edgy', 'cool' and hypersexual. These characteristics are taken out of their local contexts and incorporated by market segments such as American suburban white teenagers who remain the largest consumers of Hip-Hop music. Such activities are an act of incorporation what the black feminist scholar bell hooks (1992) calls, 'eating the other'. Hybridity is then a feature of cultural globalisation and integral to the development of niche markets and the selling of 'difference' (Sharma *et al.*, 1996). In this market-based illustration, hybridity has the tendency to 'freeze-dry' culture and affirm rather than undo race categories.

Throughout his writing, Bhabha places an acute emphasis upon the shaky ambivalence of the coloniser, a prominence that underplays the forms of resistance that colonised subjects themselves invoke. In British colonial India this included a range of activities such as public marches, peace protests, anti-colonial speeches, armed resistance, hunger strikes, non-cooperation and much more besides. While Bhabha's analysis is subtle and refined, the material aspects of direct resistance and the visceral qualities of racism we find in Fanon's accounts are relatively subdued. In Bhabha's complex literary essays the subversive potential of hybridity appears to rest less with the agency of the colonised and hinge more upon the anxiety of the coloniser. However, as the support given to Mahatma Ghandi demonstrates, resistance to colonial authority by subaltern people was crucial in generating anxiety and ambivalence in the hearts and minds of colonial authorities who feared they were no longer able to subjugate the natives.

Although Bhabha's reference to the 'third space' of hybridity is clearly of geographical significance some critics have further suggested that his work could benefit from a more context-specific analysis of place and location. Pursuing this theme, the feminist post-colonial critic Chandra Mohanty (1991) has argued that the theme of domination needs to be placed within the material politics of everyday life, especially with regard to gender relations and the struggles of 'Third World' women. A challenge that Bhabha and other post-colonial writers in the West inevitably face is the task of truly transcending Western norms. As post-colonial geographers Alison Blunt and Cheryl McEwan reflect, 'One of the major dilemmas for post-colonialism is the charge that it has become institutionalised, representing the interests of a Western-based intellectual elite who speak the language of the contemporary academy, perpetuating the exclusion of the colonized and oppressed'

(2002: 4). Bhabha's continual references to English novelists who make up the literary canon, and European social theorists such as Freud, Foucault, Fanon and Derrida, are perhaps testimony to how post-colonial scholars are faced with an almost impossible task. The 're-membering' (Bhabha, 1994) and stitching together of the colonial past in a way that recovers hidden voices and experiences can never be complete. As the subaltern is ritually unable to speak (Spivak, 1993) the best that post-colonial critics may hope for is to be able to 'trouble' the category of race and the stuck together binaries of coloniser/colonised, primitive/civilised, East/West and Christian/Muslim (Sidaway, 2000). In doing so, post-colonial geographers may humbly strive to work 'between the lines' of power rather than imagining their ideas necessarily transcend the forces of colonialism and the violent histories of past occupation.

Summary

- Colonial authority is more ambivalent and less certain than we may have imagined.
- Cultural hybridity involves the mixing of different cultures (e.g. food, music, language) to produce an altogether new culture, a third space of negotiation.
- Cultural hybridity has the potential to subvert colonial authority through forms of mimicry even if this does not always occur in practice.

Doing post-colonial geographies

Throughout this chapter we have been connecting post-colonial theory with post-colonial practices. In this final section we shall explore the doing of post-colonial geographies. Here, we shall discuss examples from geographers who have used post-colonial ideas and applied them to their objects of inquiry, including Jackson's (1988) cultural history of the Notting Hill Street Carnival (see Box 11.5). James Sidaway (2000) alludes to the significance of this generative practice where, 'at their best and most radical, post-colonial geographies will not only be alert to the continued fact of imperialism, but also thoroughly uncontainable in terms of disturbing and disrupting established assumptions, frames and methods'. Cultural geographers and those working in the field of cultural studies have quickly embraced the creative potential of this post-colonial thinking. This has given rise to multi-layered accounts of popular culture, media, geo-politics, commodities, tourism, food, urban spaces and contemporary diasporas, all of which can be connected to a wider set of post-colonial relations. For example, in our opening illustration of Raffles bar we gestured towards the potential of using post-colonial insights to develop critiques of tourism, its economies and cultures. Through a selection of examples, we will consider how post-colonial ideas can materialise in geographical practice (see McEwan, 2003).

Over the years Derek Gregory has provided meticulous accounts of Said's work and exploited the geographical potential in his thinking. In more recent work he deploys critical geo-politics to consider the role of the West and the relations it has forged with Afghanistan, Palestine and Iraq; dynamics that can only be envisioned through a reckoning with *The Colonial Present* (2004), the title of his critically acclaimed book. Gregory reveals how much geo-political conflict today carries with it the echo of the colonial past as it comes to resound in the present. Global conflict and security remain dark exemplars of the colonial present when we consider the role of US detention centres in Guantánamo Bay, Cuba, the spectacle of torture and humiliation enacted upon the bodies of Iraqi soldiers in Abu Grahib, or the barbaric practice of 'water boarding' inflicted by Western troops upon Afghan soldiers and civilians to elicit information. These violations often contravene human rights, taking place in 'out of the way' places, black spots and unmarked areas in what Giorgio Agamben (1998) regards as 'spaces of exception'. Such violent geographies remind us that the 'colonial present' and the visceral encounters between East and West are enduring. This seemingly invisible 'ever-presence' is neatly captured by Anne McClintock's remark that 'colonialism returns at the moment of its disappearance' (1995: 11). In other words, it is precisely when we forget about colonial power that it is often at its most pervasive.

A potent geographical illustration of how the past permeates the present can be found in Anthony King's (1997) detailed post-colonial exposition of the suburban bungalow. King takes this seemingly quintessential English artefact to instead reveal how the bungalow is derived from a Bengali term generated at the confluence of colonial interactions. The emergence of these low-rise buildings throughout the Western world leaves King to conclude 'That the bungalow is both a product, and a symbol of a complex yet inter-related world' (p. 263). By tracing the alternative histories and geographies of such everyday material objects, King is involved in a critical post-colonial deconstruction of the bungalow that displaces it from its symbolic place in the English and Western imaginary. Similar historical geographies could be undertaken of any number of objects or practices as post-colonial accounts of the city and suburbs have shown (Driver, 1998; Nayak, 2010). A subaltern perspective is seen in the work of Rana P. Singh (2009a, 2009b) who argues that the spiritual and religious understandings of environmentalism in India can productively penetrate the field of development geographies.

Box 11.5
Carnival as post-colonial protest

Carnival represents a rich example of the way in which the colonial past and post-colonial present fuse together. Peter Jackson's (1988) early analysis of the Notting Hill Street Carnival in London reminds us of the complex configurations that connect local places with global events. In Trinidad, Carnival was an exclusive festival for elite white French-speaking Catholic middle-classes. After the abolition of slavery in 1834, liberated blacks used Carnival as a space of transgression to celebrate their freedom in what was an overtly political mode of resistance. However, Trinidad remained under British colonial rule for the succeeding 150 years and Carnival was denigrated as an immoral and dangerous activity. Jackson meticulously reveals how the moral panic around Carnival was to see the banning of masks and masquerade, and later in 1883 a law was passed restricting the beating of drums in public places. For Jackson, this was not an aesthetic judgement but one based on colonial anxieties linking black musicality, sexuality and fear of social unrest – concerns that bleed into the present. He documents how official sanctions against popular Carnival practices such as the use of cross-dressing, masquerade costumes, stick-fighting, foul language and so on were variously implemented. Jackson's point is that the politicisation of Carnival, its heavy policing and the resistance to this surveillance by festival-goers is nothing new: it is part of the colonial legacy through which events such as Notting Hill Street Carnival must now be situated.

The history of Notting Hill Carnival (see photo) is globally connected but locally specific. In 1958 Notting Hill along with Nottingham was an early site of 'race riots' where black communities were attacked and fought

Box 11.5 continued

Transgression at the Notting Hill Carnival, London
Source: Alamy Images/Network Photographers

back. Beginning as an 'English Fayre' in the 1950s, from the 1970s onwards the Notting Hill Carnival became a site for a variety of conflicts between different groups including people and the police, black and white, young and old, Jamaican and Trinidadian, blacks and South Asians and more recently between different neighbourhood youth gangs. As with Trinidad, Carnival has also changed in London where in the early 1970s it articulated a more distinct Trinidadian identity as youth unemployment rose and racial polarisation between blacks and whites became apparent. Towards the end of the decade second-generation British Afro-Caribbean populations increasingly aligned the event with Jamaica. Jackson goes on to discuss how the event was portrayed by the popular media as a type of joyous sunshine 'calypso Carnival', an 'explosion of West Indian colour and music'. However, a different set of racial mythologies came to surround subsequent Carnivals after violence erupted in 1976 as Jackson demonstrates through the ensuing racialised media representations. He maintains that since this time the symbolic meaning of Carnival in Britain follows its Caribbean precursor as part of a threatening culture in which authority can be overturned. Jackson regards Notting Hill as 'an intensely spatial event' from which to explore the intersection of culture and politics in the creation of a specific geography of protest.

Indeed, since Jackson's informative essay on Notting Hill other racial encodings of Carnival have taken place. It has been variously associated with a global mosaic of music including Calypso, Reggae, Hip-Hop, Ragga and most recently Brit-Hop (the latter occasionally fusing Cockney rhyming with Black British patois). The Notting Hill Carnival has also been composed as a site for different racial antagonisms viscerally traced in the murder of Abdul Bhatti, a 28-year-old South Asian man who was stabbed to death in 2000, when around 50 youths targeted Asian stall-holders. In the press reportage of that year the perpetrators are described as 'mainly black youth' engaged in acts of 'steaming', as they 'rampaged' through crowds stealing belongings. The event is described as a 'racial murder'. Today Carnival continues to be represented as an ambiguous space for conflict and conviviality. Jackson previously argued that the act of symbolic resistance seen at Caribbean and London Carnivals is inevitably expressed in the form of territoriality. In this way connecting Carnival to a post-colonial geography can be a means of wrestling with the different diasporic ambiguities that surround space and belonging. It also reminds us of the violence of authority and resistance.

Source: Jackson, 1988: 213–227

Visual methods and the post-colonial spectacle

One of the more evident ways of doing post-colonial geography is through the use of textual analysis and visual methodologies that explore how places are represented. In a valiant critique of Gap Year marketing, Kate Simpson (2004) discloses how a number of volunteer-tourist organisations produce and reproduce an Orientalist discourse that reduces the global south and its people to a set of simple descriptors: 'charming', 'lively', 'shy' or 'gracious'. What is produced is a 'fixed geography' that constructs boundaries between the global north and south in which the notion of 'development', the 'Third World' and the 'Other' is packaged through familiar colonial narratives and sold back to adventurous young consumers in the West. A further example of how geographical post-colonial perspectives can be used to deconstruct words and images can be found in Kate Manzo's (2006) visual study of development in which children appear as a recurrent motif in humanitarian aid imagery and publicity shots for African NGOs. You might wish to consider why NGOs frequently use images of children in their campaigns and what might be the consequences of this repetitive action over time. Manzo suggests that such development iconography may simultaneously be read as symbols of global humanitarianism or part of an established legacy that casts Africa as primitive, child-like and in need of rescue from paternal Western authorities. In this sense, these portraits are marked by a 'colonial ambivalence' (Bhabha, 1984) that connects contemporary images to longstanding historical relations. What such cultural accounts demonstrate is the persistence of colonial ideas in the present and the unequal relations of power that sustain these connections.

The literary post-colonial theorist Gayatri Spivak (1990) has used the geographical metaphor of 'worlding' to describe the fashion in which the globe and its events are compartmentalised into discrete segments. She suggests this process of 'worlding' is a discursive 'setting apart' imbued with enormous assumptions and ethnocentric prejudices. If one looks carefully this act of 'worlding' can often be seen in news coverage of international affairs, the rich descriptions used to brand 'exotic' cuisine on supermarket shelves, or the glossy images found in tourist brochures of seemingly distant lands. There are then numerous possibilities for a critical post-colonial geography using visual and textual analysis to deconstruct, subvert or open out the norms and assumptions that underlie dominant geographical representations. It is through this process of 'worlding' that the Middle East, India, Africa and a host of developing nations are geographically uncoupled from the 'West' and designated 'Other'. For Franz Fanon (1970) these are powerful acts of 'splitting' that penetrate deep into the Western psyche as they come to inform our ideas on 'development', 'civilisation' and what is characterised as the 'Third World'. In doing so these post-colonial relations co-construct who we are and how we come to understand ourselves through these mutually defining terms (Said, 1995).

While textual and visual analysis has enabled theoretically rich and insightful post-colonial geographies to transpire there remain some limits to how these approaches are currently being used in the discipline. To begin with there is a tendency for many post-colonial geographers to focus upon text, symbols and images while saying little about what Stuart Hall (1993) terms the 'relations of representation'. Instead Hall suggests that representations should be placed within a wider 'circuit of culture' (Johnson, 1986) in which processes of artistic production and ownership, use of media technologies, practices of distribution, government legislation and regulation, as well as different forms of audience consumption across the globe, influence how an image is 'read' and understood (see Chapter 5). For Hall and other cultural studies writers, meaning does not reside within the image but is contingently being made by and through the processes described. It is not the images themselves that are then significant but the wider 'politics of representation' that sustains them and enables them to present in particular ways. By engaging with the field of representation in this way, geographers can develop more penetrating accounts of the global south that reach beyond the purely textual. This potential can be seen when Clive Barnett (2006: 155) reminds us that, 'what is most distinctive about post-colonial theory is that it is less interested in reading representations as evidence of other sorts of practice, and more concerned with the actual work that systems of textual representation do in the world'.

Second, while post-colonial visual methodologies disrupt the seamless flow of the colonial representations that saturate images of 'development' or the 'Third World', there is little attempt to offer alternatives from which to counter these forms. This is not all that surprising given the problematic of representation itself as a field of analysis. If we take seriously Hall's assertion that communication is always a mediated event, where the message 'encoded' (produced) may differ from what is actually 'decoded' (consumed), then we are again drawn to the complex processes that intervene between 'text' and 'reader'. When it comes to the production and consumption of visual imagery, meaning is an inherently unstable affair subject to material processes, geographical context and the play of identity. Third, while the development of a visual post-colonial geography can make known enduring relations of power, these methods still struggle to enable the subaltern to speak. By imposing one repertory of meaning over another – the critical 'post-colonial' over the naïve assemblage of 'colonial' iconography – the voices of those photographed continue to be indecipherable. They remain as silent figures unable to enunciate their existence. For the feminist post-colonial critic Gayatri Spivak (1993) – whose ideas on who speaks for whom were discussed previously – this raises a much bigger question about the act of representation and the problem of speaking for silenced Others. She terms this tendency to speak for subjugated others 'epistemic violence'. The power of whiteness to continually narrate the existence of racialised Others is a violent reminder of the privilege that comes with being located as the architects and inheritors of modernity.

Post-colonial economic geographies

While cultural geographers are increasingly reflexive about the nature of their work and its place in post-colonial relations, there has been a near reluctance from economic geographers to 'scrutinise' the post-colonial conditions of their analysis (Pollard *et al.*, 2009). This is all the more peculiar given the potential to consider the 'flows' of workers, finance, raw materials, commodities, music and media through post-colonial networks of power that remain enduring (Nayak and Kehily, 2007). We are encouraged by Sidaway's (2000) suggestion that the post-colonial challenge need not lead to paralysis or close down debate; rather it 'opens up layers of questions' (p. 607) to reflexively rethink the major tropes of economic geography. There are now signs that at least some economic geographers are grappling with post-colonial ideas and putting them into practice. For example, Michael Samers (1998) identifies an 'automobile diaspora' that has clustered around the Renault factory in France, made up of Algerian migrant workers whose biographies intimately connect to the French colonial legacy. In their work on Islamic banking and finance, Pollard and Samers (2007) draw upon the ideas of Dipesh Chakrabarty to reconsider or 'provincialise' their understandings of global finance as Western-centred. Here, normative and hegemonic economic practices and knowledge production come to be displaced. In order to show how economic geographers can work with the cultural to produce post-colonial accounts we might turn to two examples of the everyday material commodities – food and drink.

Danny Miller (1998) produces a post-colonial portrait of global consumption through his account of Coca-Cola and its take-up in Trinidad, a West Indian island that was formerly under British colonial rule and part of the Atlantic 'slave triangle' that connected Europe, Africa and America. Miller documents Coke's arrival to Trinidad as coinciding with American troops based on the island during World War II and its development into a staple commodity. He identifies how Coke has a unique place in Trinidadian culture as a consequence of the different post-colonial histories and geographies that make up the population. Coke is used in two ways; first as a mixer for alcoholic drinks, and second as a soft drink. Trinidad's national drink is rum, an elixir 'inextricably linked with the sugar cane fields from which modern Trinidad developed' (Miller, 1998: 34–35). These spirits are never consumed without mixers, enabling Coke to become a Trinidadian 'meta-commodity', that is an essential part of its consumer culture. However, the 'tradition' of Coke and its cultural meanings are further complicated through its diverse post-colonial population who ascribe ethnic attributes to the choice of drink and how it is consumed by different populations including East Indians, Africans and Chinese Trinidadians. Miller's argument rests with the complex way in which the meanings of a global 'meta-commodity' such as Coke are rearticulated on the ground through competing practices and local associations. As Miller's case study reveals, Coke is hardly 'the real thing' but an effect of complex post-colonial negotiations between production, consumption and various imaginings of 'tradition'.

Another example of doing post-colonial economic geography can be seen in the study of Caribbean 'cross-over' food where Ian Cook and Michelle Harrison (2003) investigate the case of Jamaican hot pepper sauces. These geographers compare two Jamaican-based manufacturers, 'Grace, Kennedy and Co.', a large Western-style firm that caters to the UK ethnic minority and 'Third World' consumers, and 'Walkerswood Caribbean Foods', a small grass-roots former collective who explicitly attempt a market 'cross-over' into the Western 'mainstream' and seek to arrest colonial inequalities in the village where they are based. In tracing the unique biographies of each of the commodities the authors demonstrate the value of re-materialising post-colonial geographies. They do so by considering the 'scale-jumping and boundary-crossing' potential of hot pepper sauce in the global market place and the different post-colonial entanglements of power each company find themselves located through. By exploring the hidden relations that lie beneath commodities Cook and Harrison (2003: 311–312) suggest, 'These re-materialized 'post-colonial' geographies are, therefore, multi-locale, connective and full of ambivalences, double meanings and glimpses of anti-colonial activities.' What a post-colonial exposition of commodity chains and practices reveals, then, is a highly contradictory set of relations that complicate understandings of economy and culture that are removed from historical and geographical context. Economic geographers can produce critical accounts of commodities if they 'follow the thing' (Cook, 2004) through its post-colonial routes by tracing the uneven geographies and forms of exploitation that may mark the journey of goods such as tea, coffee, tobacco, chocolate or exotic fruit from the point of production through to consumption. There is a host of unexploited post-colonial economic geography studies yet to be completed on everything from the global outsourcing of labour to Indian call centres, the use of Filipino domestic workers in Western households, or critical expositions that might ask how fair is 'fair trade'? Using the post-colonial theories we have outlined here could offer opportunities for enriching economic geography, challenging its universalism and displacing at least some of its master logos.

Conclusions

This chapter has considered the historical roots and critical ideas generated by the term post-colonial. Here we saw how the post-colonial moment does not mean an end to colonial legacies but is a period in which the past and the present intersect and occasionally collide as the lines of power become known. To develop a critical sense of the term we turned to the writings of key post-colonial theorists such as Frantz Fanon, Edward Said, Homi Bhabha and Gayatri Spivak. This provided us with a 'post-colonial framework' from which we could start to interpret the fraught and unequal relationships between the colonial 'core' and its 'peripheral' Others.

Having outlined the critical and historical components of post-colonialism, we then considered the doing of post-colonial geography. Here we saw how the colonial present is made material in built architecture, memorials, processes of commoditisation and so on. Post-colonial geography was particularly helpful in connecting these objects up to their colonial histories and resituating them within a new geographical imagination. Throughout we encountered numerous possibilities for post-colonial geographies including work on critical geo-politics as well as research on tourism, gap years and media representations of the global south in film and advertising. Finally, we suggested that research on post-colonial economic geographies remains relatively undeveloped and a critical focus on the colonial interconnections that may surround transnational migration, trade, the global outsourcing of labour and the circuits of particular commodity chains may develop more relational and embedded accounts of the economy. By unveiling such post-colonial histories and geographies we might begin to entertain the possibility of alternative modernities and find new ways of doing geography differently.

Summary

Post-colonial geographies:

- Speak back to the West and challenge our assumptions, knowledge and practices.
- Adopt an anti-colonial position that exposes how the spectre of colonialism continues to haunt the present.
- Demonstrate the unequal relationships between the global north and south.
- Can be used to rethink mundane material objects such as buildings or food.
- Offer reflexive and positioned accounts that seek to provide an alternative means of doing geography.

Further reading

Blunt, A. and McEwan, C. (eds) (2002) *Postcolonial Geographies*, London: Continuum.

Crush, J. (1994) Postcolonialism, Decolonization and Geography, in A. Godlewska and N. Smith (eds), *Geography and Empire*, Oxford: Blackwell, pp. 336–337.

Robinson, J. (2003) Postcolonialising Geography: Tactics and pitfalls. *Singapore Journal of Tropical Geography*, 24(3): 273–289.

Sidaway, J. (2000) Postcolonial Geographies: An exploratory essay. *Progress in Human Geography*, 24(4): 591–612.

Emotions, embodiment and lived geographies

Learning objectives

In this chapter we will:

- Demonstrate how theories of 'non-representation' challenge the representational mould of cultural geography and so mark out the limits of representation.
- Understand how an engagement with bodies, feeling and affect may develop more sensual understandings of the world.
- Explore how the idea of performance may be used to enliven geographic approaches to landscapes and everyday events.
- Identify the problematic of performance in human geography.

Introduction

It's 12:37 am inside Thunderdrome and the DJ who has been playing an eclectic mix of popular house tracks seamlessly alters the tone. The atmosphere thickens as a heavy bass note kicks in, throbbing through my solar-plexis and inflicting a pronounced, rhythmical bobbing of heads among the crowd. Without speaking we find ourselves magnetized and pulled like iron-filings to the centre of the arena where a steely crescendo of metallic techno beats increases in intensity. A plume of dry-ice causes our skin to tingle. The dance-floor is energized and we are transported to an ethereal elsewhere where our inhibitions dissipate and our limbs flow like liquid. We seem to dissolve into the sonic experience and become one with the moment. As the pulsating rhythm reaches a peak an electric-blue strobe suddenly splits the night like

Clubbing
Source: Fotolia.com/Adam Radosavlievic

> a finger of lightning. The capillaries in my pupils make rapid adjustments to the sensation of blinding white light as deliriously we embark upon a hi-energy journey into euphoric trance. It is some hours later that we are left sweaty, exhausted and elated. Bliss.
>
> (Extract from one of the author's *Clubbing Diary*)

The diary entry above captures something of the immediacy of dance. But only something. So much of the experience of going out clubbing is beyond words (see photo). Explaining how music pulls us out of our thoughts, unconsciously makes our bodies move in syncopated rhythm and enlivens our sensations would be difficult to comprehend to someone who has not experienced such an event. However much effort you put into explaining the feeling of clubbing, even with the use of photographs, film and rich description the account would inevitably be limited. Far better, you might think, to take that person out to experience the sensory worlds of club culture first-hand. We will return to the example of clubbing in the following section when we consider Ben Malbon's (1999) attempt to distil the essence of this experience into words.

A main thrust of our argument concerns the way in which embodied understandings of human relations have the capacity to 'shake up' and move beyond the more familiar scripts we discussed in Chapter 5 related to the 'cultural turn'. In doing so, these experimental approaches offer a sharp critique of representation, discourse and language. An abiding feature within these new accounts is a concern with performance, practice and affect.

To illustrate the transforming field of cultural geography this chapter will begin by considering the shift from representations to forms of post-representation, or what

Nigel Thrift (1996, 2000, 2003, 2004a, 2004b) has termed **non-representational theory**. Here, we suggest that the slide from representation to performance is not a complete displacement as the phrase 'non-representation' may imply. Rather, as we will go on to discover, representation and presentation are implicated in a dialectic relationship, feeding off and through one another. Recently, geographers have been especially interested in how presentations, events and practices offer new ways of interpreting the world through embodied actions such as that of clubbing. The chapter begins by explaining how theories of non-representation challenge what is sometimes seen as the existing orthodoxy of conventional representational forms (Thrift, 2004). We then offer an explanation of feeling and affect as they have emerged as popular terms for geographers to reach out to when discussing the significance of that which cannot be represented. To develop this way of thinking we turn to some geographical examples of performance in action, or what Lorimer (2005) styles, 'more-than-representational' geographies. Our focus here is upon *landscape* and how an events-based approach can open us up to the unfolding of place while shedding light upon hitherto unseen or taken-for-granted actions and performances. The chapter concludes by considering the 'problematic of performance' and the limits of theories of non-representation. We suggest that the rush to enact the present should not be at the expense of past representations and seemingly durable vectors of power. Throughout this chapter representation and presentation remain at the heart of our analysis. A critical question we would like to press is whether we are encountering a 'crisis of representation' affecting how we think about the social world and conduct geographical research. And if so, what possibilities are there for encountering the world 'otherwise'?

A crisis of representation? Cultural geography and non-representational theory

The cultural turn in human geography has offered critical engagements with everyday processes and activity (see Chapter 5). At the same time it has been seen to prioritise textual accounts of the immediate world with an over-reliance upon representational modes of inquiry. The privileging of discursive approaches through interviews, focus groups and the analysis of documents has led some geographers to articulate how 'that which is close up – the habitual and largely unreflective bodily practices by which we go on – is not being given adequate consideration' (Conradson, 2003: 1984). Such remarks reflect a growing concern that we have become 'prisoners of language' and that representation by its very nature always falls short. This suggests that 'The world is more excessive than we can theorise' (Dewsbury *et al.*, 2002: 437). Engaging with this excess means appreciating that when we conduct a research interview, large-scale survey, ethnographic inquiry or other methodological procedure we are always missing something – the tone in which a particular phrase was used, a reassuring gesture of welcome, a cold look of disdain, or an unfinished sentence foreclosed by a shrug of the shoulders and a downcast gaze. These brief examples reveal how a discursive reliance upon text and transcripts may serve to fetishise the word.

It is timely here to return to our opening illustration of clubbing. Reading the passage again, you may be struck by the range of emotions and sensations generated. Yet words barely do justice to the raw immediacy of the experience and offer at best a vapid translation of thick bodily encounters. If we were to try to capture the immaterial sensations of clubbing through the written word, we would soon discover that much worldly experience is simply 'felt' and beyond representation. It is this flawed attempt to convey the vitality of clubbing that leaves Ben Malbon suitably defeated by the limits of language. In what was to become a pulsating account of the clubbing experience Malbon (1999: xii–xiii) initially appears deflated by the task at hand:

> How can words – simple, linear words on a page – evoke this delirious maelstrom of movement and elation? Again and again I arrive at this point . . . I simply cannot describe it any further. How can I convey the deep, thundering bass which is felt more than heard? The mass of bobbing bodies: blurred, colourful, dimly-outlined and unceasingly in motion? The space itself, which fleetingly seems as though it has no edges, no end in time or space, yet at the same time only stretches as far as you can see into the lights, the black walls, the heaving of dancing masses? The sensation of dancing, of moving without thought, of moving before thought, of just letting go, letting it all out? Words fail me; words become redundant and unnecessary, words become pointless.

The impossibility of words to ever fully capture feeling and emotion leads us to suggest that much human experience is impenetrable to culture. But this does not mean that words are pointless as Malbon initially felt, or that more words are needed to better convey the feelings we can't express. As the social geographer Susan Smith (1997: 504) explains (in a quote we believe is drawn from an Elvis Costello interview in *Musician Magazine* 1983, see also Ingham *et al.*, 1999), 'Writing about music is like dancing about architecture, listening to ballet or feeling the texture of a painting – it might be helpful, but it is not the best, most direct or most appropriate way of illustrating the power of art.' Smith's point (via Costello, 1983a; and Ingham *et al.*, 1999) is that music has to be heard to be understood. Merely writing about it, which is a discursive and differently mediated mode of representation, will always result in a 'lack' – an absence that cannot be comprehended.

On the surface this may appear to leave us in a cul-de-sac. But a lesson we can take from this is a critical awareness of the *limits* of representation. As Harrison (2007) argues, much is revealed in broken words, when language falls short or when we are overwhelmed by the emotion of love, grief, respect, hope or jealousy. In a moving account of the role of tattoos in working-class culture, the sociologist Les Back (2007) encourages researchers to 'listen with the eye'. Back's evocative phrase is an attempt to go beyond words – what is simply said – and invites researchers to become attuned in the wider art of listening and the humble act of learning. For the middle-class observer the tattoo may be silently read as a symbol of working-class vulgarity. Yet Back discovers that many tattoos are imbued with deep emotional content, commemorating loved ones and marking feelings that cannot be expressed in words. This is a personal biography, literally written on the body. Listening to others in this 'fleshed out' fashion reminds us that it is often what is *not said* that is most personal and precious.

Words it seems are never enough. As Latham and Conradson (2003: 1901) declare, much human geography is 'Dominated by an obsession with the politics of representation', privileging it over feeling and practice. There is a powerful sense in these writings that representation seeks to fix, stabilise and capture the meaning of social life in a way that 'rendered inert all that ought to be most lively' (Lorimer, 2005: 84). The inability to express the sheer push of life in geographical writing means that all too often we are reduced to a parsimonious account. In these sparse depictions there is a tendency towards logic, linearity and what we are terming lifeless 'flat representation'. However precise we may be in our writing and geographical accounting of the 'field', there is always a haunting feeling that 'the world does not add up' (Dewsbury *et al.*, 2002: 437). This is a scary thought. It also raises a bigger issue – if language, text and discourse are inherently inadequate devices for categorising worldly existence then are we encountering a 'crisis of representation'? And if so, how might geographers respond to this 'crisis'?

In an attempt to reckon with everyday practices and the intangible sensations of lived experience, the geographer Nigel Thrift has offered the term 'non-representational theory' (NRT) to describe the eclectic ways in which academics are now trying to challenge the representational orthodoxy (see Box 12.1). For example, Dewsbury *et al.* (2002), in responding to the inadequacies of textual modes, offer two possible routes forward. First, they advocate *pluralism* as a means of acknowledging worldly excess even as we strive to endow our ideas, thoughts, analysis and writing with the illusion of coherence. This approach borrows from post-structuralist ideas discussed in Chapter 9, where texts are seen as containing multiple and open-ended meaning. Second, they draw attention to the *performative* aspects of representation. In doing so Dewsbury *et al.* (2002) posit that NRT points to the liveliness of representations and the performative manner in which they act as manifestations and doings of a particular kind. In reconsidering representations not as texts, signs or signifiers awaiting symbolic decoding, the focus is upon the performative dimensions of representation:

> In this sense representation is perhaps more usefully thought of as incessant presentation, continually assembling and disassembling, timing and spacing; worlding.
>
> (Dewsbury *et al.*, 2002: 438)

The 'doing' of representation was markedly apparent following the destruction of the World Trade Center in New York on September 11, 2001. The now familiar high-residue image of a second aeroplane crashing into the Twin Towers was incessantly replayed on television screens and print and electronic media as it was transmitted worldwide through global communication technologies. At a representational level the symbolic aspects of the destruction of the Twin Towers as phallic totems of Western global capital is certainly ripe for critical textual deconstruction. But if we consider such imagery not as 'dead text' awaiting our interpretation but as lively presentation, we can also reveal its performative aspects. The images serve to inform us of the global significance of the moment and, for many audiences far removed from New York City, these citations have even become the 'event'. It was almost as if the images had taken on a life of their own, generating fear, anxiety, loss and mourning. Indeed, the emotional content of war is vividly distilled where the US government and armed forces repeatedly declared how the subsequent invasion of Iraq and the quest to find Saddam Hussain was 'a battle of hearts and minds'. In a polemic piece on 9/11 and affect, Gerard Ó Tuathail (2003) explores the intensity of feelings aroused by this event and how US emotions were stirred through a layering of music, speeches, symbols, sounds and images that served to justify the 2003 'shock and awe' bombing of Baghdad. For Ó Tuathail the attacks 'spawned a blitzkrieg of multimedia cultural productions' (pp. 858–859) that are experienced culturally and through biophysical bodily capacities that generate feelings.

In their book *Afflicted Powers* (2005), the left-wing group Retort persuasively argue that 9/11 was a spectacle that struck at America through a field of images; it was the circulation and endless reproduction of images that was so harmful to US national esteem, exposing

Box 12.1

Nigel Thrift: non-representational theory and affect

Nigel Thrift is currently Vice-Chancellor at Warwick University and over the years has contributed to a series of geographical debates. He is widely credited in the discipline with developing the term 'non-representational theory' (NRT) to describe theories of practice which amplify the flow of events. Thrift regards non-representational theory as the mundane everyday practices that shape the conduct of human relations in particular sites and spaces. He emphasises that these acts are embodied where the body is the primary medium through which we come to interpret the social world. Thrift (1996) uses the example of dance to illustrate this point, showing how bodily activity cannot be explained through the rhetorical device of language. His argument is that dancing is a sensual, embodied activity which can only be understood in the moment of performance. To put it bluntly, one can inform another person endlessly about what it is like to dance – a mode of discourse or representation – but unless this activity is experienced first-hand, words will always fall short. Instead, Thrift deploys the term 'body-subject' to describe these humanistic, corporeal forms of knowing and being.

Thrift remarks how much NRT is located in an affective register arousing feeling, sensation and tactile forms of understanding. He admits 'there is no stable definition of affect' (2004a: 59), but understands it as a dialogic experience enmeshed within a network of human and non-human actors. To help us understand the different ways in which scholars have considered the affective domain Thrift traces at least four traditions which inform us of the origins of the work. We might summarise these affective approaches as:

1 *Embodied*: a different means of thinking and being in the world that relates to bodily practices and experiences. This involves an appreciation that emotions are mainly beyond the realm of representation.
2 *Psychoanalytic*: concerned with the unconscious and unspoken, in particular inner-drives related to desire, anxiety, fear or hatred.
3 *Spiritual*: involving the conjoining of the mind, spirit and soul. This stems from a broader religious and philosophic idea that humans are part of a universal cosmology that unites us with the natural world.
4 *Evolutionary*: it is claimed by Darwinians that certain emotional expressions conveyed in facial or vocal gestures tend to be similar across all cultures. The suggestion is that these physical ways of emoting have evolved amongst humans and animals to form a natural way of expression and communication.

It is important to regard this outline not so much as a rigid schematic but simply a set of different approaches used to understand that which exists beyond the realm of language, image and text.

Although Thrift is keen to regard non-representational theories as 'a kind of politics' (2004: 82) many critics within the discipline have questioned the moral worth of these ideas and they continue to remain a heated point of contention in geographical debate. Most controversially in a manifesto on NRT he affirms his potential to 'overturn much of the spirit and purpose of the social sciences and humanities' (2004: 81). A bold claim indeed! However, Thrift concedes that by and large his efforts have 'fallen on stony ground' (p. 81) and consequently admits to being 'frustrated with geography' (p. 83). Nevertheless his ideas have attracted some cultural geography disciples. Time will tell if these experimental approaches will leave much of a mark on the humanities and social sciences or be confined to a passing fad.

hitherto unknown vulnerabilities. For Jean Baudrillard (1994), whose work we discussed in Chapter 9, image and event are one. Baudrillard's point is that in a mediated world, reality and representation are no longer distinct but have given way to depthless simulations that render 'reality' obsolete. Rather than being a negation of power, the spectacle of signs and image is lively and active. As Latham and McCormack (2009: 253) affirm, 'images can be understood as resonant blocks of space-time: they have duration, even if they appear still'. The performative aspects of such moving and still imagery can be seen in the practices adopted by Western states in response to the threat of terrorism and people's feelings of global insecurity. This has seen attempts by the nation state to 'perform' security – the introduction of retina scans and finger-printing technologies when travelling abroad, new forms of surveillance directed towards visible and religious minorities (see Chapter 8), and a 'war on terror' that began with the invasion of Iraq and has extended into Afghanistan (see Gregory, 2004). Regarding representations as activities or 'doings' in this way enables us to understand how they operate through an affective register that can give rise to a new political geography marked out in time and space. Through such enactments representations also become events of a particular kind.

In light of this discussion we might consider non-representational theory not so much as the obliteration of representation, but rather a new means of taking representation seriously. Rather than advocating a theory of anti-representation – the non-representation originally proposed by Thrift – NRT may also be characterised by 'a firm belief in the actuality of representation' (Dewsbury *et al.*, 2002: 438). Instead of being like two separate sheets of paper, we might better consider how representation and non-representation fold into one another. This can be seen where a number of studies concerned with feelings, affect and the non-representational remain bound to certain representational forms – the use of interviews, diaries, participant observation, ethnography and photographs. Each of these techniques offer a 're-presentation' of worldly experience, the translation of activity into discourse. In musing about NRT research strategies, Ola Söderström (2005: 13) reflects how, 'practice, presentations, operations become more central than *re-presentations*'. Although some geographers have tried to experiment with performative styles of research and writing (McCormack, 2002), much human geography remains methodologically conventional.

However, as Latham (2003) has shown through the interview-diary method, familiar qualitative techniques can still be used to enhance the liveliness of our accounts. He suggests that even established research methods can be 'made to dance a little' (2003: 2000) by recognising place as a creation in process. Examples of this can be seen in experimental fieldwork conducted in the city of Berlin, where geography students consciously disrupted the rhythm and flow of everyday life by standing still in the busy stairwell of a rail underpass and filming this process, thus drawing attention to the habitual manner in which people move through these spaces almost without thinking (see Latham and McCormack, 2009). In another example of a Chinese martial art self-defence movement known as White Crane Silat, Samudra (2008) twists the familiar ethnographic term 'thick description' to discuss what he calls, 'thick participation'. He suggests that rather than fall back upon a distant observation of events, the act of encoding kinaesthetic details, describing new sensations and narrating physical training episodes can provide more lively anthropologies of action. Samudra does not wish to displace representation but find other ways of grasping bodily movement, muscle memory and the corporeal ways in which a trained body might work. In grasping the playful potential of NRT, researchers may seek to bring feelings and embodied actions into their accounts as well as the sights, sounds, smells, touch and tastes that make up much of our experiences.

NRT can then be deployed as a theoretical and methodological intervention, prising open the excessive nature of the world as active, sensual and embodied. It interprets the world as in-the-making, an accumulation of processes and on-going performances. What is interesting about this approach is its openness to the unpredictable and unknown. As Massumi explains:

> If there were no escape, no excess or remainder, no fade-out to infinity, the universe would be without potential, pure entropy, death. Actually existing, structured things live in and through that which escapes them. (2002: 35)

Instead of ruminating on a crisis of representation, NRT practitioners delight in the infinite potential, mystery and excess of the cosmos. Bare life it seems always exceeds (see Anderson, 2006). That is, the fullness of living means that any representation of friendship, love, hatred or affection is incomplete. Rather than mourn the passing away of discursive knowledge, we might just as easily celebrate the liveliness of that which defies representation. It is after all integral to our worldly experience. In this sense we might argue that NRT does not so much signal the end of representation, as offer openings for a new beginning through an engagement with what Thrift refers to as the 'immediacy of the now' (2003: 2020).

Understanding affect

In recent years a number of cultural geographers have tried to grapple with the rather abstract notion of *affect*. In doing so many have been drawn to philosophy, phenomenology, psychoanalysis, work on the sociology of emotions and the type of experience-based humanistic geographics we discussed in Chapter 3. So what do we mean by affect and how might it differ from emotion? To illustrate this we shall explore the example of music and the emotional worlds opened up through sonic experiences.

At its simplest we can understand affect as the non-conscious bodily experience of sensation. Because affects are pre-discursive – in that they are felt rather than mediated through language – they can be said to have an 'unformed and unstructured' potential (Massumi, 1987). This potential may be realised in a range of embodied activities as exemplified in the unpredictable performance of free-form jazz. Here a walk-on saxophonist may complement a musical ensemble in the live composition of an unscripted piece by connecting with the 'vibe', 'feel' and 'texture' of a live performance as it happens. Experimental jazz of this kind is not rehearsed but comes together in-the-moment as different creative energies and capacities cohere. The melodies are not thought out, but exist in an immanent state of becoming. In other words, they hang in the air as potentialities that can only be realised through the embodied dexterous finger-play of musicians attuned to the moment. What is striking about this musical form is that it is not a conscious, orchestrated sound but an assemblage of unspoken ideas, rhythm, flows, feelings and embodied motor-skills. Its potential appears to lie in the 'half-second delay' between action and mental cognition. Through musical performance, embodied knowledge is practised where 'ideas and doing work together' (Crouch, 2003: 1952) one might say, in harmony.

An aspect of affect that many writers emphasise is the manner in which it has the capacity to precede will and consciousness. Moreover, affects are not simply brought into being by a single individual but are assembled in relationships between bodies, objects, sounds, movements and encounters. Rather than residing in a subject or body, affects are composed through worldly interactions. In short, affect concerns the 'motion of emotion'

(Thien, 2005). An illustration of this transmission can be felt in the cinematic performance of 'Beatlemania' where an aura of hysteria is conjured. Mini-skirted young women are seen singing, dancing, screaming or fainting as the uncontrollable force of these affects holds sway. Such is the energy of these events that the audience became as much a focus of the spectacle as the band and music itself. For many live musicians the experience of an event is affective in that it is comprised of an intangible intersubjective coalition of forces – the bodily adrenalin of playing live, audience reaction and participation, perspiration, inspiration, movement and dance, or the throbbing vibration of a bass note. Live recordings of concerts suggest that the affective qualities of musical performance are not reducible to conscious production. In the immediacy of performance these affects erupt, spill over, rupture and disturb what is already thought to exist (Anderson, 2006). By engaging with this surplus we might begin to understand why many fans attend live events despite already owning CDs and downloaded digital recordings of the sounds and sights they are about to see performed.

The atmospheric aura of performance leads us to reflect upon the 'profound importance of affect in the unfolding of this event-ness' (Latham and Conradson, 2003: 1902). What is interesting about such performances is the way in which the affects generated may exceed bodily capacities, coming to be enacted upon other bodies, objects and things as they take on a type of trans-human quality. Affects are not tethered to an emotional human subject but can be distributed beyond this as they are constantly reassembled in the magic of performance. The atmosphere generated at a Beatles gig is a product of human and non-human agents including bodies, guitars, crowd surges, screams, stage-lights, vocal harmonies and so forth. In this way affects may amplify or diminish our experience. Such a spontaneous coming together reveals how a live music event can be thrillingly unpredictable where its affects may scatter, energise and put into motion a multiplicity of practices, actions and performances. This spilling out of affect is neatly illustrated in Ben Malbon's (1999) lively encounters with clubbing, mentioned previously. Malbon considers how clubbing may involve ecstatic and oceanic experiences; vitality and playfulness; as well as the resonance of a post-clubbing afterglow. These affects can be collectively experienced and enhanced through mutual participation. As Lorimer (2008: 552) contends, affects 'act on bodies, are produced through bodies and transmitted by bodies'. In other words affects are trans-human – working across bodies – and intersubjective – working between them. Through a constellation of these interrelations an unintended 'joint outcome' (Thrift, 1996) is tenuously formed.

The less than predictable ways in which melodies incite feeling can be seen in Anderson's (2006) study exploring how the materialities of music are bound to the routine and rhythms of everyday life. In his household interviews and observations, Anderson discovered that individuals may listen to melancholic music in times of grief and despair but in doing so may experience an immanent transcendence into hope. Feelings of recognition generated by listening to music, signalling that this is a shared experience and others have gone through this before, can enable hope to emerge from the rubble of despair. To further explore how cultural geographers are grappling with the excessive qualities of living and being, we will address the role of performance and how it is being deployed in cultural geography. This is followed by an exploration of new geographies of performance that illustrate in fine empirical detail how geographers are using these ideas of bodily affect and performativity to develop a 'more-than-representational' account of the world.

Towards 'more-than-representational' geographies

In Chapter 5 on the 'cultural turn' we discussed how 'reading the landscape' through signs, symbols, motifs and metaphors offered a new moment in British cultural geography in the 1990s. The focus on representation, which was radical at the time, is currently coming under scrutiny from a new cluster of cultural geographers. In these most recent approaches we can observe a shift from representational readings of objects, landscapes and spaces towards an engagement with action, practice and performance. Here there is a will to move beyond representation and engage with the immediacy of bodily sensations and the busy-ness of everyday life. At a methodological level these writers focus less upon what people say, as previous qualitative researchers have done, and more upon what they do. Actions can speak louder than words.

Popke's recent assessment that human geography is currently 'abuzz with passion, performance and affect, infused with a sense of playfulness and a spirit of optimism and experimentation' (2009: 81) captures this new vogue. But before this excitable bandwagon rolls away we would like to momentarily pull the hand-brake. It seems to us that Popke exaggerates the extent and impact of the turn to post-representational modes of inquiry across human geography. These ideas, although currently fashionable, remain largely the preserve of a select band of cultural geographers. Within much geographical research this is still something of a cult pursuit rather than mainstream practice. In our own institution and many others NRT is as yet not widely taught across mainstream geography curricula. We would agree that presently NRT is creating something of a stir, but this may turn out to be no more than an eddy in the backwater of cultural geography. Whether these ideas have reached their high-water mark, or will endure beyond the next decade – in the way that Marxist, feminist, post-modern and post-colonial approaches have succeeded – still remains to be seen. These latter approaches by contrast are likely to feature in explicit and subtle ways across the geography syllabus.

Having briefly contextualised the moment within which NRT is currently situated, in more restrained fashion than which excites Popke, we will now explore how these insights might influence critical thinking in the discipline. In this section we aim to trace the potential radical breach with representation offered by these accounts. However, as discussed previously, NRT does not involve a complete 'writing out' of representational forms but is a means of thinking these methods differently, a debate we shall return to towards the end of the chapter. The emerging approaches discussed offer a challenge to representational modes of analysis by instigating innovative ways of trying to interpret the emotional and sensual aspects of dwelling and being. With this in mind Haden Lorimer (2007: 89) invites us to consider, 'what is representation intended to achieve, and what *else* might be done?'

The prescient question 'what else might be done?' encourages us to reach beyond the familiar borders of cultural geography and the 'deadening effect' (Lorimer, 2005) of discourse. It asks us to consider the 'what-elseness' of human experience and look further than talk and text. In doing so, geographers have been drawn to feeling, emotion and affect as sensual ways of understanding. This does not mean forgoing representation altogether but involves a concerted opening-out towards a performative methodology that engages with events, practices and activities as they occur (see Latham, 2003; Crouch, 2003). These experimental approaches have the potential to yield what has been styled 'a more animated cultural geography' (Nayak, 2010). Following a discussion of performativity and practice, we will explore some of the imaginative ways in which geographers use performance to enliven their accounts and produce what Lorimer usefully depicts as 'more-than-representational' geographies. This includes a diverse collection of work using non-representational techniques, theories of affect, emotional geographies and embodied accounts of worldly activity. What each of these approaches share in common is a commitment to moving beyond the discursive limits of cultural representation as the sole means of interpretation. If discursive approaches looked for meaning in language, image and text, these new approaches are more concerned with the immediacy of practice and performance. For this reason the 'more-than' preface is especially apt, signalling a surplus of activity, the 'what-elseness' made open by these approaches that question whether our geographical accounts are 'dead or alive' (Thrift, 2000: 1).

In the past few years there has been a renewed interest in performativity and the way in which our identities are constituted in and through action. The work of the feminist queer theorist Judith Butler (1990, 1993) has been important here in demonstrating how identities are performed through a stylisation of the flesh that enables these identities to present as-if-real (Chapter 6). For Butler, what we might think of as identity is really a performance, a doing, an activity. What makes the fiction so compelling is how it is embodied by subjects and given the illusion of substance (see Box 12.2).

A key intervention of NRT is to implode familiar understandings of 'landscape-as-text'. This work is an explicit challenge to what was termed the 'new cultural geography' of the early 1990s which was concerned with making symbolic readings of landscape. As we saw in Chapter 5, British cultural geography offered a refreshing alternative to quantification and the cold empiricism evident in many other areas of geography at the time. Work uncovering the iconographies of landscape has been critical in breaking open these familiar ways of working, making the study of the intangible and unquantifiable acceptable. While the 'new cultural geography' (Chapter 5) along with humanistic geography (Chapter 3) acted as inadvertent handmaidens for the birth of NRT, these older practices are inevitably treated as passé by a post-representational generation of geographers. If representational approaches gave

Box 12.2
Judith Butler: performativity

Judith Butler is a feminist philosopher based at the University of California, Berkeley, US. In her work on gender she uses ideas of performance and performativity to assert that there is no pre-given identity from which actions follow but rather identities are constituted in and through action. This implies that there is no *a priori* identity, 'no doer behind the deed' (Butler, 1990: 142), but that our identities only come into being through performance. Butler claims such taken-for-granted identities as being a girl, boy, straight or gay, far from being natural, are the accumulation of successive iterative performances that appear to lend these identities the illusion of substance. In contrast to widely held assumptions that subjects produce action her insights suggest that it is actions that produce who and what a subject 'is'. Through this 'paradigm of performativity' Butler suggests that sex, the notion of being a 'man' or 'woman', is no more than an idea; a regulatory fiction sustained through repetitive, stylised performance. These enactments have a performative function as they will-to-life the notion of a coherent subject which is made manifest on the surface of the body and poses as 'real'.

primacy to signs, symbols and images, the new performative accounts turn to embodied experience, practice and doing. A limit of some of this earlier cultural geography work on landscape is that it tended to stabilise meaning, reducing relations to text and thereby draining people and places of life. The more-than-representational accounts we will now consider have the potential to usher in affective ways of understanding the landscape. They remind us that landscapes are not simply cultural texts but their materiality must be understood through the body as we encounter these environments through sights, sounds, smells, tastes, touch and other sensual experiences. These *performative* modes of address then offer a means to reinvigorate landscape by breathing life and vitality into what had once appeared 'dead space'.

In human geography the engagement with performance, practice and the vitality of our earthly surroundings has ushered in more adventurous theories and approaches to these spaces. Reaching beyond what we called 'flat representation', to engage with a living world in motion opens us up to new practices. For Thrift (2004a: 82) this entails 'inventing new relations between thought and life', in order to create a space for other things to happen. Releasing the valve on representational modes of inquiry to discharge energy, matter, life into the more-than-human worlds of our being is a means of doing geography 'otherwise'. With careful handling the transient, ephemeral and affective dimensions of everyday interactions can be used to enrich fieldwork reports and geographical accounts. In this regard affective and non-representational geographies are not only concerned with representation and meaning – although they maintain an interest in these ways of knowing – but their attention rests with the performative 'presentations', 'showings' and 'manifestations' of everyday life (Thrift, 1996: 127). To consider how performance can reanimate living relations, we will turn to a series of case accounts located in rural and urban areas that have the potential to reinvigorate landscape.

Reinvigorating landscape

A discernible area in which human geographers have sought to push the possibilities of a performative approach has been in their treatment of landscape and environment.

In particular Sarah Whatmore's (2002) work has been characterised by an attempt to rethink the material and ecological fabric of social life. Whatmore suggests that by thinking our worldly relations in 'more-than-human' ways we can arrive at different understandings of the lived environment. These insights can be used to generate what she sees as 'hybrid geographies' between nature and society. Buoyed by Thrift's declaration that the 'world is not a reflection but a continuous composition' (2003: 2021), we can now trace a new sensibility that appreciates landscape as in process, a moment of becoming. Here landscape is no longer set apart as an object which human subjects might encounter and inscribe with meaning, but instead forms a fertile part of our existence. In these performative readings human actors are not privileged above objects and the natural environment but exist as a part of a living cosmos that connects us to trees, soil, sheep, wind, rocks and rivers. The performative approaches seek to give agency to 'still life' and enable us to envision the world differently, as an unfolding event brewing with multiple possibilities.

To grasp the ways in which embodied interactions with the environment are transforming our understandings of landscape we will go on to consider some recent examples. What is interesting about these developments is that they enable us to *reanimate landscape* in lively ways beyond its familiar home in the arts and humanities. Indeed, numerous cultural geography adherents followed Carl Sauer's (2008: 103) assertion that 'Culture is the agent, the natural area is the medium, the cultural landscape the result', thereby ossifying relations between culture and nature. In geography the term landscape is weighted down with formidable traditions and accompanying disciplinary baggage. As Cresswell (2003: 269) argues, a problem with the term is, 'It is too much about the already accomplished and not enough about the processes of everyday life.' This signals a larger criticism of textual approaches and work on symbolic meaning and representation. The practices and everyday performances we shall now turn to offer an opportunity to disrupt and reinvigorate the concept of landscape. For if studies of landscape were once said to have led to the 'obliteration of practice', as Cresswell fittingly argued, this is no longer the case.

John Wylie (2002) considers the experience of ascent and elevation up Glastonbury Tor, in Somerset, South West England. The area has a spiritual history related to ancient Celtic mysticism, Christian piety and 'New Age' rituals. Rather than explore the cultural meanings and symbolism of the landscape, Wylie discusses their emergence and resonance when ascending and becoming immersed within the 'mythico-religious histories and geomantic patterns, woven around the Tor and its environs' (2002: 443). He explains how camping overnight enriched the ambience and led him to feel a sensual part of the landscape. The experience of ascent is an embodied matter from the pastoral dapple of sunlight, the mild sting of midges, the gut-wrenching vertigo, the aching limbs or the chirrup of bird-song. Wylie's argument is that it is impossible to separate out 'self' from 'landscape' to form a disembodied vision of the Tor. It is only through the act of elevation and its accompanying bodily practices that the enchantment of the Tor can be felt and understood. A further illustration of embodiment can be seen in a study in the Lake District and Peak District of England where Hannah MacPherson (2008) acted as a sight guide for blind and visually impaired walkers as she led individuals up hills, rocks, shrubbery and through wooded forestation. 'Blind walking' disrupts a common perception that ramblers use the countryside solely for visual and scenic purposes. Although touch, smell and the familiar sounds of the countryside form part of the way in which the blind and visually impaired engage with the outdoors, chatter, laughter and social interaction remain pivotal to their overall experience. In particular MacPherson reveals how laughter and humour endow the rural landscape with a temporary sonic element, used to negotiate relations between sighted guides and walkers, relieve nervousness and form bonds capable of dismantling stereotypes frequently ascribed to those with visual impairments. Here, the countryside can be understood through embodied registers and tactile engagements with the soil, air and climate (Wylie, 2002).

Accounts of this kind may help us rethink the 'physical' in physical geography by appreciating how motor-skills and bodily exertion – breathing, walking, running, climbing, crawling or swimming – are generative of a corporeal knowledge of the environment. This is an example of what Crouch (2003) terms 'grounding performance' as illustrated in his interactions with caravan campers and allotment holders who each engage with the countryside through embodied actions. These activities offer a means to disrupt the everyday by 'getting away from it all'. In doing so they tear open the fabric of daily life to generate new moods, memories, pleasures, sanctuary and solitude.

This vitalist approach, with its focus upon lived relations, is not restricted to rural landscapes and the scenic outdoors, but can equally involve an analysis of urban landscapes. In an embodied consideration of the free-running practice known as parkour (see photo), Steven John Saville (2008) produces an insightful and energetic account of these performances and how they can enable us to rethink our daily places of habitation. Saville argues that parkour has spatially transformative powers that prevent a simple closure of place through envisioning our surroundings as a composition in the making. Instead he and other practitioners known as *traceurs* are constantly refiguring the built environment by opening it out towards new hitherto unimagined, if not always attainable, mobilities. In this way Saville explores how benches, pavements, ledges, trees, walls and lamp-posts become charged through the practice of parkour where acrobatic engagements give seemingly inanimate objects a new 'feel'. This 'stretching out' of everyday material objects beyond their intended use informs us of the plasticity of place and how practice can make us see the world anew.

Here we can make a distinction between places and landscapes as simply social constructs or also as a way of 'being-in-the-world' (Cresswell, 2004). In Saville's tactile engagement with parkour, everyday architectural features come alive as new mobile possibilities unfold. This is a 'more than' social approach as, 'Parkour is full of events, where the world expands and shakes with intensity' (2008: 905). A similar comparison can be made with the dynamism of skateboarding in which everyday objects are given new life as the architecture of the city is reworked through the kinetic connections between bodies, boards and wheels (Borden, 2001). Like skateboarding, parkour requires training and repetitive practice, with *traceurs* constantly trying to extend their bodily capacities. This is seen when Saville discusses the pre-cognitive impulses that flow through the body prior to making a difficult leap, spin, roll or movement. He feels filled with fear, dread and sickness that the jump cannot be achieved and will result in a fall. But as confidence grows with practice and each successful manoeuvre, possibilities expand and fear can transform into an excitement that propels the body forward. Saville claims that before any linguistic understanding is formed the body is already undertaking a sensory reading of the environment using past

A precision jump in parkour
Source: Fotolia.com/Art Jazz

associations and tacit understandings of texture, solidity and surface. In the flow of action the body is carrying out numerous calibrations that we barely have time to consciously register as we arc towards the challenge of new movements. This is part of what we might consider to be a kinaesthetic knowledge that is informed by movement, practice, doing and an overall openness to the unplanned potentiality of becoming.

Although Saville provides a lively account of a spectacular youth practice, geographers have also attuned their radar to the types of mundane events, encounters and happenings that may occur in our daily living spaces. Such accounts include passing encounters in cafes (Laurier and Philo, 2006; Latham, 2003), commodity exchanges in farmers' markets (Slocum, 2008), bodily comportment on trains (Bissell, 2009), partying on beaches (Saldanha, 2007), meals in welfare drop-in centres (Conradson, 2003), experimental forms of rhythm and dance (McCormack, 2002), explorations of homeless spaces in the city (Cloke *et al.*, 2008), fast-food dining in burger bars (Latham, 2003), relaxing in caravan sites and allotments (Crouch, 2003), as well as rambling and climbing in rural areas (MacPherson, 2008; Wylie, 2002). These studies puncture the familiar idea of 'the city as text' (Duncan, 1990) – popular in 1990s British cultural geography (see Chapter 5) – by focusing upon the lively performances that bring these spaces to life in the hustle and bustle of the everyday.

The busyness of cityscapes is captured in Alan Latham's encounters in the thronging city of Auckland in New Zealand, where he considers 'everyday urban public culture as embodied practice' (2003: 1994). Latham explores how forms of sociability are continually produced through collective acts that are repeated in particular sites and spaces to form affective bonds of familiarity. Similarly, Laurier and Philo (2006) reflect how unexpected passing encounters in cafes may engender transient forms of conviviality to emerge or dissipate. Accepting representation as just one of many possible expressive practices, such accounts seek to understand the world as living motion replete with multiple possible geographies. The enjoined performances that momentarily fuse together to comprise 'cafe culture' suggest that these event-spaces are a collective accomplishment of a creative kind. However, we would suggest that the culture of conviviality expressed by Latham (2003) and Laurier and Philo (2006) is inevitably underwritten with relations of power. We might consider the types of mobile bodies that are welcomed into cafe culture and how a cosmopolitan 'vibe' is created by excluding or relegating others to the margin – the homeless, hobos, particular youth subcultures and seemingly 'undesirable' sections of the working-class.

These subtle forms of exclusion are enhanced through repeated performance as Arun Saldanha recounts in his provocative ethnography of racialised encounters on the Goan Trance scene in India. Saldanha explores how Goan beach parties form a space for the coming-into-being of race as it is assembled through noise, incense, dust, drugs, dance, Hippie freaks, alcohol, touristic t-shirts, motorcycles and psychedelic sonic soundscapes. It is through these intersubjective relations between human and non-human agents and actors that a Goan Trance event is summoned to life. However, the fluorescent day-glow performance of Trance does not yield a new global community. According to Saldanha, it comes to form a 'micro-fascist space' where whites can 'freak out' and local villagers and Indian tourists are once again de-centred from the main spectacle, cast adrift in the hot sands of empire. We might consider Lorimer's notion of the 'more-than-representational' to be a useful means of supplementing the more obtuse and philosophic aspects of work on affect in ways that do not write out the social but offer a more embodied, performative account of power and difference.

White lines of exclusion are also apparent in Rachel Slocum's (2008) corporeal account of the Minneapolis Farmers' Market. The site appears to offer an alternative to the 'hyper-commodified, sanitized and segregated public spaces' (p. 850) associated with US supermarket chains. Instead, using a performative approach, Slocum finds that racialised divisions are constantly forming and reforming around different spaces, commodities and vendors. She observes how the vegetables sold by the indigenous Chinese (Hmong) vendors come with soil and roots attached, so tend to be avoided by most of the white, middle-class latte drinking public. Instead, a number of market consumers opt for expensive herbs sold in hard plastic containers, thus consolidating their elite whiteness in the act of consumption (see also Jackson, 1998). Although many white customers may yet seek out an encounter with a foreign vendor selling 'exotic' produce, this frequently forms part of a wider touristic experience where they come to interpret the landscape through the gut. It is a corporeal enactment of 'eating the Other' (hooks, 1992), a palatable means of consuming difference and feeling good about oneself without really altering the dynamics of race politics. In the UK there has been a trend towards 'dirty' and 'misshapen' produce as these are markers of middle-class discernment for food that is seen as 'natural', local, rustic and organic. Through these examples of power we can begin to wrestle with the limitations of performance and the problematic of practice. In the final section we now consider emerging critiques of NRT and theories of affect.

The problem of performance

This chapter has sought to show how performance can open up new ways of approaching our environment. Here we observed a tacking movement between representation and presentation and pressed this performative potential to envision 'what else might be done' when it comes to geographical thinking and doing. Our attention was drawn to the role that feelings and affects play in sensual encounters with the world in which we come to interpret life through embodied interactions with people, objects and things. To grasp at least something of the intangible nature of experience we explored the possibilities that a 'more-than-representational' geography holds when it comes to being and belonging. Following this we examined some of the ways geographers are using performance theory in their work to recount the world 'otherwise' and engage with the immediacy of the now. Finally, in this concluding section we will consider what we might call the *problematic of performance.*

Overall non-representational approaches have been met with something of a mixed response within geography. Many are wary of yet another 'high theory' claiming to revolutionise the discipline, others see no worth in a philosophic approach they regard as devoid of politics, while most simply ignore or resent what is regarded as a niche area getting undue attention in mainstream geographical journals. Given this turbulence, what are we to make of affective geographies and the future of NRT? In many respects this is a difficult question to answer as the field is still opening up and there is little critical distance to be had. We might note, though, that despite the fervour of a few NRT enthusiasts these ideas have had little impact in revolutionising and re-envisioning the social sciences and humanities in the way Thrift had hoped. This does not mean that these ideas cannot furnish us with new and interesting ways of doing cultural geography but the grandiose aspirations of Thrift and others are unlikely to transpire.

Perhaps the most frequent charge aimed at work on performance is its alleged neglect of *power and politics.* While NRT advocates do address the political value of being open to change and doing politics differently (see Thrift, 2000, 2003, 2004a, 2004b; Popke, 2009), critics question how these ideas can be put into practice in grounded and meaningful ways. Thrift (2003) riles against those who regard his ideas as little more than fluffy academic distraction. He declares that performance theory 'wants to make things *more* political, *much more* political' (emphasis in original) and in doing so, 'expand the existing pools of alternative and corresponding forms of dissent' (p. 2021). If this is the aim of NRT projects then they fall short by this measure. As Pain (2009) has argued, this political strand of activity is more likely to be found residing within feminist work on emotional geographies or in participatory research, rather than contemporary affectual studies and philosophic approaches. The politicisation of bodies and emotions is strikingly present in critical research in race studies, to this end, 'The idea of race has long been constituted through feelings, emotions and affective dispositions' (Nayak, 2010). Post-colonial theorists recognise how the production of race categories can generate 'fear/desire' (Fanon, 1978), how daily multiculturalism is achieved through affective practices of 'conviviality' (Gilroy, 2004), or how fear is used to sketch the outline through which the figure of the asylum-seeker comes into being (Ahmed, 2000). In each of these compelling examples the *politics* of feeling and emotion is explicit.

Exploiting these creative tensions by considering how performance and practice are infused with geographies of power is at least a starting point for considering event-based understandings of a world in motion. In comparison with feminist emotional geographies and the performative anti-racist accounts described above, the declaration that NRT writing is '*much more* political' has a slightly hollow ring to it. It is also interesting how being 'political' in human geography is immediately equated with 'dissent' and radicalism. In contrast with Thrift, who claims performance as 'more political' (and presumably left-wing), and critics who dismiss the work as 'apolitical' (and presumably conservative) we suggest this is an unhelpful binary that can yet be uncoupled. Instead we would argue that theories of performance have as much to say about the exercise of power as they do resistance. This is exemplified in Ó Tuathail's (2003) interrogation of what he calls 'American affect', the nationalist feelings trumpeted after 9/11. For this reason we might rephrase Thrift's quote above to state that performativity and NRT is *no more*, or *no less* political, than routine practices of representation. In other words, it is how geographers enact performance theory in their accounts that is politically significant, not the theory itself and the moral values ascribed to it.

Some geographers also opine that much of the work on affect and emotion is guided by a **universalism** premised upon Western masculinist ideals and experiences; a criticism of humanistic geography and phenomenology more generally (see Chapter 3). For example, Thien (2005) criticises Thrift's (2004b: 58) ardent desire to affirm that his work on performance and affect is not 'nice and cuddly' or 'touchy-feely', as reflective of a masculine mode of ordering. Thrift's unending determination to assert NRT as 'more political' and his favoured use of engineering cabling metaphors may hint at some masculine anxieties here for the work to be taken seriously as 'hard science'. Certainly any discussion of performance and feeling has not come easily to what has traditionally been a masculine discipline (see Chapter 6). For Thien (2005), the turn to theories of affect may ironically be at the expense of emotion. The claim here is that the distinction between affect and emotion currently being articulated in human geography reworks a familiar gender binary of masculine/feminine. Emotional experiences as richly developed in feminist (and critical race) studies are 'emptied out' of meaning in the masculine pursuit for dense theory and philosophic explanation. In Thien's eyes, '"Affect" as a term and a concept is employed here in masculinist, technocratic and distancing ways' (2005: 452).

Although Thien's perspective is challenged by some NRT enthusiasts (see Anderson and Harrison, 2006) the issue of universalism has not fully dissipated. In a study of voluntary welfare services for the homeless, Conradson (2003) remarks how the lingering feelings of sociality generated during a meal hang in the air long after guests have eaten and left. He later goes on to claim how 'elements of its affective aura, bodily sensed and yet somehow woven into the physical spaces seemed to persist' (p. 1980). A problem with such assertions is that they rest upon the feelings (and imagination?) of the researcher concerned. Would others be similarly sensitive to these atmospheres, might we pick up on different feelings, or is this simply a projection of emotion by the author onto what is *already known* to have gone before? In other words, the social experience of previous hospitality may create a feeling of sociality in the mind of the participant that has little to do with the prevailing atmosphere. It is not simply the existence of these auras that we question here, but the universalist presentation of these 'feelings' that declare, this is like it is.

In a sophisticated and sensitive critique of dance, performativity and practice Catherine Nash (2000) considers the abiding universalism found in affectual accounts. Where dance has been a frequent visitor in NRT reports of pre-cognitive performance, Nash emphasises the durability of the past and its formative representations that shape the practice. In a critique of the performative powers of dance as advanced by Thrift, Nash reminds us of the different culturally situated configurations of power and past colonial histories that equate racialised bodies with 'natural rhythm'. For Nash, the metaphor of choreography is more appropriate then than any free-floating notion of dance as it is suggestive of intersections with power and how 'gendered and racialized bodily practices are learnt, performed and subverted' (*ibid.*). Where NRT invites us to 'live in the moment' it is vital to remember that the past may rupture and burst through into the present. The challenge is that experience is treated universally with little acknowledgement of the unequal distribution of power that enables some bodies to move freely while others are marked and held in place by past histories and geographies. Nash draws attention to 'the ways in which different material bodies are expected to do gender, class, race and ethnicity differently' (Nash, 2000: 657). Tolia-Kelly similarly reflects how many NRT accounts of emotion and affect, through their universalism, may invoke a type of 'ethnocentric encounter' by ignoring 'the political fact of different bodies having different affective capacities' (2006: 213), a theme incisively illustrated in the racially marked empirical encounters found in the work of Ahmed (2000), Saldanha (2007) and Slocum (2008) described previously.

While performative approaches do offer a sensual understanding of the world we would suggest that the tendency to celebrate the novel possibilities of NRT can present these approaches as wholly original. As our discussion of humanistic geography in Chapter 3 shows, affective relationships between neighbours, friends, family, colleagues, acquaintances and so forth are generative of an emotionally laden sense of place. Developing phenomenological insights humanistic geographers have in the past been able to show how unconscious feelings and emotions shape our connections to particular landscapes. Furthermore, a thorough engagement with the body, performance and emotional geographies is well-documented in feminist scholarship even if this work has been somewhat marginalised (see Pain, 2009). In this vein claims to *originality* are in danger of occluding the pioneering work that has gone on within the discipline such as Ley's (1974) emotionally aware account of the topography and tonality of race relations in Philadelphia as described in Chapter 8. Situating NRT alongside broader traditions of humanistic geography and phenomenology may remove the sheen of novelty from its veneer and locate it as another branch within the over-arching tree of post-modern thinking. It is possible to see these ideas as an elaboration of work on culture and representation, rather than a negation of what earlier cultural geography stands for.

As we have seen, the majority of studies using theories of performance have focused, almost obsessively, with the minutia of life. While this has offered important insights into the habits, routines and practices of our daily going about, it has also yielded a random, eclectic proliferation of studies on caravanning, dancing, allotments, cycling, boredom and even a recent attempt at 'questioning tea-spoons' (Latham and McCormack, 2009: 254)! In short, this appears to have rendered forth something akin to a geography of 'small things'. This may seem an overbearing criticism. It is certainly not our intention to dismiss the vitality of these micro-accounts nor place the everyday under erasure by ignoring its 'brute object-ness' (p. 253). Instead, what is apparent is that when NRT advocates try to 'scale up' their accounts to discuss say, the affectual nature of cities (Thrift, 2004b), some of the subtlety of performance and practice found in closely worked empirical accounts seems to dissolve into the ether of abstract generalisation. A challenge for NRT practitioners then, is to think through the difference that *scale* might make to their ideas while considering the extent to which their work can be transposed across time and place. A critical and reflexive repositioning of these ideas may more sharply clarify the use and limits of performance and the role that affectual geographies have to play in global relations, politics and everyday life.

A final criticism we would like to pose concerns the problem of representation itself. As we have argued throughout this chapter, representation and presentation do not exist as mutually exclusive practices, but can be seen as circuits through which discourse and action is continually relayed. While many adherents of NRT and performativity endeavour to achieve a radical breach with representation in terms of their thinking, more often than not a return to representational forms – interviews, diagrams, diaries, photographs – is all too evident in the doing of research and the process of write-up. Moreover, the convention of academic writing usually entails the discipline of structure, logic and coherence which may sit uneasily with the experimental uncertainty and openness of non-representational methods. Where athletes, dancers, composers, chefs, visual artists or actors may express their art in embodied ways few academics can easily forego the power of language and the written word. If we only produce personal diatribes in-the-moment our accounts may appear journalistic litanies of self-aggrandisement.

Returning to Nash, whose work critique of dance we outline above, she further argues against the tendency to locate dance movement as some pre-cognitive exercise whose impulse lies beyond the social and cultural. She intimates that 'dance is always mediated by words' (Nash, 2000: 658), citing the carefully choreographed ways in which it is taught, scripted, performed and observed. At a representational level dance is also coordinated through past traditions, genres, codes and conventions as ballet, Tango, disco, Salsa, the Foxtrot or Waltz illustrate. In this way Nash regards dance as a poor example of the pre-discursive and non-cognitive practices that Thrift alludes to, suggesting that it cannot be celebrated as libratory, un-thought-out activity. Similarly, experimental forms of writing and more conscious attempts to move beyond discourse by including photographs, video clips, scribbles and so forth do not necessarily transcend this methodological impasse, but only succeed in displacing it elsewhere. These forms still are representations of a particular kind. For Pile (2010) this is an obvious 'blind-spot' that NRT enthusiasts wilfully overlook even as they critique representational practice. As Nash (2000) elaborates a retreat from the representational devices advanced by British cultural geographers in the 1990s entails an abandonment of a whole tradition of cultural geography that has made rich use of the intersections between discourses, images, signs, symbols, texts, spaces and practices, a contribution outlined in Chapter 5. We would insist that representations can be treated as lively matter. As we saw in our example of the destruction of the Twin Towers, there is often a recursive slippage between representation and presentation. Moreover instead of regarding an interview (a popular qualitative technique in much previous cultural geography) as a partial reflection of a perceived reality it is possible to treat it as a performance, the space in which particular configurations of identity are conjured up (Nayak and Kehily, 2006). What all this tells us is that there is much that lies between the hyphen of re-presentation.

Conclusions

This chapter has explored how affective and embodied geographies can be used to deliver more sensual accounts of everyday living. We have suggested that representational accounts confined to text, words and symbolic meanings may at times fall short. This is partly because the fullness of experience can never be reduced to mere words and sparse representational codes. Second, cultural geographies that focus on representations, signs and symbols may overlook the materiality of things and simply reduce the physical properties of objects to 'dead matter'. As our examples of the circulation of 9/11 imagery, or the streetscapes of parkour reveal, there is potential to rethink material objects in more active ways through a network of relations that bring together actors, objects and processes in unpredictable ways so they may comprise an 'event' or 'happening'. What these new approaches suggest is that these relations are temporal and incomplete, being recomposed in the midst of life.

A third reason why discursive accounts may appear overly textual for performative cultural geographers is that so much of life is made up of 'non-representational' feelings including embodied gestures, day-dreams, collective sentiments, routine habits, ticks, winks, broken sentences, verbal slippages and a spectrum of indecipherable emotions. Even if we were vigilant enough to engage with some of these ephemeral feelings the problem with immaterial sensations is that they cannot adequately be conveyed in the written word, as our opening example of clubbing illustrates. Illuminating the affective, unconscious, spiritual, biophysical, chemical, tactile, sonorous and sensual aspects of our daily habitation is then a valuable and lively component of emotional and affective geographies. By turning to some of the recent advances in cultural geography concerned with non-representational theory, affect and performance we have tried to show how these ideas might enliven the discipline. Through a sensitivity to feelings, auras, bodily sensations and the like we might begin to see how the world can yet be differently arranged (Lorimer, 2007).

While many of the approaches discussed in this chapter are still quite fresh – Popke (2009: 88) remarks how we are only just 'learning to see differently' – we have gestured towards our early reflections about how this field is opening up. Where many essays are consciously written against the representational modes familiar in earlier British cultural geography, and are coming to self-define as 'non-representational', we suggest that this work might do better to place itself within, rather than against, the 'cultural turn'. To this extent we have been keen not to erase representation as it is central to much human geography and part and parcel of our modern worldly existence. We would contend that healing the fracture between representation and presentation affords more possibilities for a contingent and open-ended geography than an outward rejection of discourse, text, cultural signs and representations. As Jackson (2008) has recently argued, the ideas of social construction familiar to 1990s cultural geography were only ever a starting point used to prise open what geographical practice might be and yet become.

We closed the chapter by exploring the problematic of performance in non-representational geographies. Here we touched upon some of the more knotted aspects of NRT which at present are largely unresolved. These criticisms, or rather interruptions into the field, emphasised a tendency towards universalism, the ambiguous relationship to politics, the emergence of a philosophic masculine elitism, the frequent micro-scale of inquiry and the thorny question of representation itself. It is, after all, through questions, concerns and reflections that sharper critical thinking develops. Indeed, work on emotions, embodiment and non-representational practices has deepened the subfield of cultural geography and expanded our geographical imaginings. By exploring what lies between the gap of 're-presentation' we anticipate that more productive dialogues can be had between representation, presentation and performance.

Summary

- Theories of non-representation offer a challenge to the cultural turn where work is governed by a discursive order of words, images and language.
- Non-representational geographies approach the world as an unfolding event by focusing upon performance, practice and action.
- Non-representational theories are also concerned with pre-cognitive feelings and sensations and the unpredictable affects generated between bodies, spaces, objects and things.
- More-than-representational geographies offer a means of engaging with the surplus of feelings that cannot be reduced to words.
- Criticisms of performance and practice highlight the limitations of any new cultural geography that seeks to eschew representation altogether.

Further reading

Latham, A. (2003) Research, Performance, and Doing Human Geography: Some reflections on the diary-photograph, diary interview method. *Environment and Planning A*, 35: 1993–2017.

Lorimer, H. (2005) Cultural Geography: The busyness of being more-than-representational. *Progress in Human Geography*, 29(1): 83–94.

Nash, C. (2000) Performativity in practice: some recent work in cultural geography. *Progress in Human Geography*, 24(4): 653–664.

Pain, R. (2009) Globalized Fear? Towards an emotional geopolitics. *Progress in Human Geography*, 33: 1–21.

Saville, S.J. (2008) Playing with Fear: Parkour and the mobility of emotion. *Social and Cultural Geography*, 9(8): 891–914.

Thrift, N. (2003) Performance and . . . *Environment and Planning A*, 35: 2019–2024.

Wylie, J. (2002) An Essay on Ascending Glastonbury Tor. *Geoforum*, 33: 441–454.

Glossary

Affect – refers to immaterial sensations (light, colour, mood, atmosphere, etc.) that may exist outside of the body but nevertheless exert an influence upon our activity, thinking and overall disposition.

Alienation – a Marxist term referring to the emotional and psychic experience of feeling entirely divorced from the wage labour we undertake.

Alternative modernities – the assertion that other nations, particularly those in the global south, might have their own histories of modernisation that are now eclipsed by and judged against white Western norms.

Anti-racism – political practices that look to confront, challenge and undermine racism.

Bourgeoisie – a Marxist term referring to what was then the elite middle-classes.

Bricolage – a French term deployed by post-modernists to refer to the collage-like fashion of sticking different materials together as a canvas for a new form of expression.

Capitalism – the current system of monetary exchange for goods.

Colonialism – the forced occupation of a land, its resources and people by another nation state which rules over it.

Commodity fetishism – the dreams and desires ascribed to material goods that lend them a special quality.

Cosmopolitanism – an open and globally outward-looking attitude to difference, particularly concerning sexual diversity, ethnicity, multiculture, etc.

Cultural hybridity – a term used by the post-colonial critic Homi K. Bhabha to refer to the fusing together of different national pastimes to form a 'third space' from which a new culture can arise, often expressed through emergent identifications and new practices (e.g. black Britishness, chicken tikka massala, UK garage music, etc.).

Cultural turn – a movement across the arts, social sciences and humanities recognising that we live in an increasingly cultural world and that discourse, images, brands, signs, slogans and consumerism now complicate superficial explanations of political economy.

Deconstruction – a technique for disassembling texts, images or words in order to see how they are put together and given meaning.

Diaspora – a term originally used to refer to the forced movements of people but increasingly linked with today's complex migrations.

Discourse – forms of written, spoken and visual communication that over time may give rise to accepted norms and values.

Economic base – the fundamental organising principle in a given society be it a feudal or capitalist state.

Environmental determinism – the idea that the natural environment determines human behaviour.

Epistemology – the theory of knowledge.

Essentialism – a popular but much discredited belief that identity is biologically determined by sex, 'race' or nationality and that there exists a true 'inner self'.

Ethnicity – an anthropological term referring to the cultural habits, customs and practices of a particular community (e.g. diet, religious rituals, kinship patterns, gift giving practices, familial roles, etc.).

Ethnography – anthropological research method committed to describing culture through observations, interactions, conversations, actions and 'thick description' of fieldwork experiences.

Existentialism – a philosophical approach that seeks to understand human existence through studies of experience, awareness and the construction of meaning.

False consciousness – for Marxists this is a dominant belief that actually runs counter to one's social class interests in the long term.

Feminism – philosophies and political movements that advocate equal rights for women and seek to redress the imbalance of power between men and women.

Genre – a specific category or niche area pertaining to music, literature, film or art (e.g. heavy metal, crime fiction, science fiction, Brit-art, horror, etc.).

Geo-politics – an academic and political approach that seeks to identify spatial characteristics of world politics.

Grand narratives – or 'meta-narratives' seek to provide an over-arching explanation of the world in its totality (e.g. Christianity, evolution, Marxism, etc.).

Hegemony – a term deployed by Antonio Gramsci to refer to the exercise of power achieved by dominant groups through negotiation, conflict and consensus.

Heteronormativity– the tendency within media, culture and wider society to present heterosexuality as normal.

Historical materialism – the Marxist belief that economic inequalities can be reproduced over time and across generations.

Humanism – philosophical reflections on what it means to be human.

Hyper-reality – a term used by post-modern philosophers to refer to the way in which mass consumption and new technology in late-modern society has seen reality and fantasy fold into one another so that they are no longer distinct.

Ideology – a set of political beliefs or world view (e.g. capitalism, fascism, etc.).

Imperialism – the exercise of power over a group of people or territory ruled at a distance.

Individualisation – a consequence of social change in contemporary society that leads people to create their own lives and style their own identities.

Jouissance – a French term referring to ecstatic or even orgasmic pleasure that translates as bliss.

Means of production – a Marxist term referring to the technologies required to transform raw materials into commodities from which to generate profit.

Meta-narratives – see grand narratives.

Modernity/modernism – in Western thought this is the period after the Enlightenment associated with the birth of science and rationalism.

New cultural geography – a term associated with the rise of a British cultural geography in the 1990s that included some humanities research on landscape and the emerging critical writing on identity, representation and discourse.

New economic geography – refers to economic work that is working on the cultural turn and therefore engaging with representations, discourse and theory.

New Social Movements – political movements and interest groups that have developed in the post-socialist aftermath that are no longer restricted to a strict politics of socialism.

Non-representational theory – a phrase coined by Nigel Thrift to refer to analytic approaches that seek to challenge or move beyond discourse including certain performative, psychoanalytic, spiritual, sensual, pre-cognitive and kinaesthetic forms of knowledge.

Objectivity – a philosophical proposition that suggests that objects exist independent of individual consciousness.

Occidentalism – the study of the West, often undertaken by those who lie outside of its boundaries.

Ontology – a philosophy of how things are and how they exist in reality.

Orientalism – a specialist field of study related to the East, but according to Edward Said it is also a discourse of subjugation which renders the East inferior and through which the 'West' has come to know itself.

Paradigm – a theoretical framework or approach grounded in its own knowledge base.

Participatory methods – an approach that aims to empower research subjects by enabling them to have differing levels of participation in the overall process of research.

Pastiche – a knowing and self-aware form of imitation achieved through combining several recognisable elements of a genre.

Performativity – relates to a philosophical position that suggests speech, text and actions do not report on an external reality but rather bring this reality into being.

Phenomenology – an approach that focuses on the human experience of phenomena, such as materials, institutions and means of communication, in order to understand social orders.

Political ecology – a political reading of the relationships humans have with the natural environment and its resources, often designed on exploitation.

Positionality – a recognition of the significance of the subjectivity of the researcher and researched in the process of empirical research.

Positivism – a broad philosophy that makes the claim that the only worthwhile knowledge is knowledge gathered through repeated observation that may be tested and verified.

Post-colonialism – the period after colonial rule when a nation state becomes independent, and also a theoretical practice used to expose and critique on-going colonial relations of power.

Post-industrial – the period from around the mid-1970s when industrial production, extraction and manufacture began to decline, and later when the labour market became transformed by services and information technologies.

Post-structuralism – coming after structuralism and closely aligned with post-modernism, a theoretical approach appreciating the openness of meaning, the ambivalence of texts and the multiple and fragmented production of identity.

Pragmatism – a perspective that seeks to ground philosophical debates in everyday lives and decision making.

Proletariat – a Marxist term referring to what was then the mass of working-class subjects.

Psychoanalysis – a method from which to understand the subconscious.

Qualitative methods – in-depth approaches that may draw upon interviews, focus groups, biographies or ethnography to develop detailed explanations of social processes.

Quantitative methods – approaches using numerical data and statistical analysis often derived from or generative of large data sets.

Race – a fictitious concept premised on a belief in immutable biological or cultural differences between members of the human population.

Reflexivity – a type of self-questioning that seeks to account for one's own positionality in research and our role in the production of meaning.

Relativism – a belief that any one truth or experience is equivalent to another, opening us up to the possibility that all arguments are equally valid.

Representation – the active mediation of experience through text, words, symbols, musical notes, photographs, cinematic images and so forth.

Semiotics – derived from structural linguistics, a method for carrying out the study of signs, symbols, text and images.

Signs – the social codes used to express meaning.

Simulacra – the French philosopher Jean Baudrillard deploys this term to refer to the precision with which post-modern simulations not only mimic reality, but also have come to stand in for that perceived as 'reality'.

Simulation – the imitation of a perceived reality.

Structures of feeling – a phrase used by cultural studies writer Raymond Williams to describe the way in which particular institutions (collieries, churches, pubs, working-men's clubs, etc.) give rise to collective understandings and emotions that are actively lived and felt.

Subaltern – a term once used to refer to colonised South Asian people, subaltern theory now also refers to those ideas that have arisen outside of the authority of the West.

Subculture – a term used to refer to either marginalised, underground or oppositional groups who differ from and even resist the cultural mainstream.

Subjectivity – usually used in contrast to *objectivity*, this refers to an individual's own beliefs, wishes and perspectives.

Spatial division of labour – the way in which labour processes may be segmented across time and space.

Superstructure – the arenas in which ideologies are produced and struggled over including the law, education, media, church or family.

Typologies – fixed forms of classification.

Universalism – a belief in timeless laws or truths that apply to all and remain unaffected by geographical context or circumstance.

Whiteness – an identity and contingent marker of race privilege in late-modernity.

Working-class consciousness – a Marxist term reflecting a collective sense of one's economic place as a worker from which revolutionary change can be generated.

References

Abbott, D. (2006) Disrupting the Whiteness of Fieldwork in Geography. *Singapore Journal of Tropical Geography*, 27(3): 326–341.

Achebe, C. (1977) An Image of Africa: Racism in Conrad's 'Heart of Darkness'. *Massachusetts Review*, 18: 251–261.

Ackerman, E. (1945) Geographic Training: Wartime research, and immediate professional objectives. *Annals of the Association of American Geographers*, 35: 121–143.

Agamben, G. (1998) *Homo Sacer: Sovereign Power and Bare Life*, Stanford, CA: Stanford University Press.

Ageman, J. and Spooner, R. (1997) Ethnicity and the Rural Environment, in P. Cloke and J. Little (eds), *Contested Countryside Cultures: Otherness, Marginalization, and Rurality*, London: Routledge, pp. 197–217.

Agnew, J. (2002) *Making Political Geography*, London: Arnold.

Ahmed, S. (2000) *Strange Encounters: Embodied 'Others' in Postcoloniality*, London: Routledge.

Alexander, C. (2000) *The Asian Gang: Ethnicity, Identity, Masculinity*, Oxford: Berg.

Amin, A. (2002) Ethnicity and the Multicultural City: Living with diversity. *Environment and Planning A*, 34: 959–980.

Anderson, B. (2006) Becoming and Being Hopeful: Towards a theory of affect. *Environment and Planning D: Society and Space*, 24: 733–752.

Anderson, B. and Harrison, P. (2006) Questioning Affect and Emotion. *Area*, 38(3): 333–335.

Anderson, J. (2010) *Understanding Cultural Biography: Places and Traces*, Abingdon: Routledge.

Anderson, K. (1987) Chinatown as an Ideal: The power of place and institutional practice in the making of a racial category. *Annuls, Association of American Geographers*, 77: 580–598.

Anderson, K. (1988) Cultural Hegemony and the Race Definition Process in Chinatown, Vancouver: 1880–1980. *Environment and Planning D: Society and Space*, 6: 127–149.

Anderson, K. (1991) *Vancouver's Chinatown: Racial Discourse in Canada 1875–1980*, Montreal: McGill-Queens University Press.

Anderson, K. (1993a) Constructing Geographies: 'Race', place and the making of Sydney's Aboriginal Redfern, in P. Jackson and J. Penrose (eds), *Constructions of Race, Place and Nation*, London: UCL Press.

Anderson, K. (1993b) Place Narratives and the Origins of the Aboriginal Settlement in Inner Sydney, 1972–1973. *Journal of Historical Geography*, 9(3): 314–335.

Anderson, K. (1999 [1992]) Introduction, in K. Anderson and F. Gale (eds), *Cultural Geographies*, Melbourne: Longman.

Anderson, K. (2007) *Race and the Crisis of Humanism*, London: Routledge.

Anderson, K. and Gale, F. (eds) (1999 [1992]) *Cultural Geographies*, Melbourne: Longman.

Anderson, K. Domosh, M., Pole, S. and Thrift, N. (eds) (2003) *Handbook of Cultural Geography*, London: Sage.

Andrew, C.M. and Kanya-Forstner, A.S. (1988) Centre and Periphery in the Making of the Second French Empire 1815–1930. *Journal of Imperial and Commonwealth History*, 16: 9–34.

Appardurai, A. (1990) Disjuncture and Difference in the Global Economy. *Theory, Culture and Society*, 7: 295–310.

Appiah, A. (1992) *In My Father's House: Africa in the Philosophy of Culture*, London: Methuen.

Ashcroft, B., Griffiths, G. and Tiffin, H. (1989) *The Empire Writes Back: Theory and Practice in Post-colonial Literatures*, London: Routledge.

Association of University Teachers (AUT) (2005) Ethnicity and the Use of Discretionary Pay in UK HE. *AUT Research*, October.

Atkinson, D., Jackson, P., Sibley, D. and Washbourne, N. (eds) (2005) *Cultural Geography: A Critical Dictionary of Key Concepts*, London: I.B. Tauris.

Back, L. (1996) *New Ethnicities and Urban Culture: Racisms and Multiculture in Young Lives*, London: Routledge.

Back, L. (2007) *The Art of Listening*, Oxford: Berg.

Barker, M. (1989) *Comics: Ideology, Power and the Critics*, Manchester: Manchester University Press.

Barnes, T.J. (1994) Probable Writing: Derrida, deconstruction, and the quantitative revolution in human geography. *Environment and Planning A*, 26: 1021–1040.

Barnes, T.J. (2001a) Lives Lived and Lives Told: Biographies of geography's quantitative revolution. *Environment and Planning D: Society and Space*, 19: 409–429.

Barnes, T.J. (2001b) Retheorizing Economic Geography: From the quantitative revolution to the 'cultural turn'. *Annals of the Association of American Geographers*, 91(3): 546–565.

Barnes, T. (2009) Quantitative Revolution, in R. Johnston, D. Gregory, G. Pratt, M. Watts and S. Whatmore (eds), *The Dictionary of Human Geography*, Oxford: Wiley-Blackwell, pp. 611–612.

Barnes, T. (2010) Taking the Pulse From the Dead: History and philosophy of geography 2008–2009. *Progress in Human Geography*, 34(5): 668–677.

Barnes, T. and Duncan, J. (1992) *Writing Worlds: Discourse, Text and Metaphor in the Representation of Landscape*, London: Routledge.

Barnes, T. and Farish, M. (2006) Between Regions: Science, Militarism, and American Geography from World War to Cold War. *Annals of the Association of American Geographers*, 96: 807–826.

Barnes, T. and Gregory, D. (1997) *Reading Human Geography: The Poetics and Politics of Inquiry*, Chichester: John Wiley.

Barnett, C. (2004) *The Pentagon's New Map: War and Peace in the Twenty-first Century*, New York: G.P. Putnam's Sons.

Barnett, C. (2006) Postcolonialism, Space, Textuality and Power, in S. Aitken and G. Valentine (eds), *Approaches to Human Geography*, London: Sage, pp. 147–160.

Barrow, J. (1831) Royal Geographical Society. *The Journal of the Royal Geographical Society*, 1, at http://en.wikisource.org/wiki/Journal_of_the_Royal_Geographical_Society_of_London/Volume_1.

Barthes, R. (1972) *Mythologies*, London, Cape.

Baudrillard, J. (1983 [1981]) Simulacra and Simulations, in P. Brooker (ed.), *Modernism/Postmodernism*, Harlow: Longman.

Baudrillard, J. (1992) *Simulation*, New York: Columbia University Press.

Baudrillard, J. (1994 [1981]) *Simulation and Simulacra*, Ann Arbor, MI: University of Michigan Press.

Baudrillard, J. (1995) *The Gulf War Did Not Take Place*, Bloomington, IN: Indiana University Press.

Bauman, Z. (1988) *Freedom*, Milton Keynes: Open University Press.

Bauman, Z. (1989) *Modernity and the Holocaust*, Cambridge: Polity Press.

Bauman, Z. (2000) *Liquid Modernity*, Cambridge: Polity Press.

BBC News (2009) *Rift Flares after US Episcopal Church Elects Gay Bishop*, at http://news.bbc.co.uk/1/hi/8397653.stm.

Beck, U. (1992) *Risk Society: Towards a New Modernity*, London: Sage.

Bell, D. (1991) Insignificant Others: Lesbian and gay geographies. *Area*, 23: 323–329.

Bell, D. (1995) [Screw]ing Geography (censor's version). *Environment and Planning D: Society and Space*, 13: 127–131.

Bell, D. (2007) Fucking Geography, Again, in K. Browne, J. Lim and G. Brown (eds), *Geographies of Sexualities: Theory, Politics and Practice*, London: Ashgate, pp. 81–88.

Bell, D. and Valentine, G. (eds) (1995) *Mapping Desire: Geographies of Sexualities*, London: Routledge.

Bell, M. and McEwan, C. (1996) The Admission of Women Fellows to the Royal Geographical Society, 1892–1914: The controversy and the outcome. *The Geographical Journal*, 16(3): 295–312.

Berg, L. (1993) Between Modernism and Postmodernism. *Progress in Human Geography*, 17(4): 490–507.

Berry, B. and Garrison, W. (1958) The Functional Bases of the Central Place Hierarchy. *Economic Geography*, 34: 145–154.

Berry, B., Griffith, D. and Tiefelsdorf, M. (2008) From Spatial Analysis to Geospatial Science. *Geographical Analysis*, 40(3): 229–238.

Bhabha, H. (1984) Of Mimicry and Man: The ambivalence of colonial discourse. *October*, 28: 125–133.

Bhabha, H. (1990) *Nation and Narration*, London: Routledge.

Bhabha, H. (1994) *The Location of Culture*, London: Routledge.

Bialasiewicz, L., Campbell, D., Elden, S., Graham, S., Jeffrey, A. and Williams, A. (2007) Performing Security: The imaginative geographies of current US strategy. *Political Geography*, 26: 405–422.

Billig, M. (1995) *Banal Nationalism*, London: Sage.

Binnie, J. (1997) Coming Out of Geography: Towards a queer epistemology. *Environment and Planning D: Society and Space*, 15: 223–237.

Binnie, J. and Valentine, G. (1999) Geographies of Sexuality, A review of progress. *Progress in Human Geography*, 23: 175–187.

Bissell, D. (2009) Conceptualising Differently-mobile Passengers: Geographies of everyday encumbrance in the railway station. *Social and Cultural Geography*, 10(2): 173–195.

Blaut, J.M. (1970) Geographic Models of Imperialism. *Anitpode*, 2(1): 65–85.

Blouet, B. (2004) The Imperial Vision of Halford Mackinder. *The Geographical Journal*, 170: 322–329.

Blunt, A. and McEwan, C. (eds) (2002) *Postcolonial Geographies*, London: Continuum.

Blunt, A. and Wills, J. (2000) *Dissident Geographies: An Introduction to Radical Ideas and Practice*, Harlow: Prentice Hall.

Blunt, A., Gruffudd, P., May, J., Ogborn, M. and Pinder, D. (2003) *Cultural Geography in Practice*, New York: Edward Arnold.

Bonnett, A. (1993) Forever 'White'? Challenges and alternatives to a 'racial' monolith. *Journal of Ethics and Migration Studies*, 20(1): 173–180.

Bonnett, A. (1996) Construction of 'Race', Place and Discipline: Geographies of 'racial' identity and racism. *Ethnic and Racial Studies*, 19(4): 864–883.

Bonnett, A. (1997) Geography, 'Race' and Whiteness: Invisible traditions and current challenges. *Area*, 29(3): 193–199.

Bonnett, A. (2000a) *White Identities: Historical and International Perspectives*, Harlow: Pearson.

Bonnett, A. (2000b) *Anti-racism*, London: Routledge.

Bonnett, A. (2004) *The Idea of the West: Culture, Politics and History*, Basingstoke: Palgrave.

Bonnett, A. (2008) *What is Geography?*, London: Sage.

Borden, I. (2001) *Skateboarding, Space and the City: Architecture and the Body*, Oxford: Borg.

Boserup, E. (1970) *Woman's Role in Economic Development*, New York: St Martin's Press.

Bourdieu, P. (1984) *Distinction: A Social Critique of the Judgement of Taste*, London: Routledge.

Bourdieu, P. (1987) What Makes Social Class? On the theoretical and practical existence of groups. *Berkeley Journal of Sociology*, 32: 1–18.

Bourdieu, P. (1989) Social Space and Symbolic Power. *Sociological Theory*, 7: 14–25.

Bourdieu, P. (1995) *The Field of Cultural Production*, Cambridge: Polity Press.

Bowman, I. (1928) *The New World: Problems in Political Geography*, London, Calcutta, Sydney: George C. Harrap.

Bowman, I. (1942) Geography Vs. Geopolitics. *Geographical Review*, 32(4): 646–658.

Brah, A. (1996) *Cartographies of Diaspora: Contesting Identities*, London: Routledge.

Breitbart, M. (1981) Petr Kroptkin, the Anarchist Geographer, in D.R. Stoddart (ed.), *Geography, Ideology and Social Concern*, Oxford: Blackwell, pp. 134–153.

Brown, G. (2007a) Mutinous Eruptions: Autonomous spaces of radical queer activism. *Environment and Planning A*, 39: 2685–2698.

Brown, G. (2007b) Autonomy, Affinity and Play in the Spaces of Radical Queer Activism, in K. Browne, J. Lim and G. Brown (eds), *Geographies of Sexualities: Theory, Politics and Practice*, London: Ashgate, pp. 195–206.

Browne, K. (2004) Genderism and the Bathroom Problem: (Re)materialising sexed sites, (re)creating sexed bodies. *Gender, Place and Culture*, 11: 331–346.

Browne, K. (2007) Drag Queens and Drab Dykes: Deploying and deploring feminities, in K. Browne, J. Lim and G. Brown (eds), *Geographies of Sexualities: Theory, Politics and Practice*, London: Ashgate, pp. 113–124.

Browne, K., Lim, J. and Brown, G. (1985) (eds) (2007) *Geographies of Sexualities: Theory, Politics and Practice*, London: Ashgate.

Bunge, W. (1969) The First Years of the Detroit Geographical Expedition: A personal report. Detroit, MI: Society for Human Exploration.

Bunge, W. (1971) *Fitzgerald: Geography of a Revolution*, Cambridge, MA: Schlenkman.

Bunge, W. (1973) The Geography of Human Survival. *Annals of the Association of American Geographers*, 63: 275–295.

Bunge, W. (1979) Fred K. Shaefer and the Science of Geography. *Annals of the Association of American Geographers*, 69: 128–132.

Burgess, J.A. (1985) News from Nowhere: The press, the riots, and the myth of the inner city, in J.A. Burgess and J.R. Gold (eds), *Geography, the Media and Popular Culture*, London: Croom Helm, pp. 192–228.

Burgess, J.A. and Gold, J.R. (eds) (1985) *Geography, the Media and Popular Culture*, London: Croom Helm.

Burton, I. (1963) The Quantitative Revolution and Theoretical Geography. *The Canadian Geographer* 7: 151–162.

Butler, J. (1990) *Gender Trouble: Feminism and the Subversion of Identity*, London: Routledge.

Butler, J. (1993) *Bodies that Matter: On the Discursive Limits of Sex*, New York: Routledge.

Butler, J. (1997) Gender Trouble, Feminist Theory and Psychoanalytic Discourse, in L. McDowell and J. Sharp (eds), *Space, Gender, Knowledge: Feminist Readings*, London: Arnold, pp. 247–261.

Cahill, C., Sultana, F. and Pain, R. (eds), Participatory Ethics. *ACME: An International E-journal for Critical Geographies*, 6(3).

Carter, J., Fenton, S. and Madood, T. (1999) Ethnicity and Employment in Higher Education, Policy Studies Institute, September.

Castells, M. (1977) *The Urban Question: A Marxist Approach*, London: Edward Arnold.

Castells, M. (1983) *The City and the Grassroots: A Cross-Cultural Theory of Urban Social Movements*. Berkeley, CA: University of California Press.

Castree, N. (1999) Geography and the Renewal of Marxian Political Economy. *Transactions of the Institute of British Geographers*, 24(2): 137–158.

Castree, N. (2004) The Geographical Lives of Commodities: Problems of analysis and critique. *Social and Cultural Geography*, 5(1): 25–36.

Caute, D. (1970) *Fanon*, London: Fontana.

Center for Geographical Analysis (2010) The Center for Geographical Analysis website, Harvard University, at http://gis.harvard.edu/icb/icb.do.

Centre for Contemporary Cultural Studies (CCCS) (1981) *Unpopular Education: Schooling and Social Democracy in England Since 1944*, London: Hutchinson.

Centre for Contemporary Cultural Studies (CCCS) (1982) *The Empire Strikes Back: Race and Racism in 70s Britain*, London: Hutchinson.

Centre for Contemporary Cultural Studies (CCCS) (Women's Studies Group) (1978) *Women Take Issue: Aspects of Women's Subordination*, London: Hutchinson.

Chakrabarty, D. (2000) *Provincializing Europe: Postcolonial Thoughts and Historical Difference*, Princeton, NJ: Princeton University Press.

Charlesworth, S. (2000) *A Phenomenology of Working-class Experience*, Cambridge: Cambridge University Press.

Chatterton, P. and Hollands, R. (2003) *Urban Nightscapes: Youth Spaces, Pleasure Spaces and Corporate Power*, London: Routledge.

Chorley, R. and Haggett, P. (eds) (1967) *Models in Geography*, London: Methuen.

Clarke, D. (ed.) (1997) *The Cinematic City*, London: Routledge.

Clarke, D. (2006) Postmodern Geographies, in S. Aitken and G. Valentine (eds), *Approaches to Human Geography*, London: Sage, pp. 107–121.

Clayton, D. (2003) Critical Imperial and Colonial Geographies, in K. Anderson, M. Domosh, S. Pile and N. Thrift (eds), *Handbook of Cultural Geography*, London: Sage, pp. 354–368.

Cloke, P., Philo, C. and Saddler, D. (1991) *Approaching Human Geography: An Introduction to Contemporary Theoretical Debates*, London: Paul Chapman.

Cloke, P. and Johnston, R. (2005) Deconstructing Human Geography's Binaries, in P. Cloke and R. Johnston (eds), *Spaces of Geographical Thought: Deconstructing Human Geography's Binaries*, London: Sage, pp. 1–20.

Cloke, P., Cook, I., Crang, P., Goodwin, M., Painter, J. and Philo, C. (2004) *Practising Human Geography*, London: Sage.

Cloke, P., May, J. and Johnsen, S. (2008) Performativity and Affect in the Homeless City. *Environment and Planning D: Society and Space*, 26(2): 241–263.

Cohen, P. (1972) Subcultural Conflict and Working Class Community. *Working Papers in Cultural Studies*, 2, Birmingham: University of Birmingham.

Cohen, P. and Robins, D. (1978) *Knuckle Sandwich: Growing up in the Working Class City*, Harmondsworth: Penguin.

Conrad, J. (1924a) Geography and Some Explorers. *National Geographic*, 45: 241–256.

Conrad, J. (1924b) *Heart of Darkness*, New York: Doubleday, Page & Co.

Conradson, D. (2003) Doing Organisational Space: Practices of voluntary welfare in the city. *Environment and Planning A*, 35: 1975–1992.

Cook, I. (2004) Follow the Thing: Papaya. *Antipode*, 36(4): 642–664.

Cook, I. and Harrison, M. (2003) Cross Over Food: Re-materializing postcolonial geographies. *Transactions of the Institute of British Geographers*, 28: 296–317.

Cook, I., Crouch, D., Naylor, S. and Ryan, J.R. (eds) (2000) *Cultural Turns/Geographical Turns: Perspectives on Cultural Geography*, Harlow: Prentice Hall.

Coombes, B. and Barber, K. (2005) Environmental Determinism in Holocene Research: Causality or coincidence? *Area*, 37: 303–311.

Cosgrove, D. (2008) Geography is Everywhere: Culture and Symbolism, in T. Oakes and P.L. Price (eds), *The Cultural Geography Reader*, London: Routledge.

Cosgrove, D. and Daniels, S. (1988) *The Iconography of Landscape*, Cambridge: Cambridge University Press.

Cosgrove, D. and Jackson, P. (1987) New Directions in Cultural Geography. *Area*, 19: 95–101.

Crang, M. (1998) *Cultural Geography: An Inroduction*, London: Routledge.

Crang, M. (2002) Qualitative Methods: The new orthodoxy? *Progress in Human Geography*, 26: 647–655.

Crang, P. (1994) 'It's Showtime': On the workplace practices of display in a restaurant in south-east England. *Environment and Planning D: Society and Space*, 12: 675–704.

Crawford, J. (1863) On the Connection between Ethnology and Physical Geography. *Transactions of the Ethnological Society of London*, 2: 4–23.

Cresswell, T. (2003) Landscape and the Obliteration of Practice, in K. Anderson, M. Domosh, S. Pile and N. Thrift (eds), *Handbook of Cultural Geography*, London: Sage, pp. 269–281.

Cresswell, T. (2004) *Place: A Short Introduction*, Oxford: Blackwell.

Crouch, D. (2003) Spacing, Performing, and Becoming: Tangles in the mundane. *Environment and Planning A*, 35: 1945–1960.

Crush, J. (1994) Postcolonialism, Decolonization and Geography, in A. Godlewska and N. Smith (eds), *Geography and Empire*, Oxford: Blackwell, pp. 336–337.

Dalby, S. (1996) Writing Critical Geopolitics: Campbell, Ó Tuathail, Reynolds and dissident skepticism. *Political Geography* 15(6/7): 655–660.

Dalby, S. (2008) Imperialism, Domination, Culture: The continued relevance of critical geopolitics. *Geopolitics*, 13: 413–436.

Daniels, S. and Rycroft, S. (1993) Mapping the Modern City: Alan Sillitoe's Nottingham novels. *Transactions of the Institute of British Geographers*, NS18: 460–480.

Darby, W.J. (2000) *Landscape and Identity: Geographies of Nation and Class in England*, Oxford: Berg.

Darwin, C. (1859) *On the Origins of Species*, London: John Murray.

Datta, D. (2008) Stay Braced for a 300-point Breakfast. *Business Standard*, 8 December.

Davis, M.L. (1990) *City of Quartz: Excavating the Future in Los Angeles*, New York: Verso.

Davis, M. (1991) *City of Quartz*, London: Verso.

de Beauvoir, S. (1953) *The Second Sex*, New York: Alfred A. Knopf.

Dear, M. (1999) A Real Differentiation and Post-modern Human Geography, in D. Gregory and R. Walford (eds), *Horizons in Human Geography*, London: Macmillan Palgrave, pp. 67–96.

Dear, M. and Flusty, S. (1998) Postmodern Urbanism. *Annals of the Association of American Geographers*, 88(1): 50–72.

Delaney, D. (2002) The Space that Race Makes. *The Professional Geographer*, 54(1): 6–14.

Del Casino Jr, V.J. (2007) Health/Sexuality/Geography, in K. Browne, J. Lim and G. Brown (eds), *Geographies of Sexualities: Theory, Politics and Practice*, London: Ashgate, pp. 39–52.

Del Casino Jr, V.J. (2009) *Social Geography*, Oxford: Blackwell.

Deleuze, G. and Guattari, F. (1987) *A Thousand Plateaus*, Minneapolis, MN: University of Minnesota Press.

Derrida, J. (1976) *Of Grammatology*, Baltimore, NJ: Johns Hopkins University Press.

Derrida, J. (1991) *A Derrida Reader: Between the Blinds*, P. Kamuf (ed.), New York: Columbia University Press.

Dewsbury, J.D., Harrison, P., Rose, M. and Wylie, J. (2002) Introduction: Enacting geographies. *Geoforum*, 33: 437–440.

Dickens, C. (2003 [1838]) *Oliver Twist, or, The Parish Boy's Progress*, Harmondsworth: Penguin.

Dittmer, J. (2005) Captain America's Empire: Reflections on identity, popular culture, and post-9/11 geopolitics. *Annals of the Association of American Geographers*, 95: 626–643.

Dittmer, J. and Dodds, K. (2008) Popular Geopolitics Past and Future: Fandom, identities and audiences. *Geopolitics*, 13: 437–457.

Dodds, K. (1996) The 1982 Falklands War and a Critical Geopolitical Eye: Steve Bell and the if ... cartoons. *Political Geography*, 15: 571–592.

Dodds, K. (2003) Licensed to Stereotype: Geopolitics, James Bond and the spectre of Balkanism. *Geopolitics*, 8: 125–156.

Dodds, K. (2006) Popular Geopolitics and Audience Dispositions: James Bond and the internet movie database (IMDb). *Transactions of the Institute of British Geographers*, 31: 116–130.

Dodds, K. and Elden, S. (2008) Thinking Ahead: David Cameron, the Henry Jackson Society and British neo-conservatism. *The British Journal of Politics and International Relations*, 10(3): 347–363.

Dodds, K. and Sidaway, J.D. (2004) Halford Mackinder and the 'Geographical Pivot of History': A centennial retrospective. *The Geographical Journal*, 170: 292–297.

Dodman, D. (2007) A Place or a People? Social and cultural geographies of the Anglophone Caribbean. *Social and Cultural Geography*, 8(1): 143–150.

Doel, M. (1993) Proverbs for Paranoids: Writing geography on hollowed ground. *Transactions of the Institute of British Geographers*, 18(3): 377–394.

Doel, M. (1999) *Poststructuralist Geographies: The Diabolical Art of Spatial Science*, Edinburgh: Edinburgh University Press.

Doel, M.A. (2001) Qualified Quantitative Geography. *Environment and Planning D: Society and Space*, 19: 555–572.

Dorling, D. (2010) Our Divided Nation. *New Statesman*, 14 June.

Dorling, D. and Coles, P. (2009) Featured Graphic: Wars, massacres, and atrocities of the 20th century. *Environment and Planning A*, 41: 1779–1780.

Dorling, D. and Pritchard, J. (2010) The Geography of Poverty, Inequality and Wealth in the UK and Abroad: Because enough is never enough. *Applied Spatial Analysis and Policy*, 3: 81–106.

Dorling, D. and Thomas, B. (2004) *People and Places: A 2001 Census Atlas of the UK, Bristol-Policy.*

Dowler, L. and Sharp, J. (2001) A Feminist Geopolitics? *Space and Polity*, 5: 165–176.

Driver, F. (1992) Geography's Empire: Histories of geographical knowledge. *Environment and Planning D: Society and Space*, 10: 23–40.

Driver, F. (2001) *Geography Militant: Cultures of Exploration and Empire*, Oxford: Blackwell.

du Gay, P. (ed.) (1997) *Production of Culture/Cultures of Production*, London: Open University Press.

Duncan, J. (1980) The Superorganic in American Cultural Geography. *Annals of the Association of American Geographers*, 70: 181–198.

Duncan, J. (1990) *The City as Text: The Politics of Landscape Interpretation in the Kandyan Kingdom*, Cambridge: Cambridge University Press.

Duncan, J. (1993) Landscapes of Self/Landscapes of the Other(s): Cultural geography, 1991–92. *Progress in Human Geography*, 17: 367–377.

Dwyer, C. (1993) Construction of Muslim Identity and the Contesting of Power: The debate over Muslim schools in the United Kingdom, in P. Jackson and J. Penrose (eds), *Construction of 'Race', Place and Nation*, London: UCL Press, pp. 143–159.

Dwyer, C. (1999) Negotiations of Femininity and Identity for Young British Muslim Women, in N. Laurie, C. Dwyer, S. Holloway and F. Smith (eds), *Geographies of New Femininities*, Harlow: Pearson, pp. 135–152.

Dwyer, C. and Bressey, C. (eds) (2008) *New Geographies of Race and Racism*, Aldershot: Ashgate.

Dwyer, O. and Jones III, J. (2000) White Socio-spatial Epistemology. *Social and Cultural Geography*, 1(2): 209–222.

Dyck, I. (1990) Space, Time and Renogotiating Motherhood: An exploration of the domestic workplace. *Environment and Planning D: Society and Space*, 8: 459–483.

Eagleton, T. (1996) *Literary Theory: An Introduction*, 2nd edn, Oxford: Blackwell.

Elden, S. (2004) *Understanding Henri Lefebvre: Theory and the Possible*, London: Continuum.

Elden, S. (2006) *Speaking Against Number: Heidegger, Language and the Politics of Calculation*, Edinburgh: Edinburgh University Press.

Elden, S. (2009) *Terror and Territoriality: The Spatial Extent of Sovereignty*, Minneapolis, MN: University of Minnesota Press.

Entrikin, J.N. (1976) Contemporary Humanism in Geography. *Annals of the Association of American Geographers*, 66: 615–632.

Entrikin, N. and Tepple, J. (2006) Humanism and Democratic Place-making, in S. Aitken and G. Valentine (eds), *Approaches to Human Geography*, London: Sage, pp. 30–41.

Escobar, A. (1995) *Encountering Development*, Princeton, NJ: Princeton University Press.

Fanon, F. (1965) *A Dying Colonialism*, London: Penguin.

Fanon, F. (1968 [1961]) *The Wretched of the Earth*, New York: Grove Press.

Fanon, F. (1978 [1952]) *Black Skin, White Masks*, New York: Grove Press.

Fausto-Sterling, A. (1985) *Myths of Gender: Biological Theories about Women and Men*, New York: Basic Books.

Featherstone, M. (1998 [1991]) *Consumer Culture and Postmodernism*, London: Sage.

Flusty, S. (2005) Postmodernism, in D. Atkinson, P. Jackson, D. Sibley and N. Washbourne (eds), *Cultural Geography: A Critical Dictionary of Key Concepts*, London: I.B. Tauris, pp. 169–174.

Fortier, A. (2007) Too Close for Comfort: Loving the Neighbour and the Management of Multicultural Intimacies. *Environment and Planning D: Society and Space*, 25: 104–119.

Fotheringham, A.S. (2006) Quantification, Evidence and Positivism, in S. Aitken and G. Valentine (eds), *Approaches to Human Geography*, London: Sage, pp. 237–250.

Foucault, M. (1976) *The History of Sexuality*, vol. 1, trans. R. Hurley, Harmondsworth: Penguin.

Foucault, M. (1978) *The History of Sexuality: An Introduction*, New York: Vintage.

Frankenberg, R. (1994) *White Women, Race Matters: The Social Construction of Whiteness*, Minnesota, MN: Minneapolis University Press.

Frenkel, S. (1992) Geography, Empire and Environmental Determinism. *Geographical Review*, 8(2): 143–153.

Freud, S. (1962) *Three Essays on the Theory of Sexuality*, New York: Basic Books.

Freud, S. (1990 [1920]) *Beyond the Pleasure Principle*, New York: W.W. Norton.

Garland, J. and Chakrobarti, N. (2006) 'Race', Space and Place: Examining identity and cultures of exclusion in rural England. *Ethnicities*, 6(2): 159–177.

Gauthier, H.L. and Taaffe, E.J. (2002) Three 20th Century 'Revolutions' in American Geography. *Urban Geography*, 23: 503–527.

Geertz, C. (1993) *The Interpretation of Cultures: Selected Essays*, London: Fontana.

Gibson-Graham, J.K. (2005) Building Community Economies: Women and the politics of place, in W. Harcourt and A. Escobar (eds), *Women and the Politics of Place*, Bloomfield, CT: Kumarian Press, pp. 130–157.

Gibson-Graham, J.K. (2008) Diverse Economies: Performative practices for 'other worlds'. *Progress in Human Geography*, 32(5): 613–632.

Giddens, A. (1991) *Modernity and Self Identity: Self and Identity in the Late Modern Age*, Cambridge: Polity Press.

Gilroy, P. (1987) *There Ain't No Black in the Union Jack*, London: Hutchinson.

Gilroy, P. (1992) *The Black Atlantic: Modernity and Double Consciousness*, Cambridge, MA: Harvard University Press.

Gilroy, P. (2004) *After Empire, Melancholia or Convivial Culture?*, London: Routledge.

Godlewska, A. and Smith, N. (eds) (1994) *Geography and Empire*, Oxford: Blackwell.

Golledge, R.G. (2006) Textbooks that Moved Generations, in R.J. Chorley and P. Haggett (eds), *Models in Geography*, London: Methuen.

Gould, P. (1979) Geography 1957–1977: The augean period. *Annals of the Association of American Geographers*, 69: 139–151.

Gray, C. (1977) *The Geopolitics of the Nuclear Era: Heartland, Rimlands, and the Technological Revolution*, New York: Crane, Russak.

Gregory, D. (1981) Human Agency and Human Geography. *Transactions of the Institute of British Geographers*, 6(1): 1–18.

Gregory, D. (1989) A Real Differentiation and Postmodern Human Geography, in D. Gregory and R. Walford (eds), *Horizons in Human Geography*, London: Macmillan, pp. 67–96.

Gregory, D. (2000) Postcolonialism, in R.J. Johnston, D. Gregory, G. Pratt and M. Watts (eds), *Dictionary of Human Geography*, 4th edn, Oxford: Blackwell.

Gregory, D. (2001) Cultures of Travel and Spatial Formations of Knowledge. *Erdkunde*, 54(4): 297–319.

Gregory, D. (2004) *The Colonial Present: Afghanistan, Palestine, Iraq*, Malden, MA: Blackwell.

Gregson, N. (2003) Reclaiming 'the Social' in Social and Cultural Geography, in K. Anderson, M. Domosh, S. Pile and N. Thrift (eds), *Handbook of Cultural Geography*, London: Sage, pp. 43–47.

Gregson, N. (2007) *Living with Things: Ridding, Accommodation, Dwelling*, Oxford: Sean Kingston Publishing.

Gregson, N. and Crewe, L. (1997) The Bargain, the Knowledge and the Spectacle: Making sense of consumption in the space of the car boot sale. *Environment and Planning D: Society and Space*, 15: 87–112.

Gregson, N., Brooks, K. and Crewe, L. (2000) Narratives of Consumption and the Body in the Space of the Charity/Shop, in P. Jackson, M. Lowe, D. Miller and F. Mort (eds), *Commercial Cultures: Economies, Practices, Spaces*, Oxford: Berg, pp. 101–121.

Gregson, N., Kothari, U., Cream, J., Dwyer, C., Holloway, S., Mander, A. and Rose, G. (1997) Conclusions, in Women and Geography Study Group (ed.), *Feminist Geographies: Explorations in Diversity and Difference*, London: Longman, pp. 191–200.

Griffin, C. (1985) *Typical Girls? Young Women from School to the Job Market*, London: Routledge and Kegan Paul.

Gruffudd, P. (1994) Back to the Land: Historiography, rurality and the nation in interwar Wales. *Transactions of the Institute of British Geographers*, 19: 61–77.

Haggett, P. (2008) The Local Shape of Revolution: Reflections on quantitative geography at Cambridge in the 1950s and 1960s. *Geographical Analysis*, 40: 336–352.

Hall, S. (1980 [1973]) Encoding/Decoding, in S. Hall, D. Hobson, A. Lowe and P. Willis (eds), *Culture, Media, Language*, London: Hutchinson, pp. 107–116.

Hall, M. (1989) Private Experiences in the Public Realm: Lesbians in organisations, in J. Hearn, D. Sheppard, P. Tancred-Sherriff and G. Burrell (eds), *The Sexuality of Organisation*, London: Sage, pp. 125–138.

Hall, S. (1993) New Ethnicities, in A. Rattansi and J. Donald (eds), *Race, Culture and Difference*, London: Sage, pp. 252–260.

Hall, S. (1996) When was the 'Postcolonial'? Thinking at the limit, in I. Chambers and L. Curtis (eds), *The Postcolonial Question: Common Skies Divided Horizons*, London: Routledge, pp. 246–260.

Hall, S. and Jefferson, T. (1975) *Resistance through Rituals: Youth Subcultures in Post-war Britain*, London: Hutchinson in association with CCCS.

Hall, S., Critcher, C., Jefferson, T., Clarke, J. and Roberts, B. (1978) *Policing the Crisis: Mugging, the State and Law and Order*, London and Basingstoke: Macmillan.

Hall, S., du Gay, R., Janes, L., Mackay, H. and Negus, K. (1997) *Doing Cultural Studies: The Story of the Sony Walkman*, London: Sage.

Hamnett, C. (2003) Contemporary Human Geography: Fiddling while Rome burns? *Geoforum*, 34: 1–3.

Hampson, R. (1995) Introduction, in J. Conrad (ed.), *The Heart of Darkness*, London: Penguin, pp. xi–xlv.

Haraway, D.J. (1991) *Simians, Cyborgs and Women: The Reinvention of Nature*, London: Routledge.

Haraway, D. (1997) Situated Knowledges: The science question in feminism and the priviledge of partial perspective, in L. McDowell and J. Sharp (eds), *Space, Gender and Knowledge Feminist Readings*, London: Arnold, pp. 53–72.

Harrison, P. (2007) 'How Shall I Say it …?' Relating the nonrelational. *Environment and Planning A*, 39: 590–608.

Harvey, D. (1969) *Explanation in Geography*, London: Edward Arnold.

Harvey, D. (1972) Revolutionary and Counter-revolutionary Theory in Geography and the Problem of Ghetto Formation. *Antipode*, 4: 1–13.

Harvey, D. (1973) *Social Justice and the City, Geographies of Justice and Social Transformation*, Athens, GA: University of Georgia Press.

Harvey, D. (1982) *Limits of Capital*, London: Verso.

Harvey, D. (1984) In the History and Present Condition of Geography: An historical materialist manifesto. *The Professional Geographer*, 36: 1–11.

Harvey, D. (1985) *The Urbanization of Capital*, Oxford: Basil Blackwell.

Harvey, D. (1989a) *The Condition of Postmodernity: An Enquiry into the Origins of Cultural Change*, Oxford: Blackwell.

Harvey, D. (1989b) *The Urban Experience*, Baltimore, NJ: Johns Hopkins University Press.

Harvey, D. (1995 [1989]) *The Condition of Postmodernity: An Enquiry into the Theory of Capital*, Oxford: Blackwell.

Harvey, D. (2003) *The New Imperialism*, Oxford: Oxford University Press.

Harvey, D. (2007) *A Brief History of Neo-liberalism*, Oxford: Oxford University Press.

Hebdige, D. (1979) *Subculture: The Meaning of Style*, London: Methuen.

Heffernan, M. (1994a) A State Scholarship: The political geography of French international science during the nineteenth century. *Transactions of the Institute of British Geographers*, 19: 21–45.

Heffernan, M. (1994b) The Science of Empire: The French geographical movement and the forms of French imperialism, c.1870–c.1920, in A. Godlewska, and N. Smith (eds), *Geography and Empire*, Oxford: Blackwell, pp. 92–114.

Heffernan, M. (2000) Balancing Visions: Gerard Ó Tuathail's critical geopolitics. *Political Geography*, 19: 347–352.

Heffernan, M. (2005) Geography, Empire and National Revolution in Vichy France. *Political Geography*, 24: 731–758.

Hoggart, R. (1957) *The Uses of Literacy*, London: Chatto and Windus.

Holdar, S. (1992) The Ideal State and the Power of Geography: The life-work of Rudolf Kjellén. *Political Geography*, 11: 307–323.

Holloway, S.R. (2000) Identity, Contingency and the Urban Geography of 'Race'. *Social and Cultural Geography*, 1(2): 197–208.

hooks, b. (1992) *Black Looks: Race and Representation*, Cambridge, MA: South End Press.

Hopkins, P. (2006) Youthful Muslim Masculinities: Gender and generational relations. *Transactions for the Institute of British Geographers*, 31: 337–352.

Hopkins, P. (2007) Global Events, National Politics, Local Lives: Young Muslim men in Scotland. *Environment and Planning A*, 39(5): 1119–1133.

Howell, P. (2007) Foucault, Sexuality, Geography, in J. Crampton and S. Elden (eds), *Foucault and Geography: Space, Knowledge, Power*, Aldershot: Ashgate, pp. 291–315.

Hubbard, P. (2000) Desire/Disgust: Mapping the moral contours of heterosexuality. *Progress in Human Geography*, 24: 191–217.

Hubbard, P. (2001) Sex Zones: Intimacy, citizenship and public space. *Sexualities*, 4: 51–71.

Hubbard, P. (2002) Sexing the Self: Geographies of engagement and encounter. *Social & Cultural Geography*, 3: 365–381.

Hughes, A. (2007) Geographies of Exchange and Circulation: Flows and networks of knowledgeable capitalism. *Progress in Human Geography*, 31: 527–535.

Hughes, A. and Reimer, S. (eds) (2004) *Geographies of Commodity Chains*, London: Routledge.

Huntington, E. (1915) *Civilization and Climate*, New Haven, CT: Yale University Press.

Huntington, E. (1924) *The Character of Races as Influenced by Physical Environment, Natural Selection and Historical Development*, New York: C. Scribner's Sons.

Huntington, S. (1996) *The Clash of Civilizations and the Remaking of World Order*, New York: Simon & Schuster.

Hyndman, J. (2006) Beyond Either/Or: A Feminist Analysis of September 11th, in G. Ó Tuathail, S. Dalby and P. Routledge (eds), *The Geopolitics Reader*, 2nd edn, London: Routledge, pp. 276–280.

Ingham, J., Purvis, M. and Clarke, D.B. (1999) Hearing Places, Making Spaces: Sonorous geographies, ephemeral rhythms, and the Blackburn warehouse parties. *Environment and Planning D: Society and Space*, 17: 283–305.

Inwood, J. and Martin, D. (2008) Whitewash: White priviledge and racialized landscapes at the University of Georgia. *Social and Cultural Geography*, 9(4): 373–395.

Jackson, P. (1980) A Plea for Cultural Geography. *Area*, 12: 110–113.

Jackson, P. (ed.) (1987) *Race and Racism: Essays in Social Geography*, London: Allen and Unwin.

Jackson, P. (1988) Street Life: The politics of carnival. *Environment and Planning D: Society and Space*, 6: 213–227.

Jackson, P. (1993) Policing Difference: Race and crime in metropolitan Toronto, in P. Jackson and J. Penrose (eds), *Constructions of Race, Place and Nation*, London: UCL Press, pp. 181–200.

Jackson, P. (1995 [1989]) *Maps of Meaning: An Introduction to Cultural Geography*, London: Routledge.

Jackson, P. (1996) The Idea of Culture: A response to Don Mitchell. *Transactions of the Institute of British Geographers*, 21: 572–582.

Jackson, P. (1997) Geography and the Cultural Turn. *Scottish Geographical Magazine*, 113(3): 186–188.

Jackson, P. (1998) Constructions of Whiteness in the Geographical Imagination. *Area*, 30(2): 99–106.

Jackson, P. (2000) Rematerializing Social and Cultural Geography. *Social and Cultural Geography*, 1(1): 9–14.

Jackson, P. (2008) Afterword: New geographies of race and racism, in C. Dwyer and C. Bressey (eds), *New Geographies of Racism*, Aldershot: Ashgate, pp. 297–304.

Jackson, P. and Penrose, J. (eds) (1993) *Constructions of Race, Place and Nation*, London: UCL Press.

Jackson, P. and Smith, S. (eds) (1981) *Social Interaction and Ethnic Segregation*, London: Academic Press.

Jacobs, J. (1993) 'Shake 'im this Country': The mapping of the Aboriginal sacred in Australia – the case of Coronation Hill, in P. Jackson and S. Smith (eds), *Constructions of Race, Place and Nation*, London: UCL Press, pp. 100–108.

Jacobs, J. (1996) *Edge of Empire: Postcolonialism and the City*, London: Routledge.

James, C.L.R. (1973) *Modern Politics* (reprinted 1973 [1960]) Detroit, MI: Bewick.

James, L. (1994) *The Rise and Fall of the British Empire*. London: Little, Brown and Company.

Jameson, F. (1984) Postmodernism, or, the Cultural Logic of Late Capitalism. *New Left Review*, 146: 53–92.

Jarvis, H., Cloke, J. and Kantor, P. (2009) *Cities and Gender*, London and New York: Routledge.

Jay, E. (1992) *Keep Them in Birmingham*, London: Commission for Racial Equality.

Jeffrey, A. (2009) Containers of Fate: Problematic states and paradoxical sovereignty, in A. Ingram and K. Podds (eds), *Spaces of Security and Insecurity: Geographies of the War on Terror*, Aldershot: Ashgate, pp. 43–64.

Jencks, C. (1993) *Heteropolis: Los Angeles, the Riots and the Strange Beauty of Hetero-Architecture*, London: Academy Editions.

Johnson, R. (1986) What is Cultural Studies Anyway? *Social Text*, 16: 38–80.

Johnson, R., McLennan, G. and Schwarz, B. (eds) (1982) *Making Histories: Studies in History Writing and Politics*, London: Hutchinson.

Johnston, R. (2000) On Disciplinary History and Textbooks: Or where has spatial analysis gone? *Australian Geographical Studies*, 38: 125–137.

Johnston, R. (2009) Central Place Theory, in R. Johnston, D. Gregory, G. Pratt, M. Watts and S. Whatmore (eds), *The Dictionary of Human Geography*, Oxford: Blackwell, pp. 72–74.

Johnston, R.J. and Sidaway, J.D. (1979) *Geography and Geographers*, London: Arnold.

Johnston, R.J. and Sidaway, J. (2004) *Geograpy and Geographers Anglo-American Human Geography Since 1945*, 6th edn, London: Hodder Arnold.

Jones, S. (1988) *Black Culture, White Youth: The Reggae Tradition from JA to UK*, Basingstoke: Macmillan.

Katz, C. (2008) Bad Elements: Katrina and the scoured landscape of social reproduction. *Gender, Place and Culture*, 15(1): 15–29.

Kearns, G. (2004) The Political Pivot of Geography. *The Geographical Journal*, 170(4): 337–346.

Keith, M. (1991) Knowing Your Place: The imagined geographies of racial subordination, in C. Philo (ed.), *New Words, New Worlds: Reconceptualising Social and Cultural Geography*, Lampeter: Social and Cultural Geography Group, pp. 178–192.

Keith, M. (1993) *Race, Riots and Policing: Lore and Disorder in a Multi-racist Society*, London: UCL Press.

Kesby, M. (2007) Spatialising Participatory Approaches: The contribution of geography to a mature debate. *Environment and Planning A*, 39: 2813–2831.

King, A. (1997 [1984]) *The Bungalow: The Production of a Global Culture*, London: Routledge and Kegan Paul.

King, A. (2003) Cultures and Spaces of Postcolonial Knowledges, in K. Anderson, M. Domosh, S. Pile and S. Thrift (eds), *Handbook of Cultural Geography*, London: Sage, pp. 381–398.

Kinsman, P. (1995) Landscape, Race and National Identity: The photography of Ingrid Pollard. *Area*, 27(4): 300–310.

Kipfer, S. (2007), Fanon and Space: Colonization, urbanization, and liberation from the colonial to the global city. *Environment and Planning D: Society and Space*, 25(4): 701–726.

Kirby, S. and Hay, I. (1997) (Hetero)sexing Space: Gay men and 'straight' space in Adelaide, South Australia. *The Professional Geographer*, 49(3): 295–305.

Kitchin, R. (2006) Positivistic Geographies and Spatial Science, in S. Aitken and G. Valentine (eds), *Approaches to Human Geography*, London: Sage, pp. 20–29.

Kitchin, R. and Lysaght, K. (2003) Heterosexism and the Geographies of Everyday Life in Belfast, Northern Ireland. *Environment and Planning A*, 35: 489–510.

Kitchin, R. and Thrift, N. (eds) (2009) *The International Encyclopedia of Human Geography*, Amsterdam: Elsevier.

Klein, N. (2000) *No Logo*, London: Harper Collins.

Knopp, L. (1987) Social Theory, Social Movements and Public Policy: Recent accomplishments of the gay and lesbian movements in Minneapolis. *International Journal of Urban and Regional Research*, 11: 243–261.

Knopp, L. (1995) Sexuality and Urban Space: a Framework for analysis, in D. Bell and G. Valentine (eds), *Mapping Desire. Geographies of Sexualities*, London: Routledge, pp. 49–61.

Kobayashi, A. (1994) Colouring the Field: Gender, 'race' and the politics of fieldwork. *Professional Geographer*, 46(1): 73–80.

Kobayashi, A. (2003) The Construction of Geographical Knowledge: Racialization, spatialization, in K. Anderson, M. Domosh, S. Pile and N. Thrift, (eds), *Handbook of Cultural Geography*, London: Sage, pp. 544–556.

Kobayashi, A. and Peake, L. (2000) Racism out of Place: Thoughts on whiteness and an anti-racist geography in the new millennium. *Annals of the Association of American Geographers*, 9(2): 392–403.

Kong, L.L.L. (1997) A 'New' Cultural Geography? Debates about invention and reinvention. *Scottish Geographical Magazine*, 113(3): 177–185.

Kropotkin, P. (1885) What Geography Ought to Be. *The Nineteenth Century*, 18: 940–956.

Kropotkin, P. (1914) *Mutual Aid*, London: William Heinemann.

Kuhn, T. (1996) *The Structure of Scientific Revolutions*, Chicago, IL: University of Chicago Press.

Kuus, M. (2007) *Geopolitics Reframed: Security and Identity in Europe's Eastern Enlargement*, New York and Basingstoke: Palgrave Macmillan.

Kwan, M.-P. (2002) Feminist Visualization: Re-envisioning GIS as a method in feminist geographic research. *Annals of the Association of American Geographers*, 92: 645–661.

Kwan, M.-P. (2008) From Oral Histories to Visual Naratives: Representing the post-September 11 experiences of the Muslim women in the USA. *Social & Cultural Geography*, 9(6): 653–669.

Lambert, D. (2001) Liminal Figures: Poor whites, freedmen, and racial reinscription in colonial Barbados. *Environment and Planning D: Society and Space*, 19(3): 335–350.

Lash, S. and Urry, J. (1999) *Economies of Signs and Space*, London: Sage.

Latham, A. (2003) Research, Performance, and Doing Human Geography: Some reflections on the diary-photograph, diary interview method. *Environment and Planning A*, 35: 1993–2017.

Latham, A. and Conradson, D. (2003) Guest Editorial. *Environment and Planning A*, 35: 1901–1906.

Latham, A. and McCormack, D. (2009) Thinking with Images in Non-representational Cities: Vignettes from Berlin. *Area*, 41(3): 252–262.

Laurie, N. (1999) The Shifting Geographies of Femininity and Emergency Work in Peru, in N. Laurie, C.D.S. Holloway and F. Smith (eds), *Geographies of New Femininities*, New York: Pearson Education, pp. 67–90.

Laurier, E. and Philo, C. (2006) Possible Geographies: A passing encounter in a cafe. *Area*, 38(4): 353–363.

Lebrun, A. (1998) *Complexity of Masculine Identity in the Cruising Bar: The Cruiser as Macho or Loser*, Annual Meeting of the Association of American Geographers, Boston, MA.

Lefebvre, H. (1979) Space Social Product and Use Value, in J.W. Freiberg (ed. and trans.), *Critical Sociology: European Perspectives*, New York: Irvington.

Lefebvre, H. (1991) *The Production of Space*, Chichester: Wiley/Blackwell.

Lévi Strauss, C. (1987) *Introduction to the Work of Marcel Mauss*, London: Routledge.

Lewis, R. (1996) *Gendering Orientalism: Race, Femininity and Representation*, London: Routledge.

Ley, D. (1974) *The Black Inner City as Frontier Outpost: Images and Behavior of a Philadelphia Neighborhood*, Washington, DC, Association of American Geographers.

Ley, D. (1978) Social Geography and Social Action, in D. Ley and M. Samuels (eds), *Humanistic Geography Prospects and Problems*, London: Croom Helm, pp. 41–57.

Ley, D. (1987) Styles of the Times: Liberal and neo-conservative landscapes in inner Vancouver, 1968–1986. *Journal of Historical Geography*, 13(1): 40–56.

Ley, D. and Cybriwsky, R. (1974) Urban Graffiti as Territorial Markers. *Annals of the Association of American Geographers*, 64(4): 491–505.

Ley, D. and Samuels, M. (1978) Introduction: Contexts of modern humanism in geography, in D. Ley and M. Samuels (eds), *Humanistic Geography Prospects and Problems*, London: Croom Helm, pp. 1–21.

Livingstone, D. (1992) *The Geographical Tradition*, Oxford: Blackwell.

Livingstone, D. (2003) *Putting Science in its Place: Geographies of Scientific Knowledge*, Chicago, IL: University of Chicago Press.

Loomba, A. (1998) *Colonialism/Postcolonialism*, London: Routledge.

Lorimer, H. (2005) Cultural Geography: The busyness of being more-than-representational. *Progress in Human Geography*, 29(1): 83–94.

Lorimer, H. (2007) Cultural Geography: Worldly shapes, differently arranged. *Progress in Human Geography*, 31(1): 89–100.

Lorimer, H. (2008) Cultural Geography: Non-representational conditions and concerns. *Progress in Human Geography*, 32(4): 551–559.

Lyod, B. and Rowntree, L.R. (1978) Radical Feminists and Gay Men in San Francisco: Social space in dispersed communities, in D. Lanegran and R. Palm (eds), *An Invitation to Geography*, New York: McGraw-Hill, pp. 78–88.

Lyotard, J.F. (1984) *The Postmodern Condition*, Minneapolis, MN: Minnesota University Press.

Lyotard, J.F. (1992) *The Postmodern Explained to Children: Correspondence 1982–1985*, London: Turnaround.

Mackinder, H. (1887) On the Scope and Methods of Geography. *Proceedings of the Royal Geographical Society and Monthly Record of Geography*, 9: 141–174.

Mackinder, H. (1904) The Geographical Pivot of History. *The Geographical Journal*, 23(4): 421–444.

Mackinder, H. (1919) *Democratic Ideals and Reality*, London: Constable and Co.

MacKinnon, C. (1987) *Feminism Unmodified: Discourses on Life and Law*, Cambridge MA: Harvard University Press.

MacPherson, H. (2008) 'I Don't Know Why They Call it the Lake District They Might as Well Call it the Rock District!' The workings of humour and laughter in research with members of visually impaired walking groups. *Environment and Planning D: Society and Space*, 26: 1080–1095.

Maddrell, A. (2009) *Complex Locations: Women's Geographical Work in the UK, 1850–1970*, Oxford: Wiley Blackwell.

Madge, C. (1993) Boundary Disputes: Comments on Sidaway (1992). *Area*, 25: 294–299.

Madge, C., Raghuram, P., Skelton, T., Willis, K. and Williams, J. (1997) Methods and Methodologies in Feminist Research, in Women and Geography Study Group (ed.), *Feminist Geographies: Diversity and Difference*, London: Longman, pp. 86–111.

Mahtani, M. (2004) Mapping Race and Gender in the Academy: The experiences of women of colour faculty and graduate students in Britain, the US and Canada. *Journal of Geography in Higher Education*, 28(1): 91–99.

Mahtani, M. (2006) Challenging the Ivory Tower: Proposing anti-racist geographies within the academy. *Gender, Place and Culture*, 13(1): 21–25.

Malbon, B. (1999) *Clubbing: Dancing, Ecstasy and Vitality*, London: Routledge.

Manzo, K. (2008) Imaging Humanitarianism: NGO identity and the iconography of childhood. *Antipode*, 40(4): 632–657.

Marx, K. (1867) *Das Kapital*, Hamburg: Otto Meisner.

Marx, K. (1959 [1859]) *A Contribution to the Critique of the Political Economy*, Moscow: Progress.

Marx, K. (1968) The Eighteenth Brumaire of Louis Bonaparte, in K. Marx and F. Engles, *Selected Works*, London: Lawrence and Wishart, pp. 95–180.

Marx, K. and Engles, F. (1848) *The Communist Manifesto*, Harlo: The Echo Library.

Marx, K. and Engles, F. (1968) *The German Ideology*, New York: International Publishers.

Massey, D. (1996) A Global Sense of Place, in S. Daniels and R. Lee (eds), *Exploring Human Geography: A Reader*, London: Arnold, Chapter 11.

Massumi, B. (1987) Notes on the Translation and Acknowledgements, in G. Deleuze and F. Guattari (eds), *A Thousand Plateaus*, Minneapolis, MN: University of Minnesota Press.

Massumi, B. (2002) *Parables for the Virtual: Movement, Affect and Sensation*, Durham, NC: Duke University Press.

Matless, D. (1998) *Landscape and Englishness*, London: Reaktion.

Matless, D., Leyshon, A. and Reville, G. (1998) *The Place of Music*, New York: The Guilford Press.

May, J. (1996) Globalization and the Politics of Place: Place and identity in an inner London neighbourhood. *Transactions of the Institute of British Geographers*, 21: 194–215.

McClintock, A. (1995) *Imperial Leather: Race, Gender, and Sexuality in the Colonial Contest*, London: Routledge.

McCormack, D.P. (2002) A Paper with an Interest in Rhythm. *Geoforum*, 33: 469–485.

McDowell, L. (1979) Women in British Geography. *Area*, 11: 151–154.

McDowell, L. (1983) Towards an Understanding of the Gender Division of Urban Space. *Environment and Planning D: Society and Space*, 1: 59–72.

McDowell, L. (1992) Doing Gender: Feminism, feminists and research methods in human geography. *Transactions of the Institute of British Geographers*, 17: 399–416.

McDowell, L. (1995) Body Work: Heterosexual gender performances in city workplaces, in D. Bell and G. Valentine (eds), *Mapping Desire: Geographies of Sexualities*, London: Routledge, pp. 67–87.

McDowell, L. (1997) *Capital Culture*, Oxford: Blackwell.

McDowell, L. (2000) Learning to Serve? Employment aspirations and attitudes of young working-class men in an era of labour market restructuring. *Gender, Place and Culture*, 7(4): 389–416.

McDowell, L. (2003) *Redundant Masculinities? Employment, Change and White Working-class Youth*, Oxford: Blackwell.

McEwan, C. (2003) Material Geographies and Postcolonialism. *Singapore Journal of Tropical Geography*, 24(3): 340–355.

McEwan, C. (2005) Geography, Culture and Global Change, in P. Daniels, M. Bradshaw, D. Shaw and J. Sidaway (eds), *Human Geography, Issues for the 21st Century*, Harlow: Pearson. pp. 265–283.

McGuinness, M. (2000) Geography Matters?: Whiteness and contemporary geography. *Area*, 32(2): 225–230.

McLafferty, S. (2002) Mapping Women's Worlds: Knowledge, power and the bounds of GIS. *Gender, Place and Culture*, 9: 263–269.

McRobbie, A. (1978) *Working Class Girls and the Culture of Femininity, Women Take Issue Centre for Contemporary Cultural Studies*, London: Hutchinson.

McRobbie, A. (1981) 'Just Like a *Jackie* Story', in A. McRobbie and T. McCabe (eds), *Feminism for Girls: An Adventure Story*, London: Routledge and Kegan Paul, pp. 113–128.

McRobbie, A. (1991) *Feminism and Youth Culture: From* Jackie *to* Just Seventeen, London: Macmillan.

Megoran, N. (2004) Revisiting the 'Pivot': The influence of Halford Mackinder on analysis of Uzbekistan's international relations. *The Geographical Journal*, 170(4): 347–358.

Megoran, N. (2008) Militarism, Realism, Just War, or Nonviolence Critical Geopolitics and the Problem of Normativity. *Geopolitics*, 13: 473–497.

Miles, R. and Phizacklea, A. (1979) *Racism and Political Action in Britain*, London: Routledge and Kegan Paul.

Miller, D. (1998) Coca-Cola: A black sweet drink from Trinidad, in D. Miller (ed.), *Material Cultures*, London: University College London Press, pp. 169–188.

Miller, D. (2008) *The Comfort of Things*, Cambridge: Polity.

Miller, R. (1983) The Hoover in the Garden: Middle-class women, and suburbanization, 1850–1920. *Environment and Planning D: Society and Space*, 1: 73–87.

Mitchell, D. (1995) There's No Such Thing as Culture: Towards a reconceptualization of the idea of culture in geography. *Transactions for the Institute of British Geographers*, 20: 102–116.

Mitchell, D. (2000) Dead Labour and the Political Economy of Landscape – California Living, California Dying, in K. Anderson, M. Domosh, S. Pile and N. Thrift (eds), *Handbook of Cultural Geography*, London: Sage, pp. 223–248.

Mitchell, D. (2005 [2000]) *Cultural Geography: A Critical Introduction*, Oxford: Blackwell.

Mohanty, C. (1991) Under Western Eyes: Feminist scholarship and colonial discourses, in C.T. Mohanty, A. Russo and L. Torres (eds), *Third World Women and the Politics of Feminism*, Indiana, IN: Indiana University Press, pp. 51–80.

Monk, J. and Hanson, S. (1982) On Not Excluding Half the Human in Human Geography. *The Professional Geographer*, 34: 11–23.

Morgan, J. and Lambert, D. (2003) *Theory into Practice: Place, 'Race' and Teaching Geography*, Sheffield: Geographical Association.

Morin, K. (2004) Edward W. Said, in P. Hubbard, R. Kitchen and G. Valentine (eds), *Key Thinkers on Space and Place*, London: Sage, pp. 237–244.

Morley, D. (1980) *The Nationwide Audience: Structure and Decoding*, London: British Film Institute.

Morley, D. and Robins, K. (eds) (2001) *British Cultural Studies: Geography, Nationality, and Identity*, Oxford: Oxford University Press.

Morrill, R. (2005) Hägerstrand and the 'Quantitative Revolution': A personal appreciation. *Progress in Human Geography*, 29: 333–335.

Murphy, D.T. (1997) *The Heroic Earth: Geopolitical Thought in Weimar Germany, 1918–1933*, Kent, OH: The Kent State University Press.

Nash, C. (2000) Performativity in Practice: Some recent work in cultural geography. *Progress in Human Geography*, 24(4): 653–664.

Nash, C. (2003) Cultural Geography: Anti-racist geographies. *Progress in Human Geography*, 27(5): 637–648.

Nast, H. (1998) Unsexy Geographies. *Gender, Place and Culture*, 5: 191–206.

Nayak, A. (2003a) *Race, Place and Globalization: Youth Cultures in a Changing World*, Oxford: Berg.

Nayak, A. (2003b) Last of the 'Real Geordies'? White masculinities and the subcultural response to de-industrialization. *Environment and Planning D: Society and Space*, 21(1): 7–25.

Nayak, A. (2003c) 'Through Children's Eyes': Childhood, place and the fear of crime. *Geoforum*, 34: 303–315.

Nayak, A. (2008) Young People's Geographies of Racism and Anti-racism: The case of north east England, in C. Dwyer and C. Bressey (eds), *New Geographies of Race and Racism*, Aldershot: Ashgate, pp. 269–282.

Nayak, A. (2010) Race, Affect, and Emotion: Young people, racism, and graffiti in the postcolonial English suburbs. *Environment and Planning A*, 42(10): 2370–2392.

Nayak, A. and Kehily, M.J. (2007) *Gender, Youth and Culture: Young Masculinities and Femininities*, Basingstoke: Palgrave.

Nayak, A. and Kehily, M.J. (2008) Gender Undone: Subversion, regulation and embodiment in the work of Judith Butler. *British Journal of Sociology of Education*, 27: 459–472.

Neal, S. (2002) Rural Landscapes, Representations and Racism: Examining multicultural citizenship and policy-making in the English countryside. *Ethnic and Racial Studies*, 25: 442–461.

Nold, C. (2004) Introduction: Emotional cartography technologies of the self, in C. Nold (ed.), *Emotional Cartographies, Creative Commons*, at http://www.emotionalcartographies.net, pp. 3–14.

Nussbaum, M. (1999) Professor of Parody. *The New Republic*, 22 February.

Oakes, T.S. and Price, P.L. (2008) *The Cultural Geography Reader*, Abingdon: Routledge.

Oakley, A. (1981) Interviewing Women: A contradiction in terms, in H. Roberts (ed.), *Doing Feminist Research*, London: Routledge and Kegan Paul, pp. 31–61.

Ogborn, M. (1998) The Capacities of the State: Charles Davenant and the management of the Excise, 1683–1698. *Journal of Historical Geography*, 24: 289–312.

Ó Tuathail, G. (1986) The Language and Nature of the 'New' Geopolitics: The case of US–El Salvador relations. *Political Geography Quarterly*, 5: 73–85.

Ó Tuathail, G. (1993) The Effacement of Place? US foreign policy and the spatiality of the Gulf crisis. *Antipode*, 25(1): 4–31.

Ó Tuathail, G. (1994) The Critical Reading/Writing of Geopolitics: Re-reading/writing Wittfogel, Bowman and Lacoste. *Progress in Human Geography*, 18(3): 313–332.

Ó Tuathail, G. (1996a) *Critical Geopolitics*, London : Routledge.

Ó Tuathail, G. (1996b) An Anti-geopolitical Eye: Maggie O'Kane in Bosnia 1992–1993. *Gender, Place and Culture*, 3: 171–185.

Ó Tuathail, G. (2002) Theorizing Practical Geopolitical Reasoning: The case of the United States' response to the war in Bosnia. *Political Geography*, 21: 601–628.

Ó Tuathail, G. (2003) 'Just Out Looking for a Fight': American affect and the invasion of Iraq. *Antipode* , 35(5): 856–870.

Ó Tuathail, G. (2006) General Introduction: Thinking critically about geopolitics, in G. Ó Tuathail, S. Dalby and P. Routledge (eds), *The Geopolitics Reader*, 2nd edn, Abingdon: Routledge, pp. 1–14.

Ó Tuathail, G. (2009) Placing Blame: Making sense of Beslan. *Political Geography*, 28: 4–15.

Ó Tuathail, G. and Agnew, J. (1992) Geopolitics and Discourse: Practical geopolitical reasoning in American foreign policy. *Political Geography*, 11: 190–204.

Ó Tuathail, G. Dalby, S. and Routledge, P. (eds) (2006) *The Geopolitics Reader*, 2nd edn, Abingdon: Routledge.

Pain, R. (2000) Place, Social Relations and the Fear of Crime: A review. *Progress in Human Geography*, 24: 365–387.

Pain, R. (2001) Gender, Race, Age and Fear in the City. *Urban Studies*, 38(5–6): 899–913.

Pain, R. (2004) Social Geography: Participatory research. *Progress in Human Geography*, 28(3): 652–663.

Pain, R. (2009) Globalized Fear? Towards an emotional geopolitics. *Progress in Human Geography*, 33: 1–21.

Pain, R. and Bailey, C. (2004) British Social and Cultural Geography: Beyond turns and dualisms? *Social & Cultural Geography*, 5(2): 319–329.

Painter, J. (2000) Pierre Bourdieu, in M. Crang and N. Thrift (eds), *Thinking Space*, London: Routledge.

Palmer, C. (2002) Christianity, Englishness and the Southern English Countryside: A study of the work of H.J. Massingham. *Social and Cultural Geography*, 31(1): 25–38.

Peach, C. (1965) West Indian Migration to Britain: The economic factors. *Race*, 7: 310–347.

Peach, C. (ed.) (1975a) *Urban Social Segregation*, London: Longman.

Peach, C. (1975b) Introduction: The spatial analysis of ethnicity and class, in C. Peach (ed.), *Urban Social Segregation*, London: Longman, pp. 7–17.

Peake, L. and Kobayashi, A. (2002) Policies and Practice for an Antiracist Geography at the Millennium. *The Professional Geographer*, 54(1): 50–61.

Peet, R. (2001 [1998]) *Modern Geographical Thought*, Oxford: Blackwell.

Penrose, J. and Howard, D. (2008) One Scotland, Many Cultures: The mutual constitution of anti-racism and place, in C. Dwyer and C. Bressey (eds), *New Geographies of Race and Racism*, Aldershot: Ashgate, pp. 269–282.

Phillips, D. (1987) The Rhetoric of Anti-racism in Public Housing Allocation, in P. Jackson (ed.), *Race and Racism: Essays in Social Geography*, London: Allen and Unwin, pp. 212–237.

Phillips, D. (2006) Parallel Lives? Challenging discourses of British Muslim self-segregation. *Environment and Planning D: Society and Space*, 24: 25–40.

Phillips, D., Davis, C. and Ratcliffe, P. (2007) British Urban Narratives of Urban Space. *Transactions of the Institute of British Geographers*, 32(2): 217–234.

Phillips, R. (1997) *Mapping Men and Empire: A Geography of Adventure*, London: Routledge.

Philo, C. (ed.) (1991) *New Words, New Worlds: Reconceptualising Social and Cultural Geography*, Cambrian: St David's University College.

Philo, C. (2000) More Words, More Worlds: Reflections on the 'cultural turn' and human geography, in I. Cook, D. Crouch, S. Naylor and J.R. Ryan (eds), *Cultural Turns/Geographic Turns*, Harlow: Pearson, pp. 26–53.

Pickerill, J. (2008) The Surprising Sense of Hope. *Antipode*, 40(3): 482–487.

Pile, S. (2000 [1996]) *The Body and the City: Psychoanalysis, Space and Subjectivity*, London: Routledge.

Pile, S. (2010) Emotions and Affect in Recent Human Geography. *Transactions of the Association of British Geographers*, 35: 5–20.

Plant, S. (1996) On the Matrix: Cyberfeminist simulations, in R. Shields (ed.), *Cultures of the Internet: Virtual Space, Real Histories, Living Bodies*, London: Sage, pp. 170–183.

Pollard, I. (1989) Pastoral Interludes. *Third Text: Third World Perspectives on Contemporary Art and Culture*, 7: 41–46.

Pollard, I. (1994) *Hidden Histories: Heritage Stories: Photographic Work by Ingrid Pollard*, exhibition catalogue, Lee Valley Park: Myddleton House.

Pollard, J. and Samers, M. (2007) Islamic Banking and Finance: Post-colonial political economy and the decentring of economic geography. *Transactions of the Institute of British Geographers*, 32: 313–330.

Pollard, J., McEwan, C., Laurie, N. and Stenning, A. (2009) Economic Geography Under Postcolonial Scrutiny. *Transactions*, 34(2): 137–142.

Popke, J. (2009) Geography and Ethics: Non-representational encounters, collective responsibility and economic difference. *Progress in Human Geography*, 33(1): 81–90.

Potter, R. and Phillips, J. (2006) Both Black and Symbolically White: The 'Bajan-Brit' return migrant as post-colonial hybrid. *Ethnic and Racial Studies*, 29(5), 901–927.

Pratt, M.L. (1992) *Imperial Eyes: Travel Writing and Transculturation*, London: Routledge.

Pred, A. (2000) *Even in Sweden*, Berkeley, CA: University of California Press.

Pulido, L. (2002) Reflections on a White Discipline. *The Professional Geographer*, 54(1): 42–49.

Ratzel, F. (1897) *Politische Geographic*, Munich: Oldenburg.

Relph, E. (1976) *Place and Placelessness*, London: Pion.

Retort (2005) *Afflicted Powers*, London: Verso.

Rex, J. and Moore, R. (1967) *Race, Community and Conflict: A Study of Sparkbrook*, Oxford: Oxford University Press.

Rich, P. (1987) The Politics of 'Race Relations' in Britain and the West, in P. Jackson (ed.), *Race and Racism: Essays in Social Geography*, London: Allen and Unwin, pp. 79–97.

Robinson, J. (1996) *The Power of Apartheid: State, Power and Space in South African Cities*, London: Butterworth-Heinemann.

Robinson, J. (1998) Spaces of Democracy: Re-mapping the apartheid city. *Environment and Planning D: Society and Space*, 16(5): 533–548.

Robinson, J. (2003) Postcolonialising Geography: Tactics and pitfalls. *Singapore Journal of Tropical Geography*, 24(3): 273–289.

Rose, G. (1993) *Feminism and Geography: The Limits of Geographical Knowledge*, Cambridge: Polity Press.

Rose, G. (1997) Situating Knowledges: Positionality, reflexivities and other tactics. *Progress in Human Geography*, 21: 303–320.

Ross, E. (1919) The Climate of Liberia and its Effect on Man. *Geographical Review*, 7: 387–402.

Routledge, P. (2003a) Voices of the Dammed: Discursive resistance amidst erasure in the Narmada Valley, India. *Political Geography*, 22: 243–270.

Routledge, P. (2003b) Anti-Geopolitics, in J. Agnew, K. Mitchell and G. Toal (eds), *A Companion to Political Geography*, Oxford: Blackwell, pp. 236–248.

Routledge, P. (2006) Introduction to Part Five, in G. Ó Tuathail, S. Dalby and P. Routledge (eds), *The Geopolitics Reader*, 2nd edn, London: Routledge, pp. 233–249.

Said, E. (1993) *Culture and Imperialism*, London: Chatto and Windus.

Said, E. (1995a [1978]) *Orientalism*, New York: Vintage.

Said, E.W. (1995b) *Orientalism: Western Conceptions of the Orient*, New York: Vintage.

Saldanha, A. (2007) *Psychedelic White: Goa Trance and the Viscosity of Race*, Minneapolis, MN: University of Minnesota Press.

Sambon, L.W. (1898) Acclimatization of Europeans in Tropical Lands. *Geographical Journal*, 12: 589–606.

Samers, M. (1998) Immigration, 'Ethnic Minorities', and Social Exclusion in the European Union: A critical perspective. *Geoforum*, 29(2): 123–144.

Samudra, J. (2008) Memory in Our Body: Thick participation and the translation of kinesthetic-experience. *American Ethnologist*, 35(4): 665–681.

Samuels, M. (1978) Existentialism and Human Geography, in D. Ley and M. Samuels (eds), *Humanistic Geography Prospects and Problems*, London: Croom Helm, pp. 22–40.

Sanders, R. (2006) Social Justice and Women of Colour in Geography: Philosophical musings, trying again. *Gender, Place and Culture*, 13(1): 49–55.

Sartre, J.-P. (1948) *Being and Nothingness*, New York: Philosophical Library.

Sauer, C. (2008 [1925]) The Morphology of Landscape, in T.S. Oakes and P.L. Price (eds), *The Cultural Geography Reader*, London/New York: Routledge, pp. 96–104.

Saville, S.J. (2008) Playing with Fear: Parkour and the mobility of emotion. *Social and Cultural Geography*, 9(8): 891–914.

Schaefer, F.K. (1953) Exceptionalism in Geography: A methodological examination. *Annals of the Association of American Geographers*, 43: 226–245.

Schuurman, N. (2009) The New Brave New World: Geography, GIS, and the emergence of ubiquitous mapping and data. *Environment and Planning D: Society and Space*, 27: 571–580.

Seamon, D. (1979) *A Geography of the Lifeworld*, London: Croom Helm.

Seamon, D. (2006) A Geography of Lifeworld in Retrospect: A response to Shaun Moores. *Particip@tions*, 3(2).

Seamon, D. and Sowers, J. (2009) Existentialism/existential Geography, in R. Kitchen and N. Thrift (eds), *The International Encyclopedia of Human Geography*, in Oxford: Elsevier, pp. 666–667.

Semple, E.C. (1911) *The Influences of the Geographic Environment*, London: Constable.

Sharma, S., Hutnyk, J. and Sharma, A. (1996) *Dis-orientating Rhythms: The Politics of the New Asian Dance Music*, London: Zed Books.

Sharp, J. (2000) *Condensing the Cold War: Reader's Digest and American Identity*, Minneapolis, MN: University of Minnesota Press.

Shaw, W. (2006) Decolonizing Geographies of Whiteness. *Antipode*, 38(4): 851–869.

Shurmer-Smith, P. (ed.) (2002) *Doing Cultural Geography*, London: Sage.

Sibley, D. (1995) *Geographies of Exclusion*, New York: Routledge.

Sidaway, J. (2000) Postcolonial Geographies: An exploratory essay. *Progress in Human Geography*, 24(4): 591–612.

Sidaway, J.D. and Power, M. (2005) The Tears of Portugal: Empire, identity, 'race', and destiny in Portuguese geopolitical narratives. *Environment and Planning D: Society and Space*, 23: 527–554.

Simpson, A. (1993) *Xuxa: The Mega-marketing of Gender, Race, and Modernity*, Philadelphia, PA: Temple University Press.

Simpson, K. (2004) 'Doing Development': The gap year, volunteer tourists and a popular practice of development. *Journal of International Development*, 16: 681–697.

Sillitoe, A. (1958) *Saturday Night and Sunday Morning*, London: W.H. Allen.

Singh, R.P. (2009a) *Uprooting Geographical Thoughts in India*, Newcastle upon Tyne: Cambridge Scholars Publishing.

Singh, R.P. (2009b) *Geographical Thoughts in India: Snapshots and Visions for the 21st Century*, Newcastle upon Tyne: Cambridge Scholars Publishing.

Skelton, T. and Valentine, G. (1998) *Cool Places: Geographies of Youth Cultures*, London: Routledge.

Slocum, R. (2007) Whiteness, Space and Alternative Food Practices. *Geoforum*, 38: 520–533.

Slocum, R. (2008) Thinking Race Through Corporeal Feminist Theory: Divisions and intimacies at the Minneapolis Farmers' Market. *Social and Cultural Geography*, 9(8): 849–869.

Smith, N. (1990 [1984]) *Uneven Development: Nature, Capital and the Production of Space*, Oxford: Blackwell.

Smith, N. (1998) *The New Urban Frontier: Gentrification and the Revanchist City*, London: Routledge.

Smith, N. (2000) Socialising Culture, Radicalising the Social. *Social & Cultural Geography*, 1: 25–28.

Smith, N. (2003) *American Empire: Roosevelt's Geographer and the Prelude to Globalization*. Berkeley, CA: University of California Press.

Smith, N. and Godlewska, A. (eds) (1994a) *Geography and Empire*, Oxford: Blackwell.

Smith, N. and Godlewska, A. (1994b) Critical Histories of Geography, in N. Smith and A. Godlewska (eds), *Geography and Empire*, Oxford: Blackwell, pp. 1–8.

Smith, S. (1984) Practicing Humanistic Geography. *Annals of the Association of American Geographers*, 74(3): 353–374.

Smith, S. (1987) Residential Segregation: A geography of English racism, in P. Jackson (ed.), *Race and Racism: Essays in Social Geography*, London: Allen and Unwin.

Smith, S. (1997) Beyond Geography's Visible Worlds: A cultural politics of music. *Progress in Human Geography*, 21(4): 502–529.

Söderström, O. (2005) Representation, in D. Atkinson, P. Jackson, D. Sibley and N. Washbourne (eds), *Cultural Geography: A Critical Dictionary of Key Concepts*, London: I.B. Tauris, pp. 11–16.

Soja, E. (1989) *Postmodern Geographies: The Reassertion of Space in Critical Social Theory*, London: Verso.

Soja, E.W. (1995) Postmodern Urbanization: The six restructurings of Los Angeles, in S. Watson and K. Gibson (eds), *Postmodern Cities and Spaces*, Oxford: Blackwell, pp. 125–137.

Soja, E. (1996) *Thirdspace: Journeys to Los Angeles and Other Real-and-Imagined Places*, Cambridge, MA: Blackwell.

Soja, E.W. (2000) *Postmetropolis: Critical Studies of Cities and Regions*, Oxford: Blackwell.

Spivak, G.C. (1990) *The Postcolonial Critique: Interviews, Strategies, Dialogues*, New York: Routledge.

Spivak, G.C. (1993 [1988]) Can the Subaltern Speak?, in C. Nelson and L. Grossberg (eds), *Marxism and the Interpretation of Culture*, London: Macmillan, pp. 271–316.

Stacey, J. (1997) Can There be a Feminist Ethnography?, in L. McDowell and J. Sharp (eds), *Space, Gender, Knowledge: Feminist Readings*, London: Arnold, pp. 115–123.

Stanley, H.M. (1885) Inaugural Address – Delivered before the Scottish Geographical Society, Edinburgh, 3 December 1884. *The Scottish Geographical Magazine*, 1(1): 1–17.

Stanton, M. (2000) The Rack and the Web: The other city, in L. Lokko (ed.), *White Paper; Black Marks: Architecture, Race and Culture*, London: Athlone, pp. 114–145.

Stea, D. (1965) Space, Territory and Human Movements. *Landscape*, 15: 13.

Stoddart, D. (1966) Darwin's Impact on Geography. *Annals of the Association of American Geographers*, 56: 683–698.

Stoddart, D. (1991) Do We Need a Feminist Historiography of Geography? And If We Do, What Should It Be? *Transactions of the Institute of British Geographers*, 16: 484–748.

Taylor, I., Evans, K. and Fraser, P. (1996) *A Tale of Two Cities: Global Change, Local Feeling and Everyday Life in the North of England, A Study of Manchester and Sheffield*, London: Routledge.

Tharoor, S. (2008) Bombs and Bullets Cannot Destroy India – As Long as its Gates Remain Open, *The Guardian*, 28 November, p. 42.

The Center for Geographic Analysis (2010) *Homepage*, at http://www.gis.harvard.edu/icb/icb.do.

Thien, D. (2005) After or Beyond Feeling? A consideration of affect and emotion in geography. *Area*, 37(4): 450–456.

Thrift, N. (1996) *Spatial Formations*, London: Sage.

Thrift, N. (2000) Introduction: Dead or alive?, in I. Cook, D. Crouch, S. Naylor and J.R. Ryan (eds), *Cultural Turns/Geographical Turns: Perspectives on Cultural Geography*, Harlow: Pearson, pp. 1–6.

Thrift, N. (2003) Performance and. . . . *Environment and Planning A*, 35: 2019–2024.

Thrift, N. (2004a) Summoning Life, in P. Cloke, P. Crang and M. Goodwin (eds), *Envisioning Human Geographies*, London: Arnold, pp. 81–103.

Thrift, N. (2004b) Intensities of Feeling: Towards a spatial politics of affect. *Geografiska Annaler B*, 86(1): 57–78.

Timander, L. and McLafferty, S. (1998) Breast Cancer in West Islip, NY: A spatial clustering analysis with covariates. *Social Science and Medicine*, 46: 1623–1635.

Tolia-Kelly, D. (2006) Affect – an Ethnocentric Encounter? Exploring the 'universalist' imperative of emotional/affectual geographies. *Area*, 38(2): 213–217.

Troyna, B. and Hatcher, R. (1992) *Racism in Children's Lives: A Study of Mainly-white Primary Schools*, London: Routledge.

Tuan, Y.-F. (1974) *Topophilia: A Study of Environmental Perception, Attitudes, and Values*, Englewood Cliffs, NJ: Prentice Hall.

Tuan, Y.-F. (1976) Humanistic Geography. *Annals of the Association of American Geographers*, 66: 266–276.

Tuan, Y.-F. (1998) *Escapism*, Baltimore, NJ; Johns Hopkins University Press.

Tuan, Y.-F. (1999) *Who Am I? An Autobiography of Emotion, Mind and Spirit*, Madison, WI: University of Wisconsin Press.

Turner, R.H. (1967) Introduction, in R.H. Turner (ed.), *Robert E. Park on Social Control and Collective Behaviour: Selected Papers*, Chicago, IL: University of Chicago Press, pp. ix–xlvi.

Twine, F.W. (1996) Brown Skinned White Girls: Class, culture and the construction of white identity in suburban communities. *Gender, Place & Culture*, 3(23): 205–224.

Unwin, T. (1992) *The Place of Geography*, Harlow: Longman.

Urry, J. (1990) *The Tourist Gaze: Leisure and Travel in Contemporary Societies*, London: Sage.

Valentine, G. (1993) Negotiating and Managing Multiple Sexual Identities: Lesbian time–space strategies. *Transactions of the Institute of British Geographers*, 18: 237–248.

Valentine, G. (1998) Sticks and Stones May Break My Bones: A personal geography of harassment. *Antipode*, 30: 303–332.

Watson, S. (1986) *Housing and Homelessness: A Feminist Perspective*, London: Routledge.

Watt, P. (1998) Going Out of Town: Youth, race and place in the south east of England. *Environment and Planning D: Society and Space*, 16: 687–703.

Watts, M. (2000) Colonialism, in R. Johnston, D. Gregory, G. Pratt and M. Watts (eds), *The Dictionary of Human Geography*, Oxford: Blackwell, p.11.

Watts, M. (2005) Commodities, in P. Cloke, P. Crang and M. Goodwin (eds), *Introducing Human Geographies*, 2nd edn, London: Hodder Arnold, pp. 527–546.

Webber, M.M. (1964) Culture, Territoriality and the Elastic Mile. *Regional Science Association Papers*, 13: 61–64.

Weeks, J. (1981) *Sex, Politics and Society: The Regulation of Sexuality Since 1880*, Harlow: Longman.

Weeks, J. (1991) *Against Nature: Essays on History, Sexuality, and Identity*, London: Rivers Oram.

Weightman, B.A. (1981) Commentary: Towards a geography of the gay community. *Journal of Cultural Geography*, 1: 106–112.

Whatmore, S. (2002) *Hybrid Geographies: Natures, Cultures, Spaces*, London: Sage.

Whyte, W.F. (1993 [1943]) *Street Corner Society: The Social Structure of an Italian Slum*, Chicago, IL: University of Chicago Press.

Williams, R. (1976) *Keywords*, London: Fontana.

Williams, R. (1985 [1973]) *The Country and the City*, London: Hogarth Press.

Williams, R. (1989 [1981]) *Culture*, London: Fontana.

Williams, S. (2009) My Four Mums. *The Guardian*, 4 July.

Willis, P. (1977) *Learning to Labour: How Working-class Kids get Working-class Jobs*, London: Saxon House.

Willis, P. (1978) *Profane Culture*, London: Routledge and Kegan Paul.

Willis, P. (1990) *Common Culture: Symbolic Work at Play in the Everyday Cultures of the Young*, Milton Keynes: Open University Press.

Willis, P. with Jones, S., Canaan, J. and Hurd, G. (1990) *Common Culture: Symbolic Culture at Play in the Everyday Culture of the Young*, Milton Keynes: Open University Press.

Withers, C.W.J. and Mayhew, R.A. (2002) Rethinking 'Disciplinary' History: Geography in British universities, c. 1580–1887. *Transactions of the Institute of British Geographers*, 27(1): 11–29.

Women and Geography Study Group (WGSG) (1984) *Geography and Gender: An Introduction to Feminist Geography*, London: Hutchinson.

Women and Geography Study Group (WGSG) (1997) *Feminist Geographies: Explorations in Diversity and Difference*, Harlow: Longman.

Wren, K. (2001) Cultural Racism: Something rotten in the state of Denmark? *Social & Cultural Geography*, 2(2): 141–162.

Wyly, E. (2009) Strategic Positivism. *The Professional Geographer*, 61: 310–322.

Wylie, J. (2002) An Essay on Ascending Glastonbury Tor. *Geoforum*, 33: 441–454.

Wylie, J. (2007) *Landscape*, London: Routledge.

Yeoh, B. (2003) Postcolonial Geographies of Place and Migration, in K. Anderson, M. Domosh, S. Pile and N. Thrift (eds), *Handbook of Cultural Geography*, London: Sage, pp. 369–380.

Young, R. (1990) *Colonial Desire: Hybridity in Theory, Culture and Race*, London: Routledge.

Zelinski, W. (1973) *The Cultural Geography of the United States*, Englewood Cliffs, NJ: Prentice Hall.

Zukin, S. (1988 [1982]) *Loft Living: Culture and Capital in Urban Change*, New Brunswick, NJ: Rutgers University Press.

Zukin, S. (1993) *Landscapes of Power: From Detroit to Disney World*, Berkeley, CA: University of California Press.

Index

Bold entries are listed in the Glossary